Fluid Dynamics: Beyond the Basics

Fluid Dynamics: Beyond the Basics

Jason Wood

NY RESEARCH PRESS
New York

Published by NY Research Press
118-35 Queens Blvd., Suite 400,
Forest Hills, NY 11375, USA
www.nyresearchpress.com

Fluid Dynamics: Beyond the Basics
Jason Wood

International Standard Book Number: 978-1-64725-447-6 (Hardback)

Cataloging-in-Publication Data

Fluid dynamics : beyond the basics / Jason Wood.
p. cm.
Includes bibliographical references and index.
ISBN 978-1-64725-447-6
1. Fluid dynamics. 2. Computational fluid dynamics. 3. Turbulence.
4. Fluid mechanics. I. Wood, Jason.
QC151 .E77 2023
532.05--dc23

Contents

Preface

This book has been an outcome of determined endeavour from a group of educationists in the field. The primary objective was to involve a broad spectrum of professionals from diverse cultural background involved in the field for developing new researches. The book not only targets students but also scholars pursuing higher research for further enhancement of the theoretical and practical applications of the subject.

A fluid refers to a type of liquid, gas or other substance that constantly flows due to an external force or shear stress. Fluid dynamics is a branch of fluid mechanics that studies the flow of fluids such as liquids and gases. There are two main subdisciplines of fluid dynamics, namely, hydrodynamics and aerodynamics. Fluid dynamics can be used to study phenomena such as ocean currents, plate tectonics, evolution of stars, weather patterns and blood circulation. Wind turbines, air conditioning systems, rocket engines and oil pipelines are some of the significant technological applications of fluid dynamics. A fluid dynamics problem is normally solved by calculating different properties of the fluids such as flow density, velocity, temperature and pressure as functions of space and time. This book outlines the processes and applications of fluid dynamics in detail. It is appropriate for students seeking detailed information in this area as well as for experts.

It was an honour to edit such a profound book and also a challenging task to compile and examine all the relevant data for accuracy and originality. I wish to acknowledge the efforts of the contributors for submitting such brilliant and diverse chapters in the field and for endlessly working for the completion of the book. Last, but not the least; I thank my family for being a constant source of support in all my research endeavours.

Jason Wood

1

Complex Fluids: An Overview of Numerical Methods

Marco E. Rosti, Francesco Picano, and Luca Brandt

1.1 Introduction to Complex Fluids and Rheology

We are surrounded by a variety of fluids in our everyday life. Besides water and air, it is common to deal with fluids with peculiar behaviours such as gel, mayonnaise, ketchup and toothpaste, while water, oil and other so-called *simple* (Newtonian) fluids "regularly" flow when we apply a force, the response is different for complex fluids. In some cases, we need to apply a stress larger than a certain threshold for the material to start flowing, for example, to extract toothpaste from the tube; the same paste would behave as a solid on the toothbrush when exposed only to the gravitational force. In other cases the history of past deformations has a role in the present behaviour. Rheology studies and classifies the response of different fluids and materials to an applied force, and to this end, how the macroscopic behaviour is linked to the microscopic structure of the fluid. Hence, while simple fluids made by identical molecules show a linear response to the applied forces, complex fluids with a microstructure, such as suspensions, may show a very complex response to the applied forces.

In this chapter, we introduce numerical approaches for complex fluids focusing on the way the additional stress due to the presence of a microstructure is modelled and how rigid and deformable intrusions can be simulated. We will assume the reader has a solver for the momentum and mass conservation equations, typically using a finite-difference or finite-volume representation. An alternative approach, also very popular, are Lattice–Boltzmann methods; these will not be considered here, thus the reader is referred to Refs. [1, 2].

M. E. Rosti · L. Brandt (✉)
Linné FLOW Centre and SeRC, KTH Mechanics, Stockholm, Sweden
e-mail: luca@mech.kth.se

F. Picano
Department of Industrial Engineering, University of Padova, Padua, Italy

Newtonian and Non-Newtonian Rheology

The macroscopic rheological behaviour of a viscous fluid is well characterised in a Couette flow, i.e. the flow between two parallel walls of area A and at distance b: the upper wall moving at constant (low) velocity U_0 and the lower at rest. To keep the upper wall moving at constant velocity we need to apply a force F which is proportional to the wall area: $F \propto A$. Therefore it is more general to consider the stress $\tau = F/A$ instead of the force F itself. In a *Newtonian* fluid the shear stress is proportional to the velocity of the upper wall and to the inverse of the wall distance b, i.e. $\tau \propto U_0/b$. This linear response defines Newtonian fluids, such as air, water, oil and many others. Note that, in a simple Couette flow the ratio U_0/b equals the wall-normal derivative of the velocity profile and the shear (deformation) rate: $du/dy = \dot{\gamma} = U_0/b$. Thus, for a Newtonian fluid we can express the law relating the applied force with the response, i.e. the shear stress τ with the shear rate $\dot{\gamma}$, as

$$\tau = \mu \dot{\gamma}, \tag{1.1}$$

where the proportionality coefficient μ is called dynamic viscosity with dimension $Pa\,s$ in the SI. Many Newtonian fluids exist, each with a different value of viscosity, and therefore flowing at different velocity when subject to the same stress. The viscosity coefficient of a Newtonian fluid does not depend on the shear rate, but may vary with the temperature. Indeed, the viscosity usually increases with temperature in gases, while it decreases in liquids. This behaviour is related to the effect of the temperature on the molecular structure of the fluid, but this is outside the scope of present chapter and the reader is refereed to specialised textbooks.

Fluids that exhibit a non-linear behaviour between the shear stress τ and the shear rate $\dot{\gamma}$ are called *non-Newtonian* and fluids whose response does not depend explicitly on time but only on the present shear rate are denoted *generalised Newtonian* fluids. In particular, when the shear stress increases more than linearly with the shear rate, the fluid is called *dilatant* or *shear-thickening*, whereas in the case of opposite behaviour, i.e. when the shear stress increases less than linearly with the shear rate, the fluid is called *pseudoplastic* or *shear-thinning*. Examples of typical profiles of the shear stress τ as a function of the shear rate $\dot{\gamma}$ for Newtonian, shear-thickening and shear-thinning fluids are shown in the left panel of Fig. 1.1. The ratio of the applied stress and the resulting deformation rate is the so-called apparent effective viscosity $\mu_e = \tau/\dot{\gamma}$: it increases with $\dot{\gamma}$ for shear-thickening fluids, while it reduces for shear-thinning ones, which means that the fluidity of shear-thickening fluids reduces increasing the shear rate, while the opposite is true for shear-thinning fluids. Examples of shear-thinning fluids are ketchup, mayonnaise and toothpaste, while corn-starch water mixtures and dense non-colloidal suspensions usually exhibit a shear-thickening behaviour. Note that, sometimes, the same fluids can have plastic or elastic responses depending on the flow configuration.

Complex fluids may behave as solids, with a finite deformation, when the applied stress is below a certain threshold τ_0, while for stresses above it, they start

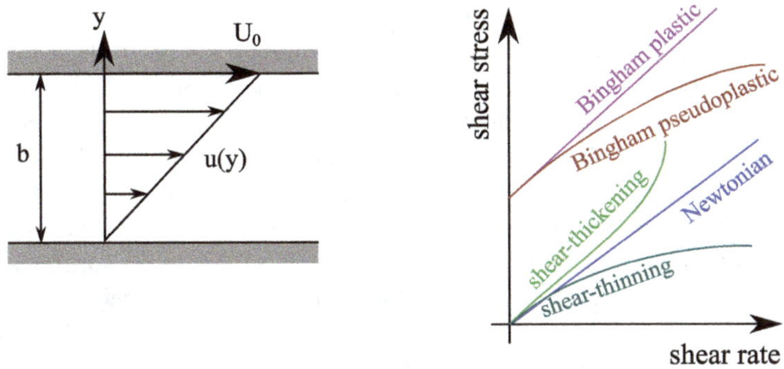

Fig. 1.1 (left) Sketch of a plane Couette flow. (right) Sketch of the shear stress τ profile as a function of the shear deformation rate $\dot{\gamma}$ for different kind of fluids

flowing as liquids. These fluids are called *yield stress* or *Bingham* fluids: when the applied stress exceeds the so-called yield stress, τ_0, these fluids can exhibit a linear relation between stress and deformation similar to Newtonian fluids or a pseudoplastic response. These macroscopic behaviours are related to changes of the microscopic structure of the fluid, and indeed these fluids are constituted by a Newtonian fluid with one or more suspended phases, such as fibres, polymers, trapped fluids (emulsions). From a qualitative point of view, the material hardly flows and deforms when the connections and interactions between the phases constituting the microstructure are intense. Changing the level of the stress τ applied on these complex fluids may either strengthen, weaken or break these interactions, thus altering their microstructure, and eventually reflecting in their non-linear rheological behaviour.

In order to describe complex fluids, we need a relation as in Eq. (1.1) between the applied stress, τ, and the deformation rate, $\dot{\gamma}$. A relation that can be used to summarise the behaviours previously described for complex fluids is the Herschel–Bulkley formula

$$\tau = \tau_0 + K\,\dot{\gamma}^n\,, \tag{1.2}$$

where τ_0 is the yield stress, n the flow index and K the fluid consistency index. A Newtonian behaviour is recovered when $\tau_0 = 0$, $n = 1$ and $K = \mu$, while values of the flow index above and below unity, $n > 1$ and $n < 1$, denote shear-thickening and shear-thinning fluids, respectively. Finally, yield-stress fluids are characterised by a finite non-zero value of the yield stress τ_0. The consistency index K measures how strong the fluid responds to the imposed deformation rate. However, the consistency index has the same dimension of a dynamic viscosity only when $n = 1$, and in general its dimension is a function of n, so that it is not possible to compare different values of K for fluids with different flow indexes n.

The fluid discussed so far are inelastic, since the stress is just a function of the present value of the deformation rate, i.e. $\tau = \tau(\dot{\gamma})$, and not on the previous history of the deformation rate (no memory effects). Another important class of non-Newtonian fluids, which cannot be described by the Herschel–Bulkley formula, is that of *viscoelastic* fluids. These materials have property similar to both a viscous liquid and an elastic solid. Indeed, the deformation is not anymore permanent, as in usual fluids, and depends on both viscous and elastic contributions. When a constant stress τ is applied the deformation of a viscoelastic fluid increases with time, but when the applied stress is removed, the fluid tends to recover its original configuration (similarly to elastic solids). Polymer solutions usually experience a viscoelastic behaviour, another culinary example being pizza dough: when softly pressed it deforms, but when the pressure is removed the original shape is recovered. However, if the dough is strongly deformed, we can rearrange it in a new stable configuration similarly to what happens in fluids. Memory and elastic effect are difficult to model, and typically require information about the microstructure deformation.

In some applications, complex fluids can be successfully modelled just by considering that their response is related to the memory of the deformation rate history; in other words, they have a time-dependent viscosity if exposed to a constant value of the shear rate. Two main kind of such fluids can be identified: *thixotropic* fluids whose effective viscosity decreases with the accumulated strain and *rheopectic* fluids, whose effective viscosity increases with the accumulated strain. A classic example of a fluid characterised by a thixotropic behaviour is painting whose apparent viscosity increases when the deformation rate reduces in order to better adhere to a surface. Rheopectic fluids are less common, and an example is the synovial fluid in our knees, whose property facilitates the absorption of shocks. Thixotropic and rheopectic fluids are usually modelled by a time-dependent viscosity, function of a scalar parameter that represents the evolution of their microstructure.

1.2 Macroscopic Approaches

1.2.1 Eulerian/Eulerian Methods

Inelastic Shear-Thinning/Thickening Fluids

Shear-thinning and shear-thickening are possibly the simplest non-Newtonian behaviours of fluids, when the viscosity μ decreases and increases under shear, i.e. $\mu = \mu(\gamma)$; these behaviours are only rarely observed in pure materials, but can often occur in suspensions. Despite its simplicity, this behaviour is able to capture the main effects induced by a microstructure in many applications. Several models have been developed to describe these fluids, an example of *shear-thinning* model being the Carreau law, usually used to describe generalised fluid where viscosity depends upon shear rate. The model is able to properly describe pseudoplastic fluid viscosity for many engineering application [3], and assumes an isotropic viscosity proportional to some power of the shear rate [4]:

$$\frac{\mu}{\mu_0} = \frac{\mu_\infty}{\mu_0} + \left(1 - \frac{\mu_\infty}{\mu_0}\right)\left[1 + (\lambda\dot{\gamma})^2\right]^{\frac{n-1}{2}}. \tag{1.3}$$

In the previous relation, μ is the viscosity, μ_0 and μ_∞ the zero and infinite shear rate viscosities, λ the relaxation time and $n < 1$ the power index; the second invariant of the strain-rate tensor $\dot{\gamma}$ can be determined as $\dot{\gamma} = \sqrt{2\, S_{ij} : S_{ij}}$, where $S_{ij} = \frac{1}{2}\left(\frac{\partial u_i}{\partial x_j} + \frac{\partial u_j}{\partial x_i}\right)$. At low shear rate ($\dot{\gamma} \ll 1/\lambda$) a Carreau fluid behaves as Newtonian, while at high shear rate ($\dot{\gamma} \gg 1/\lambda$) as a power-law fluid. For *shear-thickening* fluid a simple power-law model is frequently used,

$$\frac{\mu}{\mu_0} = \mathcal{M}\dot{\gamma}^{n-1}, \tag{1.4}$$

which reproduces a monotonic increase of the viscosity with the local shear rate for $n > 1$. The constant $\mathcal{M}$ is called the consistency index and indicates the slope of the viscosity profile. More details on the Carreau and power-law models can be found in Ref. [4].

From a numerical point of view, implementation of a shear dependent viscosity is often straightforward; however, high variations of viscosity may result in significantly time step constraint when explicit schemes are used, and disrupt the solution technique usually used to solve implicitly the viscous terms in the momentum equation. Indeed, the diffusive term cannot be reduced to a constant coefficient Laplace operator since the viscosity is now a function of space. Dodd and Ferrante [5] have introduced a splitting operator technique able to overcome this drawback, initially derived for the pressure Poisson equation; however, this splitting approach can easily be extended to the Helmholtz equation resulting from an implicit (or semi-implicit) integration of the diffusive terms as well. In particular, the viscosity is split in a constant part and in a space-varying component, i.e. $\mu(\boldsymbol{x}) = \mu_0 + \mu'(\boldsymbol{x})$, and the resulting diffusive term split consequently in a constant coefficients operator that can be treated implicitly, and in a variable coefficients operator which can be treated explicitly.

Viscoelastic Fluids

Viscoelasticity is the property of materials that exhibit both viscous and elastic characteristics when undergoing deformation. Unlike purely elastic substances, a viscoelastic substance has an elastic component and a viscous component, and the latter gives the substance a strain rate dependence on time. Viscoelastic materials have been often modelled in the past as linear combinations of springs and dashpots; famous examples are the Maxwell model, represented by a purely viscous damper

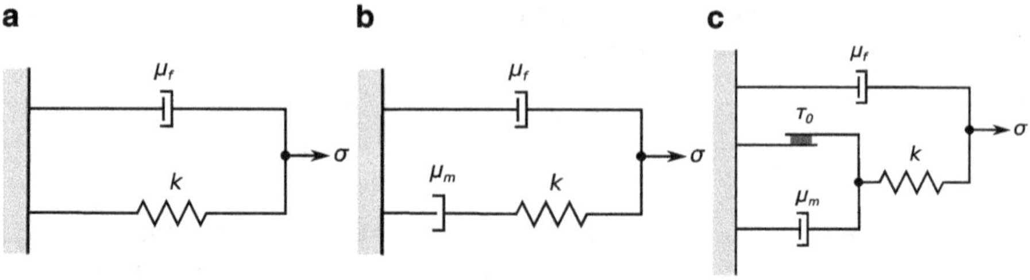

Fig. 1.2 Sketch of the mechanical model of the (**a**) Kelvin–Voigt model, (**b**) Oldroyd-B viscoelastic model and of the (**c**) elastoviscoplastic fluid proposed by Saramito

and a purely elastic spring connected in series, the Kelvin–Voigt model (Fig. 1.2a), made by a Newtonian damper and a Hookean elastic spring connected in parallel, and the standard linear solid model, which combines the Maxwell model and a Hookean spring in parallel. In 1950 Oldroyd proposed a famous viscoelastic model [6], often called Oldroyd-B model (Fig. 1.2b), where the fluid is assumed to consist of dumbbells, beads connected elastic springs. In a frame-independent form, it can be expressed in terms of the upper-convected derivative of the stress tensor

$$\lambda\left(\frac{\partial\tau_{ij}}{\partial t}+u_k\frac{\partial\tau_{ij}}{\partial x_k}-\tau_{kj}\frac{\partial u_i}{\partial x_k}+\tau_{ik}\frac{\partial u_j}{\partial x_k}\right)+\tau_{ij}=2\eta_m S_{ij}, \tag{1.5}$$

where τ_{ij} is the stress tensor, λ the relaxation time, η_m the material viscosity and S_{ij} the rate of strain tensor. Although the model provides good approximations of viscoelastic fluids in shear flow, it has an unphysical singularity in extensional flow, where the dumbbells are infinitely stretched [7]. In order to overcome this problem, the finite elastic non-linear elastic (FENE) model has been proposed; it consists of a sequence of beads with non-linear springs, with forces governed by the inverse Langevin function. Subsequently, the finite elastic non-linear extensibility-Peterlin (FENE-P) model has been developed, by extending the dumbbell version of the FENE model and assuming the Peterlin statistical closure for the restoring force. The model is suited for numerical simulations, since it removes the need of statistical averaging at each grid point at any instant in time, and because the polymer suspension is treated as a continuum and its dynamics represented by an evolution equation of the phase-averaged configuration tensor C_{ij}, a symmetric second-order tensor defined as $C_{ij} =< q_i q_j >$, where q_i are the components of the end-to-end vector for a polymer molecule. The evolution of the polymer conformation is governed by the balance of stretching and restoring forces in an Eulerian framework, such that the transport equation for the conformation tensor can be expressed as

$$\frac{\partial C_{ij}}{\partial t}+u_k\frac{\partial C_{ij}}{\partial x_k}=C_{kj}\frac{\partial u_i}{\partial x_k}+C_{ik}\frac{\partial u_j}{\partial x_k}-\tau_{ij}, \tag{1.6}$$

where τ_{ij} is the polymeric stress tensor, defined as

$$\tau_{ij} = \frac{1}{\lambda}\left(\frac{C_{ij}}{1-\frac{C_{kk}}{L^2}} - \delta_{ij}\right), \tag{1.7}$$

with L the maximum polymer extensibility, δ_{ij} the Kronecker delta and λ the polymer relaxation time. A non-dimensional number can be defined based on the polymer relaxation time λ, which is usually called Weissenberg number We, and is defined as

$$We = \frac{\lambda U^{ref}}{L^{ref}}. \tag{1.8}$$

The previous transport equation is a balance between the advection of the configuration tensor on the left-hand side, and the stretching and relaxation of the polymer, represented by the first two terms and the last one on the right-hand side, respectively. Polymer stresses result from the action of polymer molecules to keep their configuration close to the highest entropic state, i.e. the coiled configuration (see Refs. [3, 8]). The polymer stress is then added to the momentum equation, the Navier–Stokes equation for an incompressible flow in the case of polymer solutions.

The numerical solution of Eq. (1.6) is cumbersome, and many researchers showed that the numerical solution of a viscoelastic fluid is unstable, especially in the case of high Weissenberg numbers, since any disturbance amplifies over time [9–11]. Indeed, the numerical solution of this equation can easily diverge and lead to the numerical breakdown since it is an advection equation without any diffusion term [12]. One of the earliest solution to this problem has been to introduce a global *artificial diffusivity* (AD) to the transport equation of the conformation tensor [11, 13, 14] by adding to the right-hand side of Eq. (1.6) the term $k\frac{\partial^2 C_{ij}}{\partial x_k \partial x_k}$, where k is a coefficient. Subsequently, global AD was replaced by local AD, where the diffusion is applied only to locations where the tensor C_{ij} experiences a loss of positiveness. Recently, researchers started to use high-order *weighted essentially non-oscillatory* (WENO) schemes [15] for the advection terms in the equation. WENO scheme are non-linear finite-volume or finite-difference methods which can numerically approximate solutions of hyperbolic conservation laws and other convection dominated problems with high-order accuracy in smooth regions and essentially non-oscillatory transition for solution discontinuities. Apart from that, the governing differential equations can be solved on a staggered grid using a second-order central finite-difference scheme. This methodology has been proved to work properly by Sugiyama et al. [16] and also successfully used in Refs. [17–19]. A comprehensive review on the properties of different numerical schemes for the advection terms is reported by Min et al. [9].

An alternative methodology to overcome such problems is the so-called *log-representation of the conformation tensor* that ensures the positive-definiteness of the tensor C_{ij}, even at high Weissenberg number [20–23]; this consists in solving equivalent transport equations for $A_{ij} = \log C_{ij}$, instead of the ones for the conformation tensor C_{ij}. Following the notation used in Ref. [23], we write $\boldsymbol{A} = \log \boldsymbol{C} = \boldsymbol{R} \log \boldsymbol{D} \boldsymbol{R}^T$, where $\boldsymbol{D}$ is a diagonal matrix containing the eigenvalues of $\boldsymbol{C}$ and $\boldsymbol{R}$ an orthogonal matrix containing the eigenvectors of $\boldsymbol{C}$. First, we define a decompose of the velocity gradient such that $(\nabla \boldsymbol{u})^T = \boldsymbol{\Omega} + \boldsymbol{B} + \boldsymbol{N}\boldsymbol{C}^{-1}$; note that, $\boldsymbol{\Omega}$ and $\boldsymbol{N}$ are antisymmetric and that $\boldsymbol{B}$ is traceless, symmetric and commutes with $\boldsymbol{C}$. Next, we introduce four new matrices, $\widetilde{\boldsymbol{M}} = \boldsymbol{R}^T (\nabla \boldsymbol{u})^T \boldsymbol{R}$, $\widetilde{\boldsymbol{\Omega}} = \boldsymbol{R}^T \boldsymbol{\Omega} \boldsymbol{R}$, $\widetilde{\boldsymbol{B}} = \boldsymbol{R}^T \boldsymbol{B} \boldsymbol{R}$ and $\widetilde{\boldsymbol{N}} = \boldsymbol{R}^T \boldsymbol{N} \boldsymbol{R}$, and rewrite the decomposition of the velocity gradient as $\widetilde{\boldsymbol{M}} = \widetilde{\boldsymbol{\Omega}} + \widetilde{\boldsymbol{B}} + \widetilde{\boldsymbol{N}}\boldsymbol{D}^{-1}$. Note that, in order to ensure a unique decomposition $\widetilde{\boldsymbol{B}}$ is diagonal, while $\widetilde{\boldsymbol{\Omega}}$ and $\widetilde{\boldsymbol{N}}$ are antisymmetric matrices. $\widetilde{\boldsymbol{N}}$ and $\widetilde{\boldsymbol{\Omega}}$ can then be found by satisfying the equations

$$\widetilde{\boldsymbol{B}} + \frac{1}{2}\left(\widetilde{\boldsymbol{N}}\boldsymbol{D}^{-1}\boldsymbol{D}^{-1}\widetilde{\boldsymbol{N}}\right) = \frac{1}{2}\left(\widetilde{\boldsymbol{M}} + \widetilde{\boldsymbol{M}}^T\right) \tag{1.9}$$

and

$$\widetilde{\boldsymbol{\Omega}} + \frac{1}{2}\left(\widetilde{\boldsymbol{N}}\boldsymbol{D}^{-1} + \boldsymbol{D}^{-1}\widetilde{\boldsymbol{N}}\right) = \frac{1}{2}\left(\widetilde{\boldsymbol{M}} - \widetilde{\boldsymbol{M}}^T\right). \tag{1.10}$$

Finally, the original transport equation for $\boldsymbol{C}$ is rewritten into an equivalent one for $\boldsymbol{A}$

$$\frac{\partial \boldsymbol{A}}{\partial t} + (\boldsymbol{u} \cdot \nabla)\boldsymbol{A} - \left(\boldsymbol{\Omega}\boldsymbol{A} - \boldsymbol{A}\boldsymbol{\Omega}\right) - 2\boldsymbol{B} = \frac{1}{We}\left(e^{-\boldsymbol{A}} - \boldsymbol{I}\right)\frac{\alpha}{We}\left(e^{-\boldsymbol{A}} - \boldsymbol{I}\right)^2, \tag{1.11}$$

where $e^{\boldsymbol{A}} = \boldsymbol{R}\boldsymbol{D}\boldsymbol{R}^T$ and $e^{-\boldsymbol{A}} = \boldsymbol{R}\boldsymbol{D}^{-1}\boldsymbol{R}^T$.

Plastic Effects

Viscoplasticity is a theory in continuum mechanics that describes the rate-dependent inelastic behaviour of solids. Rate dependence in this context means that the deformation of the material depends on the rate at which loads are applied. The first viscoplastic rheological model based on yield stress (stress at which a material begins to deform plastically) was proposed by Schwedoff [24] as a plastic viscoelastic version of the Maxwell model:

$$\begin{cases} \dot{\varepsilon} = 0 & \text{if } \tau \leq \tau_0 \\ \lambda \dfrac{d\tau}{dt} + (\tau - \tau_0) = \eta_m \dot{\varepsilon} & \text{if } \tau > \tau_0 \end{cases}, \tag{1.12}$$

where $\dot{\varepsilon}$ is the rate of deformation, η_m the solid viscosity and τ_0 the yield stress. The previous model states that when the stress τ is less than the yield stress τ_0, the material is completely solid, and the rate of deformation is zero, while when the stress is greater than the yield value, it behaves as a fluid. Note that, at steady state, we obtain $\tau = \tau_0 + \eta_m \dot{\varepsilon}$. Bingham [25] proposed a similar model:

$$\max\left(0, \frac{|\tau| - \tau_0}{|\tau|}\right)\tau = \eta_m \dot{\varepsilon}, \tag{1.13}$$

which can be rewritten as

$$\begin{cases} \dot{\varepsilon} = 0 & \text{if } |\tau| \leq \tau_0 \\ \dfrac{|\tau| - \tau_0}{|\tau|}\tau = \eta_m \dot{\varepsilon} & \text{if } |\tau| > \tau_0 \end{cases}. \tag{1.14}$$

Bingham model is exactly equivalent to the steady case of the one proposed by Schwedoff for positive rates of deformation. In 1947, Oldroyd modified the Bingham model and proposed the following constitutive equation [26]:

$$\begin{cases} \tau = \mu \varepsilon & \text{if } |\tau| \leq \tau_0 \\ \dfrac{|\tau| - \tau_0}{|\tau|}\tau = \eta_m \dot{\varepsilon} & \text{if } |\tau| > \tau_0 \end{cases}, \tag{1.15}$$

which combines the yielding criterion with a linear Hookean elastic behaviour before yielding and a viscous behaviour after yielding. Differently from the previously described models, here when the stress is less than the yield value, the material is not completely rigid. The numerical simulation of a Bingham fluid is not a straightforward task, because of the mathematical non-smoothness of the model and the indeterminacy of the stress tensor below the yield stress threshold [27]. Two kind of solution methods has been proposed in the literature, the *regularisation approach* [28–32] and the *augmented Lagrangian* [33–40] algorithm. The former solution method consists in modifying the constitutive equation in order to avoid the numerical and mathematical complexities, while the second consists in solving the whole problem as a minimisation of a functional with a step descent Uzawa algorithm [41]. In other words, the former method consists in solving modified equations which are computationally more permissive, while the second solves the actual yield stress model, but is computationally much more expensive. Among the first category of regularised approaches, in 1987, Papanastasiou [29] developed a modified constitutive relation for Bingham plastics whose main feature is that the tracking of the yield surfaces is completely eliminated. The model assumes

$$\tau = \left[\mu + \frac{\tau_0}{|\dot{\gamma}|}\left(1 - e^{-\mathcal{M}|\dot{\gamma}|}\right)\right]\dot{\gamma}, \tag{1.16}$$

where $\mathcal{M}$ is a constant that, when chosen sufficiently big, provides a quick stress growth even at relatively low strain rates. This behaviour is consistent with materials in their practically unyielded state, i.e. plastic material that exhibits little or no deformation up to a certain level of stress determined by the yield stress. Due to the fast growing stress, this model has been sometimes used to represent fluids that exhibit extreme shear-thickening behaviour.

Motivated by experimental observations, where yield-stress fluid have an elastic response, Saramito [42, 43] combined the Bingham and Oldroyd models, and proposed a model for elastoviscoplastic fluids (Fig. 1.2c)

$$\lambda \frac{d\tau}{dt} + \max\left(0, \frac{|\tau| - \tau_0}{|\tau|}\right)\tau = \eta_m \dot{\varepsilon}, \tag{1.17}$$

where the total stress is again $\sigma = \eta\dot{\varepsilon} + \tau$. While Schwedoff proposed a rigid behaviour when $|\tau| \leq \tau_0$ and Oldroyd a change of model when reaching the yield value, Saramito assures a continuous change from a solid to a fluid behaviour of the material. The mechanical model is composed by a friction element inserted in the Oldroyd viscoelastic model: at stresses below the yield stress, the friction element remains rigid, and the whole system predicts only recoverable Kelvin–Voigt viscoelastic deformation due to a spring and a viscous element η in parallel. Note that, the elastic behaviour $\tau = \mu\varepsilon$ is expressed in differential form and that $\mu = \eta_m/\lambda$ is the elasticity of the spring. As soon as the strain energy exceeds the level required by the von Mises criterion [44], the friction element breaks allowing deformation of another viscous element (η_m), and the material is described by the Oldroyd viscoelastic model.

After expanding the time derivative in the previous equation, the general Saramito model can be written as

$$\lambda\left(\frac{\partial \tau_{ij}}{\partial t} + u_k \frac{\partial \tau_{ij}}{\partial x_k} - \tau_{kj}\frac{\partial u_i}{\partial x_k} + \tau_{ik}\frac{\partial u_j}{\partial x_k}\right) + \max\left(0, \frac{|\tau_{ij}^d| - \tau_0}{|\tau_{ij}^d|}\right)\tau_{ij} = 2\eta_m S_{ij}, \tag{1.18}$$

where $\tau_{ij}^d = \tau_{ij} - \frac{1}{N}\tau_{kk}\delta_{ij}$ is the deviatoric part of τ_{ij}, with $N = 2$ or 3 the dimension of the problem at hand, and δ_{ij} the Kronecker delta. Note that for yield stress $\tau_0 = 0$, the Oldroyd-B model is recovered. A non-dimensional number can be defined based on the field stress τ_0, which is usually called Bingham number Bn, and is defined as

$$Bn = \frac{\tau_0 L^{ref}}{\mu U^{ref}}. \tag{1.19}$$

The yield stress value of certain materials, for example, liquid metals, is a function of the temperature [45, 46]. Indeed, while in crystal solids the yielding involves bond switch in an orderly manner, in metallic glasses this should be determined by bond breakage [47, 48]. By computing separately the mechanical and thermal energies that are required for bond breakage, a simple relation between the yield stress and the temperature can be obtained: $\tau_0 = 50\rho/M\left(T_g - T\right)$, where T is the ambient temperature, ρ the density, M the molar mass and T_g the glass transition temperature. Guan et al. [49] used molecular dynamic simulations and found that the yield strength and the temperature are well correlated through a simple expression

$$\frac{T}{T_0} + \left(\frac{\tau}{\tau_0}\right)^2 = 1, \tag{1.20}$$

where T_0 and τ_0 are viscosity-dependent, normalised constants.

The numerical solution of Eq. (1.18), similarly to Eq. (1.6), may be cumbersome. The use of high-order WENO schemes for the advection terms in the equation is suggested to have high-order accuracy in smooth regions and essentially non-oscillatory transition for solution discontinuities [50, 51]. The previously discussed log-representation of the equation can be used as well.

Fluid–Structure Interaction

A fully Eulerian formulation of a fluid structure problem can be obtained with a technique similar to the one discussed in the previous sections. Indeed, we can consider fluid and solid motion governed by the conservation of momentum and the incompressibility constraint:

$$\frac{\partial u_i^f}{\partial t} + \frac{\partial u_i^f u_j^f}{\partial x_j} = \frac{1}{\rho}\frac{\partial \sigma_{ij}^f}{\partial x_j}, \tag{1.21a}$$

$$\frac{\partial u_i^f}{\partial x_i} = 0, \tag{1.21b}$$

$$\frac{\partial u_i^s}{\partial t} + \frac{\partial u_i^s u_j^s}{\partial x_j} = \frac{1}{\rho}\frac{\partial \sigma_{ij}^s}{\partial x_j}, \tag{1.21c}$$

$$\frac{\partial u_i^s}{\partial x_i} = 0, \tag{1.21d}$$

where the suffixes ^{f} and ^{s} are used to distinguish the fluid and solid phase. In the previous set of equations, σ_{ij} is the Cauchy stress tensor. The kinematic and dynamic interactions between the fluid and solid phases are determined by enforcing the continuity of the velocity and traction force at the interface between the two phases

$$u_i^f = u_i^s, \tag{1.22a}$$

$$\sigma_{ij}^f n_j = \sigma_{ij}^s n_j, \tag{1.22b}$$

where n_i denotes the normal vector. The problem at hand can be solved numerically by using the so-called one-continuum formulation [52], where only one set of equations is solved over the whole domain. This is achieved by introducing a monolithic velocity vector field u_i valid everywhere obtained by a volume averaging procedure [53, 54], i.e.

$$u_i = \left(1 - \phi^s\right) u_i^f + \phi^s u_i^s, \tag{1.23}$$

where ϕ^s is an indicator function expressing the local solid volume fraction. Thus, we can write the stress in a mixture form as

$$\sigma_{ij} = \left(1 - \phi^s\right) \sigma^f_{ij} + \phi^s \sigma^s_{ij}. \tag{1.24}$$

A fully Eulerian formulation is obtained after properly defining the fluid and solid Cauchy stress, with examples given in [16–18, 55].

1.3 Microscopic Approaches

In this section we will discuss approaches used to perform interface-resolved simulations of the intrusions defining the microstructures and thus at the origin of the non-Newtonian behaviours described above. We will consider rigid and deformable particles, as well as two-fluid systems. Indeed, recent developments in computational power and efficient numerical algorithms have allowed the scientific community to numerically resolve the microstructure of suspensions in fluids.

1.3.1 Eulerian/Lagrangian Methods

Eulerian/Lagrangian methods are often used to simulate suspension in fluids, and are also called immersed boundary methods (IBM). The main feature of this method is that the numerical grid does not need to conform to the geometry of the object, which is instead replaced by a body force distribution f that mimics the effect of the body on the fluid and restores the desired velocity boundary values on the immersed surfaces. To do that, two separate grids coexist, the Eulerian fixed grid where the flow is solved, and the Lagrangian grid representing the moving immersed boundary (see Fig. 1.3a); a singular force distribution at the Lagrangian positions is first determined and then applied to the flow equations in the Eulerian frame via a regularised Dirac delta function.

The primary advantage of the IB method is associated with the simplification of the grid generation task: indeed, grid complexity and quality are not significantly affected by the complexity of the geometry. The advantage of the IB method becomes eminently clear for flows with moving boundaries, where the process of generating a new grid at each time step is avoided, because the grid remains stationary and non-deforming. A drawback of this approach is that the grid lines are not aligned with the body surface, so in order to obtain the required resolution, higher number of grid points may be required. Many IBMs have been created so far, which differ in the way the immersed boundary force is computed [56–60]. The different methods are often grouped in two categories, continuous and direct forcing: in the first approach the forcing is incorporated into the continuous equations before discretisation, whereas in the second approach the forcing is introduced after

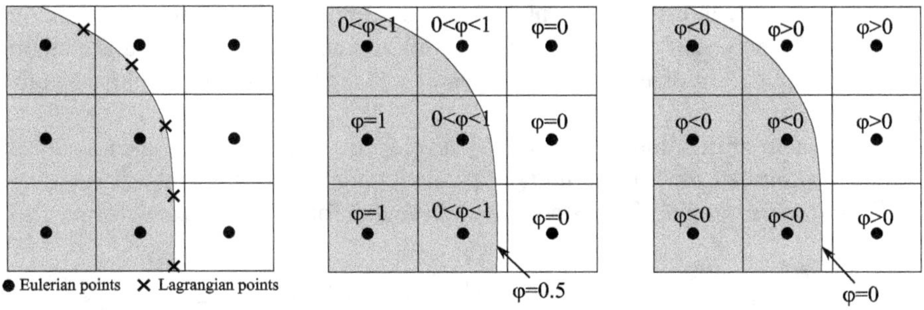

Fig. 1.3 (**a**) Sketch of an immersed surface (grey) and of the Eulerian and Lagrangian grids used in the immersed boundary method. (**b**) Sketch of the volume of fluid method. (**c**) Sketch of the level-set method

the equations are discretised. The continuous forcing approach is very attractive for flows with immersed deforming boundaries, whereas the direct one is more commonly used to simulate rigid boundaries.

The original IB method was developed by Peskin [61] for the coupled simulation of blood flow and muscle contraction in a beating heart and is generally suitable for flows with immersed elastic boundaries. The IB is represented by a set of elastic fibres and the location of these fibres is tracked in a Lagrangian fashion by a collection of massless points moving with the local fluid velocity, i.e. the coordinate $\boldsymbol{X}_k$ of the k-th Lagrangian point is governed by the equation

$$\frac{\partial \boldsymbol{X}_k}{\partial t} = \boldsymbol{u}\left(\boldsymbol{X}_k, t\right), \tag{1.25}$$

where $\boldsymbol{u}$ is the local fluid velocity. The stress (denoted by $\boldsymbol{F}$) is related to deformation of these elastic fibres by a constitutive law, such as the Hooke's law, and the effect of the IB on the surrounding fluid is captured by transmitting the fibre stress to the fluid through a localised forcing term in the momentum equations

$$\boldsymbol{f}\left(\boldsymbol{x}, t\right) = \sum_k \boldsymbol{F}_k\left(t\right) \delta\left(\left|\boldsymbol{x} - \boldsymbol{X}_k\right|\right), \tag{1.26}$$

where δ is the Dirac delta function. Because the location of the fibres does not generally coincide with the nodal points of the Cartesian grid, the forcing is distributed over a band of cells around each Lagrangian point and added on the momentum equations of the surrounding nodes. Thus, the sharp delta function is replaced by a smoother distribution function, denoted here by d, so that the forcing at any grid point $x_{i,j}$ is given by

$$\boldsymbol{f}\left(\boldsymbol{x}_{i,j}, t\right) = \sum_k \boldsymbol{F}_k\left(t\right) d\left(\left|\boldsymbol{x} - \boldsymbol{X}_k\right|\right). \tag{1.27}$$

The fibre velocity in Eq. (1.25) is also obtained through the use of the same smooth function. The choice of the distribution function d is a key ingredient in this method, and several different distribution functions have been derived and employed in the past [61–64].

In the same spirit, Goldstein et al. [65] developed another model, called feedback forcing, to simulate the flow around rigid and moving bodies, where the effect of the body on the surrounding flow is modelled through a forcing term of the form

$$\boldsymbol{F}(\boldsymbol{x},t)=\alpha\int_0^t\left[\boldsymbol{u}(\boldsymbol{x},\tau)-\boldsymbol{V}(\boldsymbol{x},\tau)\right]d\tau+\beta\left[\boldsymbol{u}(\boldsymbol{x},t)-\boldsymbol{V}(\boldsymbol{x},t)\right], \tag{1.28}$$

where the coefficients α and β are selected to best enforce the boundary condition at the immersed solid boundary, whose velocity is V. The above relation is a feedback to the velocity difference $\boldsymbol{u}-\boldsymbol{V}$ and behaves in such a way to enforce $\boldsymbol{u}=\boldsymbol{V}$ on the immersed boundary. Indeed, the first term on the right-hand side of the equation tends to annihilate the difference between $\boldsymbol{u}$ and $\boldsymbol{V}$, whereas the second term can be interpreted as the resistance opposed by the surface element to assume a velocity $\boldsymbol{u}$ different from $\boldsymbol{V}$. In an unsteady flow the magnitude of α must be large enough so that the restoring force can react with a frequency which is bigger than any frequency in the flow; however, big values of α and β render the forcing equation stiff and its time integration requires very small time steps. The method has been used to simulate flexible filaments as well [66, 67]. Even if the original intent behind Eq. (1.28) is to provide feedback control of the velocity near the surface, from a physical point of view it can also be interpreted as a damped oscillator [68] with frequency $1/(2\pi)\sqrt{\alpha}$ and damping coefficient $-\beta/\left(2\sqrt{\alpha}\right)$.

Immersed Boundary Methods for Suspensions of Rigid Particles

Uhlmann [56] proposed a computationally efficient numerical method based on the IBM to simulate suspension of rigid particles. Firstly, the surface Γ of the immersed surface delimiting the body is discretised using N markers, called Lagrangian points $\boldsymbol{X}$; note that, in general they do not correspond to the grid nodes $\boldsymbol{x}$. The solution of the incompressible Navier–Stokes is based on the fractional-step method [69]. Indeed, a simple prediction step is first performed, without taking into account the immersed object. The obtained velocity field $\boldsymbol{u}^*$ is then interpolated (with an interpolator operator $\mathcal{I}$) onto the embedded geometry Γ,

$$\boldsymbol{U}^*=\mathcal{I}\left(\boldsymbol{u}^*\right). \tag{1.29}$$

The values of $\boldsymbol{U}^*$ are used to determine a distribution of singular forces along the boundary of Γ that restore the prescribed boundary values $\boldsymbol{U}^\Gamma$ as

$$\boldsymbol{F}^*=\frac{\boldsymbol{U}^\Gamma-\boldsymbol{U}^*}{\Delta t}. \tag{1.30}$$

The force field defined over Γ is then transformed into a body force distribution applied to the Eulerian grid using a convolution operator C

$$f^* = C\left(F^*\right). \tag{1.31}$$

The momentum conservation equation is then solved again with the computed volume force field added as a source term and the time advancement step is completed with the usual solution of the pressure Poisson equation and the projection step where velocity and pressure are corrected to ensure mass conservation. Note that, this procedure is common to the most modern IB methods, and the step that defines the method is the way in which the operators $\mathcal{I}$ and C are built: in particular, the interpolation and spreading operations are based on the regularised Dirac delta function by Roma et al. [70], which extends over three grid cells in all coordinate directions.

The desired velocity $\boldsymbol{U}^{\Gamma}$ at a location $\boldsymbol{X}$ on the interface between the fluid and the immersed boundary is given by the rigid-body motion of the solid object:

$$\boldsymbol{U}^{\Gamma} = \boldsymbol{u}^{c} + \boldsymbol{\omega}^{c} \times \boldsymbol{r}, \tag{1.32}$$

where $\boldsymbol{r} = \boldsymbol{X} - \boldsymbol{x}^{c}$ is the position vector relative to the particle centroid, $\boldsymbol{u}^{c}$ is the translational velocity of the particle centroid and $\boldsymbol{\omega}^{c}$ is the angular velocity of the particle. The translational and angular velocities of a particle are described by the Newton–Euler equations, which for a sphere reduce to

$$\rho_p \mathcal{V}_p \frac{d\boldsymbol{u}^{c}}{dt} = \int_{\delta\mathcal{V}_p} \tau \cdot \boldsymbol{n}\, dA + \left(\rho_p - \rho_f\right) \mathcal{V}_p \boldsymbol{g} - \mathcal{V}_p \nabla p + \boldsymbol{F}_c, \tag{1.33}$$

and

$$I_p \frac{d\boldsymbol{\omega}^{c}}{dt} = \int_{\delta\mathcal{V}_p} \boldsymbol{r} \times (\tau \cdot \boldsymbol{n})\, dA + \boldsymbol{T}_c. \tag{1.34}$$

In the previous two equations, Eqs. (1.33) and (1.34), ρ_p is the density of the particle, $\mathcal{V}_p$ its volume ($4/3\pi R^3$ for a sphere with radius R), τ the fluid stress tensor, $\boldsymbol{n}$ the outward-pointing unit normal at the surface $\delta\mathcal{V}_p$ of the particle, $\boldsymbol{g}$ the gravitational acceleration and I_p the moment of inertia of the particle ($2/5\rho_p\mathcal{V}_p R^2$ for a solid sphere). $\boldsymbol{F}_c$ and $\boldsymbol{T}_c$ represent the force and torque acting on the particle as a result of collisions and contact with other particles or solid walls.

Breugem [59] proposed two major improvements to the method discussed above. The first is the so-called *multidirect forcing scheme*. The use of a regularised Dirac delta function for the interpolation and spreading operations results in a diffuse distribution of the IBM force around the interface and because of that, the influence region of neighbouring Lagrangian points overlaps. Eulerian grid points in the overlap region are used to enforce the boundary value multiple times; thus, the resulting forcing is perturbed and the final distribution of the IBM force may not properly enforce the desired boundary condition. The remedy for this problem is to iteratively determine the IBM forces on the relevant Eulerian grid points such that they collectively enforce the desired boundary condition at the different Lagrangian points [71, 72]. The second improvement suggested by Breugem is the *inward retraction of Lagrangian grid*. The delta function of Roma et al. [70] has a width of three grid cells, and because of that the (outer) radius of the particle actually increases from R to $R + 3\Delta x/2$; this effect results in an increase of the particle drag force, which is partially balanced by an overall permeability of the particle due to a non-perfect boundary condition. As shown by Breugem, the first inaccuracy (increase in the effective radius) is stronger than the other one, and the suggested solution is to slightly retract the Lagrangian points from the surface towards the interior of the particle [73, 74].

In Eqs. (1.33) and (1.34), $\boldsymbol{F}_c$ and $\boldsymbol{T}_c$ are the force and torque acting on the particle as a result of collisions and contact with other particles or solid walls. A recent model for particle–particle and particle–wall interactions in interface-resolved simulations of particle-laden flows is described by Costa et al. [75]. The model consists of three different interactions: long- and short-range hydrodynamic interactions, and solid–solid contact. The long-range interactions are directly obtained by the immersed boundary method, while the short-range ones are based on a lubrication model employed when the gap between particles is below the grid size, and is based on asymptotic expansions of the analytical solution for canonical lubrication interactions between spheres in the Stokes regime. Roughness effects can be accounted for as well. This correction is applied until the particles reach contact when a linear soft-sphere collision model is used. Note that, the approach described above can be extended to particles of different shapes [76], and that alternative collision models can be found in the literature, for example, Refs. [77, 78].

Front-Tracking Methods for Suspensions of Deformable Droplets

The so-called front-tracking method is an evolution of the immersed boundary method used to simulate viscous, incompressible, immiscible two-fluid systems, first developed by Unverdi [79] and Tryggvason [80]. In such multiphase problems, the density and viscosity fields of each fluid remain constant, but they are discontinuous across the interface. In order to avoid numerical diffusion or oscillations problems close to the jump, these fluid properties are not advected directly, but instead a Lagrangian grid is created to describe the boundary between the different fluids, which is then moved with the fluid velocity. Therefore, at every time step it is necessary to reset the fluid properties and to do so, an indicator function $\mathcal{F}(\boldsymbol{x})$ is also introduced, equal to 1 inside one fluid and 0 in the other one. This function is constructed from the known position of the interface and is used to evaluate the proper values of density and viscosity at each grid point:

$$\rho(\boldsymbol{x}) = \rho_1 + (\rho_2 - \rho_1)\mathcal{F}(\boldsymbol{x}) \quad \text{and} \quad \mu(\boldsymbol{x}) = \mu_1 + (\mu_2 - \mu_1)\mathcal{F}(\boldsymbol{x}), \tag{1.35}$$

where the suffixes 1 and 2 indicate the two fluids. The jump in the indicator function carried by the interface is spread to the grid points nearest to the interface, in order to ensure that the fluid properties change smoothly across the interface. This generates a grid-gradient field which is zero everywhere except near the interface and has a finite thickness of the order of the mesh size. The spreading of the jump onto the grid is done in such a way that the volume integral of the gradient is conserved, i.e. if $\mathcal{G}(\boldsymbol{x})$ is the gradient of the indicator function evaluated at a stationary grid point $\boldsymbol{x}$, and $\mathcal{D}$ is a distribution function, such as the one introduced by Peskin [81], then

$$\mathcal{G}(\boldsymbol{x}) = \sum_k \mathcal{D}(\boldsymbol{x} - \boldsymbol{X}_k)\, \boldsymbol{N}_k \Delta S_k, \tag{1.36}$$

where $\boldsymbol{N}_k$ is the unit normal vector to the interface element of area ΔS_k whose centroid is at $\boldsymbol{X}_k$. The indicator function is found everywhere by solving the following Poisson equation:

$$\nabla^2 \mathcal{I} = \nabla \cdot \mathcal{G} \tag{1.37}$$

where the right-hand side is computed by simple numerical differentiation.

Since the fluid velocities are computed on the fixed grid and the front moves with the fluid velocities, the velocity of the interface points must be found by interpolation; thus, similarly to Eq. (1.36), to interpolate the velocity on the interface Lagrangian points we use

$$\boldsymbol{U}_k = \sum_i \mathcal{D}(\boldsymbol{x}_i - \boldsymbol{X}_k)\, \boldsymbol{u}_i, \tag{1.38}$$

where the sum is now over the points on the stationary grid in the vicinity of the considered k-th Lagrangian point. Finally, the new position of the interface is found by solving a simple advection equation

$$\frac{d\boldsymbol{X}_k}{dt} = \boldsymbol{U}_k. \tag{1.39}$$

As the front moves, it deforms and stretches, and the resolution along some parts of the interface can become inadequate or overly crowded. To maintain accuracy, either additional elements must be added when the separation among points becomes too large or points must be redistributed to maintain adequate resolution.

In those calculations where the surface tension σ is needed, the magnitude of the surface tension force is obtained from the local curvature $\mathcal{K}$ of the interface: $\boldsymbol{F}_k = \sigma \mathcal{K}_k \boldsymbol{N}_k \Delta S_k$. This force is then distributed onto the grid as

$$\boldsymbol{f}(\boldsymbol{x}) = \sum_k \mathcal{D}(\boldsymbol{x} - \boldsymbol{X}_k)\, \boldsymbol{F}_k. \tag{1.40}$$

Note that, alternative ways have been proposed to calculate the surface tension without having to find the curvature, but only the local tangent defined by the Lagrangian points on the interface [80].

1.3.2 Eulerian/Eulerian Methods

When dealing with moving and deformable boundaries, an alternative approach to the IB front-tracking methods previously discussed are the so-called front-capturing methods, which are fully Eulerian and handle topology changes automatically. A strong advantage of such methods is that they are easier to parallelised and typically achieve higher efficiency than their Lagrangian counterpart. However, interactions between approaching particles and droplets are difficult to control and may depend on the resolution adopted. Eulerian interface representations include essentially the volume of fluid (VOF) [82] and level-set (LS) [83–85] methods, their variants and combination. The VOF method defines different fluids with a discontinuous colour function, and its main advantage is an intrinsic mass conservation; however, it suffers from an inaccurate computation of the interface properties, such as normals and curvatures [86, 87]. Contrary to the VOF, the LS method prescribes the interface through a (Lipschitz-)continuous function which usually takes the form of the signed distance to the interface. Thus, normals and curvatures can be readily and accurately computed, while mass loss/gain may occur since the LS function has no volume information. Furthermore, this approach requires a procedure to reshape the LS into a distance function, i.e. the reinitialisation step. More recently, researchers have started to develop coupled VOF-LS methods [88] in order to overcome the disadvantages of both the techniques.

Volume of Fluids

We introduce an indicator (or colour) function H to identify a given fluid so that $H = 1$ in the region occupied by fluid 1 and $H = 0$ in fluid 2. Considering that the fluid is transported with the flow velocity, we update H in the Eulerian framework by the following advection equation:

$$\frac{\partial H}{\partial t} + \boldsymbol{u} \cdot \nabla H = 0, \tag{1.41}$$

where $\boldsymbol{u}$ is the velocity vector. The cell-averaged value of the indicator function is defined as the volume fraction or volume of fluid (VOF) function within a cell

$$\phi = \frac{1}{\delta V} \int_{\delta V} H dV. \tag{1.42}$$

Thus, the VOF function assumes values $0 \leq \phi \leq 1$ (see Fig. 1.3b). Combining the two previous equations, we obtain the advection equation of the VOF function in the divergence form:

$$\frac{\partial \phi}{\partial t} + \nabla \cdot (\boldsymbol{u} H) = \phi \nabla \cdot \boldsymbol{u}. \tag{1.43}$$

In a conventional VOF method, the interface separating different fluids is piecewise reconstructed for each cell by straight line segments, which are then used to calculate the numerical fluxes necessary to update the VOF function. This geometric reconstruction effectively eliminates the numerical diffusion that smears out the compactness of the transition layer of the interface. Different methodologies have been proposed to accurately recover the exact surface geometry from the discretised VOF function: the simple line interface calculation (SLIC) method [89], the piecewise linear interface calculation (PLIC) [90, 91], the latter being further modified by several authors [92–96]. Another technique is the tangent of hyperbola for interface capturing (THINC) method [97]: this avoids the explicit geometric reconstruction by using a continuous sigmoid function rather than the Heaviside function, thus allowing a completely algebraic description of the interface and enabling the computation of the numerical flux. An improvement was proposed by combining the original THINC method with the first-order upwind scheme in the so-called THINC/WLIC (THINC/weighted linear interface capturing) method [98]. Recently, the method has been further developed in the multi-dimensional THINC (MTHINC) method where the fully multi-dimensional hyperbolic tangent function is used to reconstruct the interface [99, 100]. The numerical fluxes can be directly evaluated by integrating the hyperbolic tangent function which also prevents the numerical diffusion that smears out the interface transition layer. Another advantage of the method is that the normal vector, curvature and approximate delta function can be directly obtained from the derivatives of the function, thus the standard smoothing or convolution techniques used in conventional VOF methods is not required.

The unit normal vector is defined as $\boldsymbol{n} = \boldsymbol{m}/|\nabla\phi|$, being $\boldsymbol{m}$ the gradient of the VOF function, i.e. $\mathbf{m} = \nabla\phi$, which can be computed using the usual Young's approach [90, 91]. For example, in the $2D$ case, first the values of the derivative at the four cell corners indexed as $i \pm \frac{1}{2}, j \pm \frac{1}{2}$ are calculated by the VOF function on its surroundings, for example,

$$\begin{aligned} m^x_{i+\frac{1}{2},j+\frac{1}{2}} &= \frac{\phi_{i+1,j} + \phi_{i+1,j+1} - \phi_{i,j} - \phi_{i,j+1}}{\Delta x_i + \Delta x_{i+1}}, \\ m^y_{i+\frac{1}{2},j+\frac{1}{2}} &= \frac{\phi_{i,j+1} + \phi_{i+1,j+1} - \phi_{i,j} - \phi_{i+1,j}}{\Delta y_j + \Delta y_{j+1}}, \end{aligned} \tag{1.44}$$

and then averaged to find the cell-centre value

$$\begin{aligned} m^x_{i,j} &= \left(m^x_{i-\frac{1}{2},j-\frac{1}{2}} + m^x_{i-\frac{1}{2},j+\frac{1}{2}} + m^x_{i+\frac{1}{2},j-\frac{1}{2}} + m^x_{i+\frac{1}{2},j+\frac{1}{2}}\right), \\ m^y_{i,j} &= \left(m^y_{i-\frac{1}{2},j-\frac{1}{2}} + m^y_{i-\frac{1}{2},j+\frac{1}{2}} + m^y_{i+\frac{1}{2},j-\frac{1}{2}} + m^y_{i+\frac{1}{2},j+\frac{1}{2}}\right). \end{aligned} \tag{1.45}$$

The curvature k is then found by taking the divergence of the normal vector

$$k = -\nabla \cdot \boldsymbol{n}. \tag{1.46}$$

The surface tension force $\boldsymbol{f} = \sigma k \boldsymbol{n} \delta$ can be computed using the continuum surface force (CSF) model [101], where the $1D$ approximate delta function δ is directly approximated by $\delta \approx |\nabla \phi|$. Thus, we obtain

$$\boldsymbol{f} = \sigma k \boldsymbol{n} \delta \approx \sigma k \nabla \phi. \tag{1.47}$$

Finally, the mixture density and dynamic viscosity are simply averaged in terms of the VOF function (similarly to Eq. (1.35)):

$$\rho = \rho_1 \phi + \rho_2 (1 - \phi)) \qquad \text{and} \qquad \mu = \mu_1 \phi + \mu_2 (1 - \phi). \tag{1.48}$$

Due to the non-uniformity of the density, the Poisson equation used to enforce a divergence-free velocity field becomes

$$\nabla \cdot \left(\frac{1}{\rho^{n+1}} \nabla p^{n+1}\right) = \frac{1}{\Delta t} \nabla \cdot \boldsymbol{u}^*, \tag{1.49}$$

which is in an equation with variable coefficients. In order to utilise an efficient FFT-based pressure solver with constant coefficients [5, 102], we use the following splitting of the pressure term [103]:

$$\frac{1}{\rho^k} \nabla p^{n+1} \rightarrow \frac{1}{\rho_0} \nabla p^{n+1} + \left(\frac{1}{\rho^{n+1}} - \frac{1}{\rho_0}\right) \nabla \left(2p^n - p^{n-1}\right), \tag{1.50}$$

where ρ_0 is a constant density equal to the lowest density between the two phase. With this splitting, the Poisson equation can be rewritten as

$$\nabla \cdot \nabla p^{n+1} = \nabla \cdot \left[\left(1 - \frac{\rho_0}{\rho^{n+1}}\right) \nabla \left(2p^n - p^{n-1}\right)\right] + \frac{\rho_0}{\Delta t} \nabla \cdot \mathbf{u}^*. \tag{1.51}$$

Note that, the correction step of the fractional-step method needs to be modified accordingly and that the term $2p^n - p^{n-1}$ is a linear extrapolation consistent with using a second-order scheme to integrate in time the momentum equation, for example, Adams–Bashforth. The methodology can be extended to the Runge–Kutta schemes by following Ref. [104].

Level-Set Method

The level-set function ϕ approximates the signed distance from the interface, thus $\phi = 0$ denotes the interface and $\phi > 0$ or $\phi < 0$ the two different fluids separated by it (see Fig. 1.3c). The motion of the interface is governed by the following transport equation (formally similar to Eq. (1.41)):

$$\frac{\partial \phi}{\partial t} + \boldsymbol{u} \cdot \nabla \phi = 0, \tag{1.52}$$

where $\boldsymbol{u}$ is the flow velocity field. The equation is closed and allows to solve the system of equations in a fully Eulerian fashion. Notwithstanding the formal simplicity of the equation, its numerical solution is challenging. The time integration is often performed by a three-stage total variation diminishing third-order Runge–Kutta scheme [105, 106], while the advection term of the equation is usually discretised with one of the following schemes: the high-order upstream-central (HOUC) scheme [107], the weighted essentially non-oscillatory (WENO) scheme [108], the semi-Lagrangian scheme [109] or the semi-jet scheme [105]. Quantitative comparisons of these schemes in various test cases can be found in Ref. [110]. As reported by Ge et al. [106], for flows involving moderate deformations, the HOUC scheme is usually sufficient and the most efficient, while for more complex flows, the WENO or the semi-Lagrangian/jet schemes combined with grid refinement should be used. Note that, Eq. (1.52) does not need to be solved over the entire computational domain, but only near the $\phi = 0$ value, where the interface is located and the normal and curvature are needed. Thus, in the so-called narrow band approach [111, 112], the level set is computed and stored only around the interface, and fast computation and low memory usage may be achieved.

Although the level-set function is initialised to be a signed distance, this property is lost with time, causing numerical issues in the evaluation of the normal and curvature [83]. These issue requires an additional treatment in order to reshape the level-set function ϕ into a distance function, i.e. $|\nabla \phi| = 1$. This is usually performed by converting it into a time-dependent Hamilton–Jacobi equation [83]

$$\frac{\partial \phi}{\partial \mathcal{T}} + S(\phi_0)(|\nabla \phi| - 1) = 0, \tag{1.53}$$

ϕ_0 being the level-set field before redistancing, $\mathcal{T}$ a pseudo-time and $S(\phi_0)$ the mollified sign function of the original level set. When the steady state solution of the equation is reached, the zero level-set contour is unaltered, while the rest of the field has re-obtained the property of being a signed distance function. In practice, this equation is iterated only for few steps towards its steady state every certain number of time steps. Although an alternative approach exists, i.e. the fast marching method (FMM) [84], the reinitialisation procedure allows the use of high-order schemes and is easy to parallelised; thus, it has been a much more popular choice.

Finally, the unit normal vector $\boldsymbol{n}$ and the local mean curvature κ can be simply computed directly from the LS function as follows:

$$\boldsymbol{n} = \frac{\nabla\phi}{|\nabla\phi|} \qquad \text{and} \qquad \kappa = -\nabla\cdot\mathbf{n}, \tag{1.54}$$

and the body force $\mathbf{f}$ due to surface tension, included in the momentum equation of the Navier–Stokes equation, is expressed as

$$\mathbf{f} = \sigma\kappa\delta(\phi)\mathbf{n}, \tag{1.55}$$

being δ the delta function and σ the surface tension. The density and viscosity vary across the fluid interface, and can be expressed in a mixture form (similarly to Eq. (1.35)) as

$$\rho = \rho_1 H(\phi) + \rho_2\big(1 - H(\phi)\big) \qquad \text{and} \qquad \mu = \mu_1 H(\phi) + \mu_2\big(1 - H(\phi)\big), \tag{1.56}$$

where $H(\phi)$ is the regularised Heaviside function defined such that it is zero inside one fluid and unity in the other one. In order to utilise an efficient FFT-based pressure solver with constant coefficients, the technique described by Dodd and Ferrante [5, 102] and discussed in the previous section can be easily used.

We conclude by noting that, although easy to implement, CSF effectively introduces an artificial spreading of the interface. An alternative approach is the ghost fluid method (GFM) [113–115], which provides a finite-difference discretisation of the gradient operator even if the stencil includes shocks.

Phase-Field Methods

From a macroscopic point of view, the interface between two immiscible fluids can be usually assumed to be sharp, since its thickness is of the order of few nanometres. This aspect motivated the numerical methods discussed above, that, however, need to overcome the difficulties associated with evolving a discontinuous front through Eulerian fields. An alternative approach has been therefore proposed and successfully used for different applications, the so-called phase-field (or diffuse interface) method, where the interface is modelled as a thin layer across which the fluid properties continuously change avoiding discontinuous fields. The original idea can be attributed to Van der Waals who suggested that the interface thickness in a binary mixture is determined by the balance of counteracting weakly non-local terms in the free energy: these, on the one hand, provide diffusion and, on the other hand, a sharpening of the interface. Cahn and Hilliard [116] used this approach in the context of phase separation problems deriving an evolution equation for the concentration field. The thermodynamic consistency of the coupled Cahn–Hilliard/Navier–Stokes model [117] and its ability to handle topological changes are the main reasons that justify the increasing use of phase-field methods [118–122]. The fundamental variable of phase-field models is a scalar $\Phi(x_i, t)$ which represents the relative quantity of one of the two phases, and whose extreme values, $\Phi = \pm 1$, correspond to the two pure fluids. Since the two fluids are immiscible, the flow domain is essentially divided in subdomains containing only one of the two phases, separated by an extremely thin interface where the two substances mix, $-1 < \Phi < 1$. The equation describing the evolution of the phases is

$$\frac{\partial \Phi}{\partial t} + \boldsymbol{u} \cdot \nabla \Phi = \nabla \cdot (M \nabla G), \tag{1.57}$$

where $G = \delta \mathcal{F} / \delta \Phi$ is the thermodynamic chemical potential defined as the functional derivative of the free energy $\mathcal{F}$ with respect to Φ, and M is a proportionality constant called mobility. The left-hand side of Eq. (1.57) represents the convective transport of Φ, whereas the right-hand side the driving force from the chemical potential, ensuring phase separation with the exception of the thin layer constituted by the interface. The mobility M has no intuitive physical meaning and is related to the time scale of the interface dynamics. In the limit of vanishing mobility, we recover pure convection neglecting the inner interface dynamics, whereas for infinite mobility the interface will reach equilibrium immediately and do not vary.

Following the seminal work of Cahn and Hilliard [116], it is possible to write the free energy in the following form:

$$\mathcal{F} = \int \frac{3}{2\sqrt{2}} \sigma \ell \left[\frac{(\Phi^2 - 1)^2}{4\epsilon^2} + \frac{1}{2} |\nabla \Phi|^2 \right] dV, \tag{1.58}$$

with σ the surface tension and ℓ the interface thickness. The form of the first term between brackets in Eq. (1.58) guarantees the presence of two minima of $\mathcal{F}$ for the two pure fluids at $\Phi = \pm 1$. The gradient for the free energy drives therefore the system towards phase separation. However, the second term in the brackets is proportional to the square of $|\nabla \Phi|$. Hence, when reducing the interface thickness, the free energy increases because of the sharpening of the gradient of Φ. The combination of these two terms tends to create an equilibrium interface of finite thickness, mimicking what originally assumed by Van der Waals (see, for example, [120] for more details). Using Eq. (1.58), the chemical potential G can be explicitly written as

$$G = 3\sigma \left(\Phi^3 - \Phi - \ell^2 \nabla^2 \Phi \right) / \left(2\sqrt{2}\,\ell \right), \tag{1.59}$$

which can be directly inserted in Eq. (1.57) to compute the phase field. Eq. (1.57) needs to be coupled with the Navier–Stokes equations (convection) which, in turn, are influenced by the phase field because of surface tension. The surface tension across the relatively thick interface between the two phases can be expressed in terms of the free energy (and chemical potential). From the consideration that, without dissipation, the total energy of a systems needs to be conserved it is possible to find the expression for the momentum forcing arising from the interface (surface tension) [123],

$$\boldsymbol{F}_s = -\delta\mathcal{F}/\delta\boldsymbol{u} = -\phi\nabla G. \tag{1.60}$$

Considering Eqs.(1.57), (1.59) and (1.60), together with the incompressible Navier–Stokes system, we can write the complete Navier–Stokes/Cahn–Hilliard model for the immiscible fluid dynamics. In the following it is presented directly in dimensionless form:

$$\nabla\cdot\boldsymbol{u} = 0 \tag{1.61}$$

$$\frac{\partial\boldsymbol{u}}{\partial t} + \boldsymbol{u}\cdot\nabla\boldsymbol{u} = -\nabla p + \frac{1}{Re}\nabla^2\boldsymbol{u} + \frac{3}{2\sqrt{2}}\frac{1}{\mathrm{We\,Cn}}\left(\Phi^3 - \Phi - \mathrm{Cn}^2\nabla^2\Phi\right)\nabla\Phi \tag{1.62}$$

$$\frac{\partial\Phi}{\partial t} + \boldsymbol{u}\cdot\nabla\Phi = \frac{M^*}{\mathrm{Cn}}\nabla^2\left[\Phi^3 - \Phi - \mathrm{Cn}^2\nabla^2\Phi\right]. \tag{1.63}$$

Here, the equations have been made dimensionless using the typical flow scales, L for length, U for velocity, μ and ρ for dynamic viscosity and density. We have also introduced the Cahn number $\mathrm{Cn} = \ell/L$ as the dimensionless measure of the interface thickness, the Weber number, $\mathrm{We} = \rho U^2 L/\sigma$ as the ratio between convection and surface tension effects, the dimensionless mobility $M^* = 3\,M\sigma/(2\sqrt{2}UL^2)$ measuring the (dimensionless) mobility intensity and the Reynolds number $Re = \rho UL/\mu$. When using the phase-field method it is important to correctly set the Cahn Cn and mobility M^* numbers. While the other dimensionless numbers used above are classical and depend only on the macroscopic properties of the system, the values of Cn and M^* are set also by numerical considerations.

Since the physical thickness of the interface is usually on the nanometre scale, realistic values of the Cahn number for micro- and macro-scale applications may range between 10^{-3} and 10^{-10}. As the interface needs to be resolved using $4 \div 5$ control points for accurate numerical simulations, the use of the realistically tiny values of the Cahn number is unpractical in simulations and an artificial thickening of the interface is necessary. Clearly, a trade-off is necessary when deciding the numerical interface thickness; to this end, it has been proved that the dynamics is not affected by the artificial thickening if the typical interface size is well below the smallest flow length scales. In other words, the Cahn number should be small, yet it can be much larger than the real one, for example, around 10^{-2}; this is the so-called *sharp-interface limit*, [124]. In this limit, any further decrease of the Cahn number does not produce appreciable differences in the macroscopic dynamics.

More difficult is to precisely define or even measure the mobility M [123]. Moreover, it is crucial to determine how the mobility depends on the interface thickness because of its artificial magnification necessary in simulations. Recently, it has been found using matching asymptotic expansions that the optimal scaling to recover the sharp-interface limit is $M^* = \alpha\, Ca^2$ (α an order one constant), see Ref. [120] for more details.

The main advantage of phase-field methods is in their strict thermodynamic derivation and the ability to automatically handle interface topology changes, the former opening also possibilities for direct simulations of phase changes. The main drawback is represented by the high computational costs since it is important to use at least $4 \div 5$ points within the interface thickness which, in turn, needs to be smaller than the flow length scales. The Navier–Stokes/Cahn–Hilliard model has recently been adopted in several studies of immiscible fluids both in laminar and turbulent conditions, for example, Refs. [122, 125–127]. Besides the Cahn–Hilliard formulation, different phase-field methods have been proposed, based on alternative forms of the free energy, see, for example, [128] for compressible flows with phase change for cavitation problems. Similar methods exist also in the Lattice–Boltzmann framework, for more details the reader is referred to [129–131] and the references therein.

1.3.3 Other Approaches

In many applications the suspended objects are much smaller than the smallest hydrodynamical scale; in these cases the so-called point particle method can be used, in which the particle are treated as Lagrangian points moving with the local flow velocity. When the volume fraction of the particles is small enough hydrodynamic interactions and collisions among particles can be effectively neglected, while for large values of the particle-to-fluid density ratio, i.e. for significant mass loads, the momentum exchange between the two phases is still significant and must be accounted for. In general, the motion of the particles is described by a set of ordinary differential equations for the particle velocity and position, with the velocity depending on the Stokes drag and the buoyancy force [132] but also added mass and history force can become important [133, 134]; note that, the Stokes drag coefficient is sometimes corrected with empirical correlations when the particle Reynolds number does not remain small. The interested readers are referred to Refs. [135–138] for more details. Other approaches have been proposed to model the feedback of the point particle on the flow, which should be independent of the resolutions used; details can be found in Refs. [139–142].

In the recent years, various alternative methods have been proposed which are modifications and/or combinations of the methods previously discussed. A commonly used approach is the force coupling method: this is a bridge between methods for Stokes flow and for low-Reynolds number conditions and is based on a low-order, finite force multipole representation of the effect the particles have on the surrounding fluid flow. In particular, the full Navier–Stokes equations are solved with an additional force density: the force monopole corresponds to the force the particle transmits to the fluid if it were replaced by a rigid particle of mass with the same density as the fluid, while the force dipole is a combination of a symmetric stresslet and a torque that acts on the fluid: the torque is set in a similar manner as the force monopole in terms of the angular momentum of the displaced fluid, while the stresslet is chosen to ensure that the average rate of strain within the particle is zero. The aim is to create a flow outside the particle that matches the actual flow within a short distance from the surface. Note that, the fluid inside the particle volume is an active part of the simulation and satisfies the same integral moments as a rigid particle. The interested reader is referred to the detailed review in Ref. [143] and the references therein.

Another technique is the so-called volume of fluid tensorial penalty method: the method is based on the one-fluid formulation modified for dealing with particle flows. In particular, the fluid and solid phases are treated as two fluids with different rheological property and distinguished by a phase function, such as the solid volume fraction. The solid particle behaviour is recovered in the Navier–Stokes equations by properly decomposing of the stress tensor: in particular, the stress tensor is rewritten in a way to separate the compression, tearing, shearing and rotation contributions [144]. The decomposition is used to separate the stress components operating in a viscous flow and to facilitate the implementation of a numerical penalty method. In the solid phase it is imposed that the local flow admits no shearing and tearing while preserving a constant rotation; these flow constraints are implicitly transmitted to the particle sub-domain as they are solved with the fluid motion. Note that, the previous viscous penalty method is formally equivalent to choosing a viscosity much bigger than the fluid one, similarly to what previously done when discussing plastic effects. However, on a discrete point of view, the two formulations are not equivalent: when the dynamic viscosity is used to impose the solid behaviour, a first-order convergence in space is usually obtained and a rasterisation effect can be produced at the particle–fluid interface; on the other hand, when the viscous stress tensor splitting is used, a more accurate fluid–solid interface is used, thus reducing the rasterisation effect and a second-order convergence can be achieved. To solve the unsteady Navier–Stokes equations together with the incompressibility and solid constraints, the augmented Lagrangian method can be applied. The interested readers are referred to Refs. [145–147].

1.4 Conclusions

In this chapter, we have introduced present numerical methods for complex fluids. In the first part, we have discussed continuum approaches to viscoplastic and viscoelastic fluids, whereas interface-resolved methods for the simulations of suspensions of rigid and deformable particles, droplets, bubbles have been presented in the second part.

Continuum approaches require rheological models and constitutive equations for the additional stresses due to the fluid microstructure. These are derived theoretically or from experimental data; however, recently, the fast development of computational resources has enabled us to also resolve the microstructure with numerical simulations of the type described in the second part of this chapter. Although important results are continuously reported on the behaviour of viscoplastic and viscoelastic fluids, both in laminar and turbulent flows, these are somewhat restricted to relatively simple geometries. It is therefore relevant to explore the performance of the current numerical algorithms in more complex geometry and thus come closer to industrial applications and natural flows as, for example, avalanches. In addition, new and more sophisticated models are continuously proposed, also thanks to the development of new experimental techniques able to probe fluids subject to time varying shear or stress. From a numerical point of view, these will pose new challenges which we may not be able to tackle with the tools currently available. One such example is the isotropic kinematic hardening idea, based on the concept that the material yield stress builds up and evolves in time together with the flow field, where the steady state yield stress is determined via the back stress modulus (a new material parameter) and the deformation of microstructure (a hidden internal dimensionless evolution variable). Another important point for industrial applications is represented by the case where the fluid stress induces a breakage in a solid structure, i.e. hydraulic fracturing. Once a break-up occurs, the fragment behaves as a suspended particle that may contribute to additional solid breakage events. Despite its relevance in applications, the (automatic) numerical handling of this complex dynamics is still challenging.

Examples of simulations of multiphase and multicomponent flows are numerous and ever increasing, so that the approaches reported here constitute necessarily a quick and limited overview. Two directions are seen here as emerging and of potential interest. On one side the study of intrusions in a complex fluid, ubiquitous in applications, and, on the other side, the need to improve short-range near-contact interactions among particles/droplets. Indeed, when particles/bubbles are significantly larger than the colloids or macromolecules providing plastic and elastic effects, it is reasonable to model the complex fluid as intrusions in a non-Newtonian matrix, thus combining the two types of approaches presented here, denoted as macro and microscopic.

Although the approaches discussed here fully resolve the surface of the suspended phase, rigid or deformable, solid or fluid, models are necessary for the short-range interactions: (i) from a numerical point of view, the grid resolution inevitably becomes coarse when two objects approach each other, and (ii) microscale chemical and physical effects determine different interactions, for example, repulsive or attractive forces, slippage, Marangoni effects and depletion forces associated with the microstructure of the suspending phase. These become therefore truly multiscale problems, ranging from nanoscale surface interactions to millimetre/centimetre flows. Coupling of the numerical approaches mentioned here with nanoscale simulations, for example, molecular dynamics, is, therefore, a very active research area.

We would like to conclude by saying that we believe it is fundamental to always be aware of the limitations of the simulations one wishes to perform. This allows the researcher to design configurations and propose, though sometimes unphysical, numerical experiments that can unveil important physics, something which may not be possible in laboratory experiments.

References

1. S. Succi, *The Lattice Boltzmann Equation: For Complex States of Flowing Matter*, 2nd edn. (Oxford University Press, Oxford, 2018)
2. T. Kruger, H. Kusumaatmaja, A. Kuzmin, O. Shardt, G. Silva, E.M. Viggen, *The Lattice Boltzmann Method* (Springer, Berlin, 2017)
3. R.B. Bird, R.C. Armstrong, O. Hassager, *Dynamics of Polymeric Liquids. Vol. 1: Fluid Mechanics* (Wiley, Hoboken, 1987)
4. F.A. Morrison, *Understanding Rheology* (Oxford University Press, Oxford, 2001)
5. M.S. Dodd, A. Ferrante, A fast pressure-correction method for incompressible two-fluid flows. J. Comput. Phys. **273**, 416 (2014)
6. J.G. Oldroyd, in *Proceedings Royal Society London A Mathematical, Physical Engineering Sciences*, vol. 200 (The Royal Society, London, 1950), pp. 523–541
7. R.J. Poole, M.A. Alves, P.J. Oliveira, Purely elastic flow asymmetries. Phys. Rev. Lett. **99**(16), 164503 (2007)
8. Y. Doi, *Microbial Polyesters* (Wiley-Vch, Hoboken, 1990)
9. T. Min, J.Y. Yoo, H. Choi, Effect of spatial discretization schemes on numerical solutions of viscoelastic fluid flows. J. Fluid Mech. **100**(1), 27 (2001)
10. F. Dupret, J.M. Marchal, Loss of evolution in the flow of viscoelastic fluids. J. Fluid Mech. **20**, 143 (1986)
11. R. Sureshkumar, A.N. Beris, Effect of artificial stress diffusivity on the stability of numerical calculations and the flow dynamics of time-dependent viscoelastic flows. J. Fluid Mech. **60**(1), 53 (1995)
12. Y. Dubief, V.E. Terrapon, C.M. White, E.S.G. Shaqfeh, P. Moin, S.K. Lele, New answers on the interaction between polymers and vortices in turbulent flows. Flow Turbul. Combust. **74**(4), 311 (2005)
13. G. Mompean, M. Deville, Unsteady finite volume simulation of Oldroyd-B fluid through a three-dimensional planar contraction. J. Fluid Mech. **72**(2–3), 253 (1997)
14. M.A. Alves, F.T. Pinho, P.J. Oliveira, Effect of a high-resolution differencing scheme on finite-volume predictions of viscoelastic flows. J. Fluid Mech. **93**(2), 287 (2000)
15. G.S. Jiang, C.W. Shu, Efficient implementation of weighted ENO schemes. J. Comput. Phys. **126**(1), 202 (1996)
16. K. Sugiyama, S. Ii, S. Takeuchi, S. Takagi, Y. Matsumoto, A full Eulerian finite difference approach for solving fluid–structure coupling problems. J. Comput. Phys. **230**(3), 596 (2011)
17. M.E. Rosti, L. Brandt, Numerical simulation of turbulent channel flow over a viscous hyper-elastic wall. J. Fluid Mech. **830**, 708 (2017)
18. M.E. Rosti, L. Brandt, Suspensions of deformable particles in a Couette flow. J. Fluid Mech. **262**, 3 (2018)
19. A. Shahmardi, S. Zade, M.N. Ardekani, R.J. Poole, F. Lundell, M.E. Rosti, L. Brandt, Turbulent duct flow with polymers. J. Fluid Mech. **859**, 1057 (2019)
20. R. Fattal, R. Kupferman, Constitutive laws for the matrix-logarithm of the conformation tensor. J. Fluid Mech. **123**(2), 281 (2004)

21. M.A. Hulsen, R. Fattal, R. Kupferman, Flow of viscoelastic fluids past a cylinder at high Weissenberg number: stabilized simulations using matrix logarithms. J. Fluid Mech. **127**(1), 27 (2005)
22. A. Afonso, P.J. Oliveira, F.T. Pinho, M.A. Alves, The log-conformation tensor approach in the finite-volume method framework. J. Fluid Mech. **157**(1), 55 (2009)
23. M. Yang, S. Krishnan, E.S.G. Shaqfeh, Numerical simulations of the rheology of suspensions of rigid spheres at low volume fraction in a viscoelastic fluid under shear. J. Fluid Mech. **234**, 51 (2016)
24. T. Schwedoff, La rigidité des fluides, Rapports du Congrès intern. Physique **1**, 478 (1900)
25. E.C. Bingham, *Fluidity and Plasticity*, vol. 2 (McGraw-Hill, New York, 1922)
26. J.G. Oldroyd, in *Mathematical Proceedings of the Cambridge Philosophical Society*, vol. 43 (Cambridge University Press, Cambridge, 1947), pp. 100–105
27. P. Saramito, A. Wachs, Progress in numerical simulation of yield stress fluid flows. Rheol. Acta **56**, 211–230 (2016)
28. M. Bercovier, M. Engelman, A finite-element method for incompressible non-Newtonian flows. J. Comput. Phys. **36**(3), 313 (1980)
29. T.C. Papanastasiou, Flows of materials with yield. J. Rheol. **31**(5), 385 (1987)
30. E. Mitsoulis, S.S. Abdali, N.C. Markatos, Flow simulation of Herschel-Bulkley fluids through extrusion dies. Can. J. Chem. Eng. **71**(1), 147 (1993)
31. S.D.R. Wilson, A.J. Taylor, The channel entry problem for a yield stress fluid. J. Fluid Mech. **65**(2), 165 (1996)
32. E. Mitsoulis, R.R. Huilgol, Entry flows of Bingham plastics in expansions. J. Fluid Mech. **122**(1), 45 (2004)
33. M.R. Hestenes, Multiplier and gradient methods. J. Optim. Theory Appl. **4**(5), 303 (1969)
34. R.T. Rockafellar, Augmented Lagrangians and applications of the proximal point algorithm in convex programming. Math. Oper. Res. **1**(2), 97 (1976)
35. R. Glowinski, *Lectures on Numerical Methods for Non-linear Variational Problems* (Springer Science & Business Media, New York, 2008)
36. M. Fortin, R. Glowinski, in *Studies in Mathematics and its Applications*, vol. 15 (Elsevier, Amsterdam, 1983), pp. 1–46
37. R. Glowinski, P. Le Tallec, *Augmented Lagrangian and Operator-Splitting Methods in Nonlinear Mechanics* (SIAM, Philadelphia, 1989)
38. P. Saramito, N. Roquet, An adaptive finite element method for viscoplastic fluid flows in pipes. Comput. Method. Appl. Mech. Engg. **190**(40), 5391 (2001)
39. D. Vola, L. Boscardin, J.C. Latché, Laminar unsteady flows of Bingham fluids: a numerical strategy and some benchmark results. J. Comput. Phys. **187**(2), 441 (2003)
40. O. Turan, N. Chakraborty, R.J. Poole, Laminar natural convection of Bingham fluids in a square enclosure with differentially heated side walls. J. Fluid Mech. **165**(15), 901 (2010)
41. H. Uzawa, in *Studies in Linear and Nonlinear Programming*, ed. by K.J. Arrow, L. Hurwicz, H. Uzawa (Stanford University Press, Stanford, 1958)
42. P. Saramito, A new constitutive equation for elastoviscoplastic fluid flows. J. Fluid Mech. **145**(1), 1 (2007)
43. P. Saramito, A new elastoviscoplastic model based on the Herschel–Bulkley viscoplastic model. J. Fluid Mech. **158**(1), 154 (2009)
44. R. von Mises, Mechanik der festen köper im plastisch deformablen zustand. Göttin. Nachr. Math. Phys. **1**, 582 (1913)
45. C.A. Schuh, A.C. Lund, Atomistic basis for the plastic yield criterion of metallic glass. Nat. Mater. **2**(7), 449 (2003)
46. L. Wang, M.C. Liu, J.C. Huang, Y. Li, W.H. Wang, T.G. Nieh, Effect of temperature on the yield strength of a binary CuZr metallic glass: stress-induced glass transition. Intermetallics **26**, 162 (2012)
47. B. Yang, C.T. Liu, T.G. Nieh, Unified equation for the strength of bulk metallic glasses. Appl. Phys. Lett. **88**(22), 221911 (2006)

48. Y.H. Liu, C.T. Liu, W.H. Wang, A. Inoue, T. Sakurai, M.W. Chen, Thermodynamic origins of shear band formation and the universal scaling law of metallic glass strength. Phys. Rev. Lett. **103**(6), 65504 (2009)
49. P. Guan, M. Chen, T. Egami, Stress-temperature scaling for steady-state flow in metallic glasses. Phys. Rev. Lett. **104**(20), 205701 (2010)
50. M.E. Rosti, D. Izbassarov, O. Tammisola, S. Hormozi, L. Brandt, Turbulent channel flow of an elastoviscoplastic fluid. J. Fluid Mech. **853**, 488 (2018)
51. D. Izbassarov, M.E. Rosti, M.N. Ardekani, M. Sarabian, S. Hormozi, L. Brandt, O. Tammisola, Computational modeling of multiphase viscoelastic and elastoviscoplastic flows. Int. J. Numer. Methods Fluids **88**(12), 521–543 (2018)
52. G. Tryggvason, M. Sussman, M.Y. Hussaini, in *Computational Methods for Multiphase Flow*, ed. by A. Prosperetti, chap. 3 (Cambridge University Press, Cambridge, 2007)
53. S. Takeuchi, Y. Yuki, A. Ueyama, T. Kajishima, A conservative momentum-exchange algorithm for interaction problem between fluid and deformable particles. Int. J. Numer. Methods Fluids **64**(10–12), 1084 (2010)
54. M. Quintard, S. Whitaker, Transport in ordered and disordered porous media II: generalized volume averaging. Transp. Porous Media **14**(2), 179 (1994)
55. M.E. Rosti, L. Brandt, D. Mitra, Rheology of suspensions of viscoelastic spheres: deformability as an effective volume fraction. Phys. Rev. Fluids **3**(1), 12301 (2018)
56. M. Uhlmann, An immersed boundary method with direct forcing for the simulation of particulate flows. J. Comput. Phys. **209**(2), 448 (2005)
57. E.A. Fadlun, R. Verzicco, P. Orlandi, J. Mohd-Yusof, Combined immersed-boundary finite-difference methods for three-dimensional complex flow simulations. J. Comput. Phys **161**(1), 35 (2000)
58. A. Pinelli, I.Z. Naqavi, U. Piomelli, J. Favier, Immersed-boundary methods for general finite-difference and finite-volume Navier-Stokes solvers. J. Comput. Phys. **229**(24), 9073 (2010)
59. W.P. Breugem, A second-order accurate immersed boundary method for fully resolved simulations of particle-laden flows. J. Comput. Phys. **231**(13), 4469 (2012)
60. J. Favier, A. Revell, A. Pinelli, A Lattice Boltzmann - immersed boundary method to simulate the fluid interaction with moving and slender flexible objects. J. Comput. Phys. **261**, 145 (2014)
61. C.S. Peskin, Flow patterns around heart valves: a numerical method. J. Comput. Phys. **10**(2), 252 (1972)
62. R.P. Beyer, R.J. LeVeque, Analysis of a one-dimensional model for the immersed boundary method. SIAM J. Numer. Anal. **29**(2), 332 (1992)
63. M.C. Lai, C.S. Peskin, An immersed boundary method with formal second-order accuracy and reduced numerical viscosity. J. Comput. Phys. **160**(2), 705 (2000)
64. E.M. Saiki, S. Biringen, Numerical simulation of a cylinder in uniform flow: application of a virtual boundary method. J. Comput. Phys. **123**(2), 450 (1996)
65. D. Goldstein, R. Handler, L. Sirovich, Modeling a no-slip flow boundary with an external force field. J. Comput. Phys. **105**(2), 354 (1993)
66. W.X. Huang, S.J. Shin, H.J. Sung, Simulation of flexible filaments in a uniform flow by the immersed boundary method. J. Comput. Phys. **226**(2), 2206 (2007)
67. M.E. Rosti, A.A. Banaei, L. Brandt, A. Mazzino, Flexible fiber reveals the two-point statistical properties of turbulence. Phys. Rev. Lett. **121**(4), 44501 (2018)
68. G. Iaccarino, R. Verzicco, Immersed boundary technique for turbulent flow simulations. Appl. Mech. Rev. **56**(3), 331 (2003)
69. J. Kim, P. Moin, Application of a fractional-step method to incompressible Navier-Stokes equations. J. Comput. Phys. **59**(2), 308 (1985)
70. A.M. Roma, C.S. Peskin, M.J. Berger, An adaptive version of the immersed boundary method. J. Comput. Phys. **153**(2), 509 (1999)
71. K. Luo, Z. Wang, J. Fan, K. Cen, Full-scale solutions to particle-laden flows: multidirect forcing and immersed boundary method. Phys. Rev. E **76**(6), 66709 (2007)

72. S. Kriebitzsch, M. der Hoef, H. Kuipers, in *Proceedings Academy Colloquium Immersed Boundary Methods Current Status and Future Research Directions* (2010)
73. K. Hoefler, S. Schwarzer, Navier-Stokes simulation with constraint forces: finite-difference method for particle-laden flows and complex geometries. Phys. Rev. E **61**(6), 7146 (2000)
74. Z. Yu, X. Shao, A direct-forcing fictitious domain method for particulate flows. J. Comput. Phys. **227**(1), 292 (2007)
75. P. Costa, B.J. Boersma, J. Westerweel, W.P. Breugem, Collision model for fully resolved simulations of flows laden with finite-size particles. Phys. Rev. E **92**(5), 53012 (2015)
76. M.N. Ardekani, P. Costa, W.P. Breugem, L. Brandt, Numerical study of the sedimentation of spheroidal particles. Int. J. Multiphase Flow **87**, 16 (2016)
77. T. Kempe, J. Froehlich, An improved immersed boundary method with direct forcing for the simulation of particle laden flows. J. Comput. Phys. **231**(9), 3663 (2012)
78. E. Biegert, B. Vowinckel, E. Meiburg, A collision model for grain-resolving simulations of flows over dense, mobile, polydisperse granular sediment beds. J. Comput. Phys. **340**, 105 (2017)
79. S.O. Unverdi, G. Tryggvason, A front-tracking method for viscous, incompressible, multi-fluid flows. J. Comput. Phys. **100**(1), 25 (1992)
80. G. Tryggvason, B. Bunner, A. Esmaeeli, D. Juric, N. Al-Rawahi, W. Tauber, J. Han, S. Nas, Y.J. Jan, A front-tracking method for the computations of multiphase flow. J. Comput. Phys. **169**(2), 708 (2001)
81. C.S. Peskin, Numerical analysis of blood flow in the heart. J. Comput. Phys. **25**(3), 220 (1977)
82. R. Scardovelli, S. Zaleski, Direct numerical simulation of free-surface and interfacial flow. Annu. Rev. Fluid Mech. **31**(1), 567 (1999)
83. M. Sussman, P. Smereka, S. Osher, A level set approach for computing solutions to incompressible two-phase flow. J. Comput. Phys. **114**(1), 146 (1994)
84. J.A. Sethian, *Level Set Methods and Fast Marching Methods: Evolving Interfaces in Computational Geometry, Fluid Mechanics, Computer Vision, and Materials Science*, vol. 3 (Cambridge University Press, Cambridge, 1999)
85. J.A. Sethian, P. Smereka, Level set methods for fluid interfaces. Annu. Rev. Fluid Mech. **35**(1), 341 (2003)
86. M.M. Francois, S.J. Cummins, E.D. Dendy, D.B. Kothe, J.M. Sicilian, M.W. Williams, A balanced-force algorithm for continuous and sharp interfacial surface tension models within a volume tracking framework. J. Comput. Phys. **213**(1), 141 (2006)
87. S.J. Cummins, M.M. Francois, D.B. Kothe, Estimating curvature from volume fractions. Comput. Fluids **83**(6–7), 425 (2005)
88. M. Sussman, E.G. Puckett, A coupled level set and volume-of-fluid method for computing 3D and axisymmetric incompressible two-phase flows. J. Comput. Phys. **162**(2), 301 (2000)
89. W.F. Noh, P. Woodward, in *Proceedings of fifth International Conference on Numerical Methods fluid Dynamics* (Springer, Berlin, 1976), pp. 330–340
90. D.L. Youngs, Time-dependent multi-material flow with large fluid distortion. Numer. Methods Fluid Dyn. **24**, 273–285 (1982)
91. D.L. Youngs, An interface tracking method for a 3D Eulerian hydrodynamics code. Tech. Rep. 44/92, Atomic Weapons Research Establishment (1984)
92. E.G. Puckett, A.S. Almgren, J.B. Bell, D.L. Marcus, W.J. Rider, A high-order projection method for tracking fluid interfaces in variable density incompressible flows. J. Comput. Phys. **130**(2), 269 (1997)
93. W.J. Rider, D.B. Kothe, Reconstructing volume tracking. J. Comput. Phys. **141**(2), 112 (1998)
94. D.J.E. Harvie, D.F. Fletcher, A new volume of fluid advection algorithm: the stream scheme. J. Comput. Phys. **162**(1), 1 (2000)
95. E. Aulisa, S. Manservisi, R. Scardovelli, S. Zaleski, A geometrical area-preserving volume-of-fluid advection method. J. Comput. Phys. **192**(1), 355 (2003)
96. J.E. Pilliod Jr., E.G. Puckett, Second-order accurate volume-of-fluid algorithms for tracking material interfaces. J. Comput. Phys. **199**(2), 465 (2004)

97. F. Xiao, Y. Honma, T. Kono, A simple algebraic interface capturing scheme using hyperbolic tangent function. Int. J. Numer. Methods Fluids **48**(9), 1023 (2005)
98. K. Yokoi, Efficient implementation of THINC scheme: a simple and practical smoothed VOF algorithm. J. Comput. Phys. **226**(2), 1985 (2007)
99. S. Ii, K. Sugiyama, S. Takeuchi, S. Takagi, Y. Matsumoto, F. Xiao, An interface capturing method with a continuous function: the THINC method with multi-dimensional reconstruction. J. Comput. Phys. **231**(5), 2328 (2012)
100. M.E. Rosti, F. De Vita, L. Brandt, Numerical simulations of emulsions in shear flows. Acta Mech. **230**(2), 667–682 (2019)
101. J.U. Brackbill, D.B. Kothe, C. Zemach, A continuum method for modeling surface tension. J. Comput. Phys. **100**(2), 335 (1992)
102. M.S. Dodd, A. Ferrante, On the interaction of Taylor length scale size droplets and isotropic turbulence. J. Fluid Mech. **806**, 356 (2016)
103. S. Dong, J. Shen, A time-stepping scheme involving constant coefficient matrices for phase-field simulations of two-phase incompressible flows with large density ratios. J. Comput. Phys. **231**(17), 5788 (2012)
104. F. Capuano, G. Coppola, M. Chiatto, L. de Luca, Approximate projection method for the incompressible Navier–Stokes equations. AIAA J. **54**(7), 2179 (2016)
105. C.W. Shu, S. Osher, Efficient implementation of essentially non-oscillatory shock-capturing schemes. J. Comput. Phys. **77**(2), 439 (1988)
106. Z. Ge, J.C. Loiseau, O. Tammisola, L. Brandt, An efficient mass-preserving interface-correction level set/ghost fluid method for droplet suspensions under depletion forces. J. Comput. Phys. **353**, 435 (2018)
107. R.R. Nourgaliev, T.G. Theofanous, High-fidelity interface tracking in compressible flows: unlimited anchored adaptive level set. J. Comput. Phys. **224**(2), 836 (2007)
108. X.D. Liu, S. Osher, T. Chan, Weighted essentially non-oscillatory schemes. J. Comput. Phys. **115**(1), 200 (1994)
109. J. Strain, Semi-Lagrangian methods for level set equations. J. Comput. Phys. **151**(2), 498 (1999)
110. G. Velmurugan, E.M. Kolahdouz, D. Salac, Level set jet schemes for stiff advection equations: the semijet method. Comput. Method. Appl. Mech. Engg. **310**, 233 (2016)
111. M.B. Nielsen, K. Museth, Dynamic tubular grid - an efficient data structure and algorithms for high resolution level sets. J. Sci. Comput. **26**(3), 261 (2006)
112. E. Brun, A. Guittet, F. Gibou, A local level-set method using a hash table data structure. J. Comput. Phys. **231**(6), 2528 (2012)
113. R.P. Fedkiw, T. Aslam, B. Merriman, S. Osher, A non-oscillatory Eulerian approach to interfaces in multimaterial flows (the ghost fluid method). J. Comput. Phys. **152**(2), 457 (1999)
114. M. Kang, R.P. Fedkiw, X.D. Liu, A boundary condition capturing method for multiphase incompressible flow. J. Sci. Comput. **15**(3), 323 (2000)
115. X.D. Liu, T. Sideris, Convergence of the ghost fluid method for elliptic equations with interfaces. Math. Comput. **72**(244), 1731 (2003)
116. J.W. Cahn, J.E. Hilliard, Free energy of a nonuniform system. I. Interfacial free energy. J. Chem. Phys. **28**(2), 258 (1958)
117. M.E. Gurtin, Generalized Ginzburg-Landau and Cahn-Hilliard equations based on a micro-force balance. Phys. D **92**(3–4), 178 (1996)
118. V.E. Badalassi, H.D. Ceniceros, S. Banerjee, Computation of multiphase systems with phase field models. J. Comput. Phys. **190**(2), 371 (2003)
119. P. Yue, J.J. Feng, C. Liu, J. Shen, A diffuse-interface method for simulating two-phase flows of complex fluids. J. Fluid Mech. **515**, 293 (2004)
120. F. Magaletti, F. Picano, M. Chinappi, L. Marino, C.M. Casciola, The sharp-interface limit of the Cahn–Hilliard/Navier–Stokes model for binary fluids. J. Fluid Mech. **714**, 95 (2013)

121. L. Scarbolo, F. Bianco, A. Soldati, Turbulence modification by dispersion of large deformable droplets. Eur. J. Mech. B-Fluid **55**, 294 (2016)
122. Y. Wang, G. Amberg, A. Carlson, Local dissipation limits the dynamics of impacting droplets on smooth and rough substrates. Phys. Rev. Fluids **2**(3), 33602 (2017)
123. D. Jacqmin, Contact-line dynamics of a diffuse fluid interface. J. Fluid Mech. **402**, 57 (2000)
124. P. Yue, C. Zhou, J.J. Feng, Sharp-interface limit of the Cahn–Hilliard model for moving contact lines. J. Fluid Mech. **645**, 279 (2010)
125. M. Ahmadlouydarab, J.J. Feng, Motion and coalescence of sessile drops driven by substrate wetting gradient and external flow. J. Fluid Mech. **746**, 214 (2014)
126. L. Scarbolo, F. Bianco, A. Soldati, Coalescence and breakup of large droplets in turbulent channel flow. Phys. Fluids **27**(7), 73302 (2015)
127. H. Ding, B.Q. Chen, H.R. Liu, C.Y. Zhang, P. Gao, X.Y. Lu, On the contact-line pinning in cavity formation during solid–liquid impact. J. Fluid Mech. **783**, 504 (2015)
128. F. Magaletti, L. Marino, C.M. Casciola, Shock wave formation in the collapse of a vapor nanobubble. Phys. Rev. Lett. **114**(6), 64501 (2015)
129. M. Sbragaglia, R. Benzi, L. Biferale, S. Succi, K. Sugiyama, F. Toschi, Generalized lattice Boltzmann method with multirange pseudopotential. Phys. Rev. E **75**(2), 26702 (2007)
130. A.E. Komrakova, O. Shardt, D. Eskin, J.J. Derksen, Lattice Boltzmann simulations of drop deformation and breakup in shear flow. Int. J. Multiphase Flow **59**, 24 (2014)
131. A. Gupta, M. Sbragaglia, A. Scagliarini, Hybrid Lattice Boltzmann/finite difference simulations of viscoelastic multicomponent flows in confined geometries. J. Comput. Phys. **291**, 177 (2015)
132. M.R. Maxey, J.J. Riley, Equation of motion for a small rigid sphere in a nonuniform flow. Phys. Fluids **26**(4), 883 (1983)
133. S. Olivieri, F. Picano, G. Sardina, D. Iudicone, L. Brandt, The effect of the Basset history force on particle clustering in homogeneous and isotropic turbulence. Phys. Fluids **26**(4), 41704 (2014)
134. A. Daitche, On the role of the history force for inertial particles in turbulence. J. Fluid Mech. **782**, 567 (2015)
135. C. Marchioli, V. Armenio, A. Soldati, Simple and accurate scheme for fluid velocity interpolation for Eulerian–Lagrangian computation of dispersed flows in 3D curvilinear grids. Comput. Fluids **36**(7), 1187 (2007)
136. C. Marchioli, M.V. Salvetti, A. Soldati, Some issues concerning large-eddy simulation of inertial particle dispersion in turbulent bounded flows. Phys. Fluids **20**(4), 40603 (2008)
137. C. Marchioli, A. Soldati, J.G.M. Kuerten, B. Arcen, A. Taniere, G. Goldensoph, K.D. Squires, M.F. Cargnelutti, L.M. Portela, Statistics of particle dispersion in direct numerical simulations of wall-bounded turbulence: results of an international collaborative benchmark test. Int. J. Multiphase Flow **34**(9), 879 (2008)
138. A. Soldati, C. Marchioli, Physics and modelling of turbulent particle deposition and entrainment: review of a systematic study. Int. J. Multiphase Flow **35**(9), 827 (2009)
139. P. Gualtieri, F. Picano, G. Sardina, C.M. Casciola, Exact regularized point particle method for multiphase flows in the two-way coupling regime. J. Fluid Mech. **773**, 520 (2015)
140. J. Capecelatro, O. Desjardins, An Euler–Lagrange strategy for simulating particle-laden flows. J. Comput. Phys. **238**, 1 (2013)
141. G. Akiki, W.C. Moore, S. Balachandar, Pairwise-interaction extended point-particle model for particle-laden flows. J. Comput. Phys. **351**, 329 (2017)
142. M. Esmaily, J.A.K. Horwitz, A correction scheme for two-way coupled point-particle simulations on anisotropic grids. J. Comput. Phys. **375**, 960 (2018)
143. M. Maxey, Simulation methods for particulate flows and concentrated suspensions. Annu. Rev. Fluid Mech. **49**, 171 (2017)
144. J.P. Caltagirone, S. Vincent, Sur une méthode de pénalisation tensorielle pour la resolution des equations de Navier–Stokes. C. R. Acad. Sci. Paris II B **329**(8), 607 (2001)

145. J. Janela, A. Lefebvre, B. Maury, in *ESAIM: Proceedings*, vol. 14 (EDP Sciences, Les Ulis, 2005), pp. 115–123
146. T.N. Randrianarivelo, G. Pianet, S. Vincent, J.P. Caltagirone, Numerical modelling of solid particle motion using a new penalty method. Int. J. Numer. Methods Fluids **47**(10–11), 1245 (2005)
147. S. Vincent, J.C.B. De Motta, A. Sarthou, J.L. Estivalezes, O. Simonin, E. Climent, A Lagrangian VOF tensorial penalty method for the DNS of resolved particle-laden flows. J. Comput. Phys. **256**, 582 (2014)

2

Stokes Flows: Fundamental Concepts

Christopher I. Trombley and Maria L. Ekiel-Jeżewska

2.1 Introduction

Stokes flows have many applications in both physical theory and practice. For example, they have been used to describe dynamics of complex fluids in microfluidics, lab-on-chip technologies [1], medical applications [2, 3], design of innovative materials [4–6] and micro-devices—e.g. to carry drugs [7, 8] or act as fuel cells [9]—and in biological systems [10–15].

In this chapter, we discuss some fundamental properties of Stokes flows, namely: negligibility of inertial forces, reversibility and the minimum energy dissipation theorem. First we will briefly discuss how the neglecting of inertial forces simplifies the nonlinear Navier–Stokes equations to the linear Stokes equations. We then discuss two basic aspects of Stokes flows: reversibility and the minimum energy dissipation theorem. In order to bring out the nature of the three principles, we will demonstrate by example how these properties can be used to obtain conclusions about investigated fluid systems without laborious construction of analytical solutions. We then move beyond the Stokes approximation in various ways in order to see how the principles work in a general context. Finally, we conclude by discussing the logical structure of the principles as revealed by the examples considered.

C. I. Trombley · M. L. Ekiel-Jeżewska (✉)
Institute of Fundamental Technological Research, Polish Academy of Sciences, Warsaw, Poland
e-mail: mekiel@ippt.gov.pl

2.2 Navier–Stokes and Stokes Equations

2.2.1 *Navier–Stokes Equations*

We start with the general Navier–Stokes equations for an incompressible fluid. These are [16–18]

$$\rho\frac{\partial \mathbf{u}}{\partial t} + \rho(\mathbf{u}\cdot\nabla)\mathbf{u} = \mu\nabla^2\mathbf{u} - \nabla p + \mathbf{F} \tag{2.1}$$

$$\nabla\cdot\mathbf{u} = 0 \tag{2.2}$$

where ρ is the density of the fluid, $\mathbf{u}$ is the velocity field of a fluid, μ is the dynamic viscosity of the fluid, p is the fluid pressure field,[1] and $\mathbf{F}$ captures the effects of external forces. The left-hand side of this equation is the inertial forces, that is, the acceleration of a fluid element with unit volume. The right-hand side is sum of the viscous and pressure forces, $\mu\nabla^2\mathbf{u}$ and ∇p, respectively, exerted on surfaces of this fluid element, and any external body forces $\mathbf{F}$ acting on the fluid element. The second equation is based on the conservation of mass of a fluid element and achieves its simple form because of incompressibility of the fluid element.

We use non-dimensionalisation in order to capture the relative scale of the forces. Define U to be a characteristic velocity of the fluid and L to be a characteristic length scale. Other characteristic dimensional scales of the flow, for instance, a time scale $T = L/U$, can be defined implicitly from these scales. There is still some freedom when normalising pressure p and body forces $\mathbf{F}$. We choose to normalise pressure by a characteristic viscous force per unit area and $\mathbf{F}$ by a characteristic viscous force per unit volume as in [18]. Using a star to denote non-dimensionalised objects, this results in the following definitions:

$$\begin{aligned}
\mathbf{u}^* &= \frac{\mathbf{u}}{U}\\
\nabla^* &= L\nabla\\
\frac{\partial}{\partial t^*} &= \frac{L}{U}\frac{\partial}{\partial t}\\
p^* &= \frac{L}{\mu U}p\\
\mathbf{F}^* &= \frac{L^2}{\mu U}\mathbf{F}
\end{aligned} \tag{2.3}$$

[1]In the presence of a gravitational field, p is the so-called modified pressure, which takes into account also gravitational potential energy per unit fluid volume.

With the above characteristic dimensional scales, the inertial force per unit volume is estimated by $\frac{\rho U^2}{L}$ and the scale of the viscous force per unit volume is $\frac{\mu U}{L^2}$. The Reynolds number, Re, a non-dimensional number defined as the ratio of inertial and viscous forces in a fluid, takes the form

$$Re = \frac{(\rho U^2)/L}{(\mu U)/L^2} = \frac{\rho U L}{\mu} \tag{2.4}$$

The end result is the following non-dimensional version of the Navier–Stokes equation (2.1):

$$Re\left(\frac{\partial \mathbf{u}^*}{\partial t^*} + \mathbf{u}^* \cdot \nabla^* \mathbf{u}^*\right) = \nabla^{*2}\mathbf{u}^* - \nabla^* p^* + \mathbf{F}^* \tag{2.5}$$

The left-hand side is the inertial force and the right-hand side is the viscous, pressure and body forces. Flows with the same Re are hydrodynamically similar [18].

A difficulty to using Eqs. (2.1) and (2.2) (or their non-dimensional form) in the analysis of fluids is that the inertial forces are nonlinear in $\mathbf{u}$. In terms of forces, the so-called Stokes approximation can be understood as when the viscous and pressure forces dominate the inertial forces absolutely. The Reynolds number allows one to test the applicability of Stokes approximation to fluids. The Stokes approximation holds exactly in the limit as this ratio goes to zero [17–22]. For this reason, Stokes flows are often called low Reynolds number, non-inertial or viscous flows.

2.2.2 *Stokes Flows*

Taking the limit $Re \to 0$ in Eq. (2.5) one obtains the non-dimensional steady Stokes equations. In dimensional form, without external body forces, sources or sinks, they read

$$\mu \nabla^2 \mathbf{u} - \nabla p = 0 \tag{2.6}$$

$$\nabla \cdot \mathbf{u} = 0 \tag{2.7}$$

The first equation states the balance of forces in a non-accelerating fluid. The second equation is, as in Eq. (2.2), the conservation of mass for incompressible fluids.

The Stokes equations (2.6) and (2.7) must be combined with boundary conditions appropriate to the physical situation. The so-called stick or no-slip boundaries for rigid walls and at the surfaces of particles are important examples. Consider a surface S moving with local velocity $\mathbf{w}$. It has no-slip boundary condition for the fluid velocity $\mathbf{u}$ if on S one has

$$\mathbf{u}(\mathbf{r}) = \mathbf{w}(\mathbf{r}) \qquad \text{for } \mathbf{r} \in S \tag{2.8}$$

There are many other important examples of boundary conditions, such as the boundaries for a free surface [18], but we will focus on the stick boundary conditions, which are sufficient for considering inertial forces, reversibility and the minimum energy dissipation. When considering reversibility especially, one must remember that boundaries can be time dependent. This means that the boundaries move, such as in the classical Taylor–Couette experiment involving a fluid between two rotating cylinders [23]. The Stokes equations (2.6) and (2.7) can also apply to unbounded flow problems by the selection of an appropriate boundary at infinity. For instance, a fluid can be constrained to be at rest at infinity, as in case of particles settling in a quiescent fluid.

Equations (2.6)–(2.7) are linear, so that any linear combination of solutions $(\mathbf{u}_1, p_1)$ and $(\mathbf{u}_2, p_2)$ is also a solution $(\mathbf{u}_1 + \mathbf{u}_2, p_1 + p_2)$. Linearity allows for classes of solutions to be constructed. One example is the case of flow around a rigid sphere, where a complete set of elementary solutions to Eqs. (2.6) can be constructed, as done by Lamb [16]. In his families of elementary solutions, the pressure p is expanded in spherical harmonics and the velocity field $\mathbf{u}$ is written as an infinite series of solid harmonics. This concept is used in the multipole method of solving the Stokes equations for systems of particles moving in fluids [20, 24–30].

2.3 Reversibility of Fluid Flows

Because Stokes equations (2.6)–(2.7) are steady and linear, the motion they predict is reversible in time. Mathematically, it means that the reversibility transformation of any solution, that is, $(\mathbf{u}(\mathbf{x}, t), p(\mathbf{x}, t)) \longmapsto -(\mathbf{u}(\mathbf{x}, t), p(\mathbf{x}, t))$, will also give a solution. This can be checked by simple algebraic manipulation of the governing equations. G.I. Taylor explained in his film *Low Reynolds Number Flows* [31] the physical meaning of reversibility—"low Reynolds number flows are reversible when the direction of motion of the boundaries which gave rise to the flow is reversed". Actually, a reversed fluid flow can result from reversing velocity of the boundary (equal to the fluid velocity at the surfaces of particles or walls) or from reversing directions of the external and the opposite hydrodynamic forces. In the following, we will show how reversibility allows to predict symmetries of fluid flows and motion of particles in fluids.

2.3.1 Examples of Reversibility

One of the most dramatic presentations of reversibility is seen in the film mentioned above [31]. In this experiment, the volume between two transparent cylinders is filled with glycerine. Dyes are injected which form a compact coloured volume into the glycerine to help visualise the flow. The inner cylinder is rotated causing the dyes to stir and apparently mix. The inner cylinder is then rotated in the opposite direction and one sees the seemingly mixed fluids unstir themselves. This experiment demonstrates the difficulty of mixing low Reynolds number fluids, an important problem for microfluidics.

G.I. Taylor used this experiment to explain the concept of reversibility in the following way: "On reversal of the motion of the boundary, every particle retracts exactly the same path on its return journey as on the outward journey, and at every point its speed is the same fraction of the boundary speed as it was at the same point on its outward journey, so that when the boundary has returned to its original position every particle in the fluid has also done so and the original pattern of dye is reproduced" [32].

A very important consequence of reversibility in biology is that the ordinary swimming motion done by an idealised swimmer with a rigid tail could not produce forward motion in a non-inertial fluid, since any propulsion created by the swimmer when the tail moves left is exactly cancelled when it moves right, as demonstrated in [31]. This is a consequence of the "Scallop theorem" fundamental to the study of the locomotion of microscale organisms [11].

One can use reversibility to derive basic properties of solutions to the Stokes equations without finding the solutions explicitly. Take the case of a rotating, but not translating, sphere immersed in fluid governed by Eqs. (2.6) and (2.7). This situation is illustrated in Fig. 2.1. We might ask how much force such a sphere would feel. Here we mean the force exerted by the fluid on the sphere owing to stick boundary conditions on its surface. This hydrodynamic force needs to be balanced by the opposite external (non-hydrodynamic) force acting on the sphere. Through the use

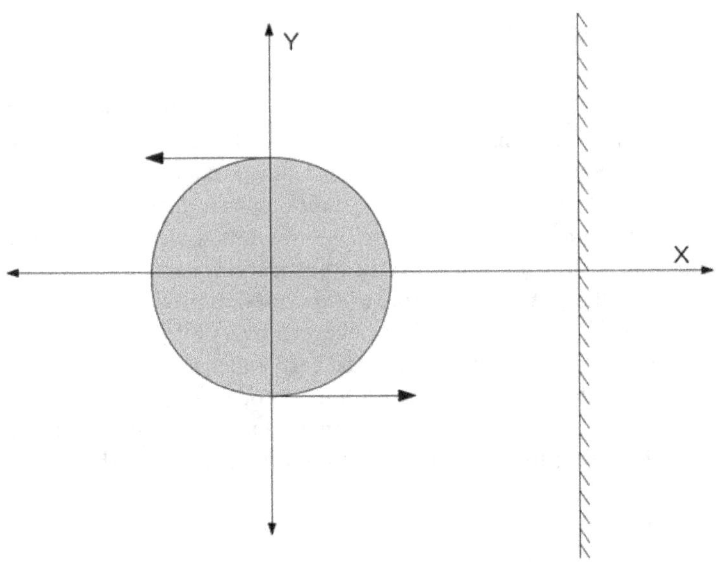

Fig. 2.1 A solid sphere rotating without translations near a solid wall. Reversibility implies that the sphere does not feel any external force perpendicular to the wall

of superposed reversibility and symmetry transformations, we can discover that in this situation the answer is that the sphere would not feel any hydrodynamic force in the direction perpendicular to the wall [33]. This can be proven by contradiction. Suppose $F_x \neq 0$. Notice that if we put the origin at the centre of the sphere, the system is symmetric for the transformation $y \longmapsto -y$. This reflection reverses the rotation of the sphere, but leaves the x-component of the force the same. Now apply the reversibility transformation. The rotation is now reversed back to the original sense. The force vector should have the opposite direction. The result is that the sphere is at the same position, has the same physical rotation, but opposite F_x. This is a contradiction. This argument shows how reversibility and symmetry arguments can be combined to put strong restrictions on Stokes flow [33, 34].

We can also apply reversibility arguments again to the case of a sphere which moves under a constant gravitational force parallel to a solid wall. Applying the time reversal, we now reverse also the direction of the sphere velocity and force. By the same argument above, i.e. by combining the time reversal with the reflection with respect to the plane $y = 0$, there will be no velocity in the direction perpendicular to the wall; the sphere will keep translating parallel to the wall [34]. This reasoning applies to the study of sedimentation of a slowly moving particle of any shape and material symmetric with respect to reflection in the plane $y = 0$ [33, 34]. We have demonstrated that reversibility has observable consequences which do not require elaborate constructions.

2.3.2 *Irreversible Trajectories in Stokes Flow*

Applying reversibility, one must take care that reversibility applies to time and forces. In particular, the paths that particles take need not be reversible in time even though the Stokes equation is reversible in time. As an example, consider the system shown in Fig. 2.2: two spheres of the same radii—one fixed and another one settling from above under gravity. For non-touching spheres, trajectories of the moving sphere centre are symmetric with respect to reflection in the plane $z = 0$. Under the time reversal, the gravitational force is reversed and the sphere centre moves backwards along the same trajectory. However, reversibility of the trajectories is broken when two spheres come so close to each other that their surfaces interact by direct forces, such as van der Waals attraction or mechanical reaction of rough surfaces at the contact [35–39]. The reason is that central direct forces are not symmetric with respect to superposition of the time reversal with reflection in the horizontal plane $z = 0$.

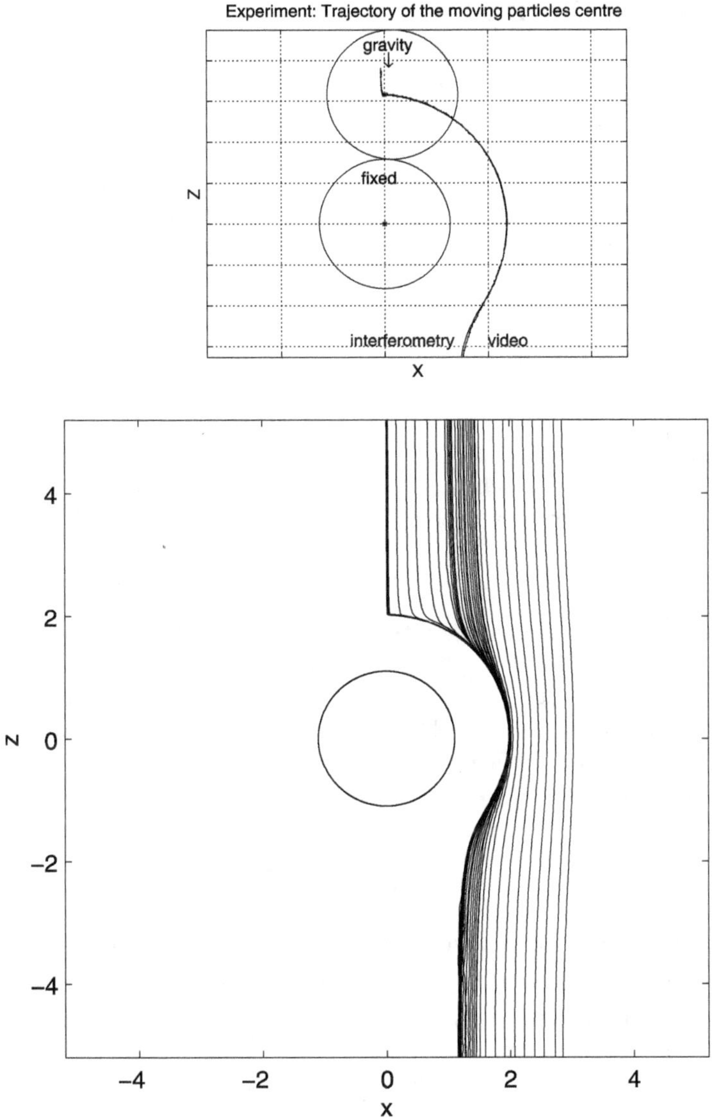

Fig. 2.2 Experimentally observed trajectories of the centre of sphere settling under gravity in a silicon oil towards another fixed sphere of the same radius. Top: reprinted by permission from Ref. [36]. Copyright Kluwer Academic Publisher (2002). Bottom: reprinted with permission from Ref. [37]. Initally, the line of the sphere centres is inclined with respect to gravity. For a large inclination, the surfaces of the spheres are always separated by a fluid, and the trajectories are reversible. However, if the initial inclination is small enough, after some time the surfaces come into contact and the resulting direct forces break the reversibility of the trajectories

2.4 Minimum Energy Dissipation Theorem

We will now give a "variational" view of Stokes flow. A solution to Stokes equations (2.6) and (2.7) is the unique divergence-free vector field that minimises the extensive energy dissipation rate (that is, the energy dissipated by the bulk of the fluid) [20]. In this section, we will state this minimum energy dissipation theorem precisely and sketch a proof. After that, we will apply it to derive "inclusion monotonicity", a principle about particles moving through Stokes flows.

2.4.1 *Statement*

Consider a fluid filling a volume V with an impermeable boundary $\partial V = S$. Let $\mathbf{u}$ be the velocity of a Stokes flow defined by Eqs. (2.6) and (2.7). Let $\mathbf{v}$ be a divergence-free vector field describing a flow in V with the same boundary conditions as $\mathbf{u}$. The minimum energy dissipation theorem is

$$\epsilon^{\mathbf{u}} \leq \epsilon^{\mathbf{v}} \tag{2.9}$$

where $\epsilon^{\mathbf{u}}$ is the extensive energy dissipation rate of the Stokes flow and $\epsilon^{\mathbf{v}}$ is the extensive energy dissipation rate of the other flow.

For an excellent discussion of how these relations for the change of internal energy over time are established physically, see section 3.4 of [18]. For now we will simply use the fact that for an incompressible fluid, the intensive energy dissipation rate (i.e. the energy dissipated per unit volume) Φ is

$$\Phi^{\mathbf{u}} = 2\mu\mathbf{e}^{\mathbf{u}} : \mathbf{e}^{\mathbf{u}} \tag{2.10}$$

$$\Phi^{\mathbf{v}} = 2\mu\mathbf{e}^{\mathbf{v}} : \mathbf{e}^{\mathbf{v}} \tag{2.11}$$

where $\mathbf{e}$ is the rate of strain tensor for the Stokes flow $\mathbf{u}$ given component-wise as $e^{\mathbf{u}}_{ij} = \frac{1}{2}(\frac{\partial u_i}{\partial x_j} + \frac{\partial u_j}{\partial x_i})$ (and similarly for $e^{\mathbf{v}}_{ij}$) and : is the double dot product. Integrating Φ over V gives the extensive energy dissipation rates ϵ, so that

$$\epsilon^{\mathbf{u}} = \int \Phi^{\mathbf{u}} dV \tag{2.12}$$

$$\epsilon^{\mathbf{v}} = \int \Phi^{\mathbf{v}} dV \tag{2.13}$$

Having thus connected the energy dissipation rate to the mechanical properties of the flow, we can now discuss the proof of Eq. (2.9). Because the minimum energy dissipation theorem is proven and discussed in many textbooks, such as [20], we will only give a brief outline. One starts by demonstrating

$$\int (e^{\mathbf{v}}_{ij} - e^{\mathbf{u}}_{ij}) e^{\mathbf{u}}_{ij} dV = 0 \tag{2.14}$$

from Green's theorem, the divergence theorem and Stokes equations (2.6) and (2.7). Then one subtracts Eq. (2.14) from the extensive energy dissipation rate for $\mathbf{v}$ and rearranges

$$2\mu \int e_{ij}^{\mathbf{v}} e_{ij}^{\mathbf{v}} dV = 2\mu \int \left(e_{ij}^{\mathbf{v}} e_{ij}^{\mathbf{v}} - (e_{ij}^{\mathbf{v}} - e_{ij}^{\mathbf{u}}) e_{ij}^{\mathbf{u}}\right) dV \tag{2.15}$$

$$= 2\mu \int \left(e_{ij}^{\mathbf{u}} e_{ij}^{\mathbf{u}} + (e_{ij}^{\mathbf{v}} - e_{ij}^{\mathbf{u}}) e_{ij}^{\mathbf{v}}\right) dV \tag{2.16}$$

$$= 2\mu \int \left(e_{ij}^{\mathbf{u}} e_{ij}^{\mathbf{u}} + (e_{ij}^{\mathbf{v}} - e_{ij}^{\mathbf{u}}) e_{ij}^{\mathbf{v}} - (e_{ij}^{\mathbf{v}} - e_{ij}^{\mathbf{u}}) e_{ij}^{\mathbf{u}}\right) dV \tag{2.17}$$

$$= 2\mu \int \left(e_{ij}^{\mathbf{u}} e_{ij}^{\mathbf{u}} + (e_{ij}^{\mathbf{v}} - e_{ij}^{\mathbf{u}})^2\right) dV \tag{2.18}$$

Which shows that $2\mu \int \left(e_{ij}^{\mathbf{v}} e_{ij}^{\mathbf{v}} - e_{ij}^{\mathbf{u}} e_{ij}^{\mathbf{u}}\right) dV \geq 0$, which by Eqs. (2.10) and (2.11) is the same as

$$\int (\Phi_{\mathbf{v}} - \Phi_{\mathbf{u}}) dV \geq 0 \tag{2.19}$$

By Eqs. (2.12) and (2.13), one sees that Eq. (2.19) is the same as Eq. (2.9), the minimum energy dissipation theorem.

2.4.2 An Application of the Minimum Energy Dissipation Theorem

One advantage of variational principles such as the minimum energy dissipation theorem is that they can be used to describe the behaviour of general rigid bodies in a Stokes flow. We will give an example through the principle of "inclusion monotonicity". If one particle is large enough to completely contain another particle, then we can compare the magnitude of the so-called drag force resulting from a Stokes flow. Inclusion monotonicity follows from the minimum energy dissipation theorem, which we will now show in a manner following [20].

Let a rigid particle 1 take up a volume V_1 with surface $\partial V_1 = S_1$ and compare with the flow around rigid particle 2 taking up a volume V_2 with a surface $\partial V_2 = S_2$. They are undergoing the same translational motion with velocity $\mathbf{w}$ without rotation. The fluid is described by Stokes equations (2.6) and (2.7). Further, the particles have no-slip boundary conditions on their surfaces. The forces the fluid flow exerts on these particles are

$$\mathbf{f}_1 = \oint \boldsymbol{\sigma}_1 \cdot \mathbf{n}_1 dS_1 \tag{2.20}$$

$$\mathbf{f}_2 = \oint \boldsymbol{\sigma}_2 \cdot \mathbf{n}_2 dS_2 \tag{2.21}$$

where $\boldsymbol{\sigma}_i$ is the fluid stress tensor and $\mathbf{n}_i$ is the normal coming out of surface of particle i. The force of the fluid on the particle has the same magnitude but opposite direction.

$$\textbf{Inclusion monotonicity principle: If} \quad V_2 \subset V_1, \quad \text{then} \quad \mathbf{f}_2 \cdot \mathbf{w} \leq \mathbf{f}_1 \cdot \mathbf{w} \tag{2.22}$$

The drag is the component of the fluid force on the particle in the direction of $\mathbf{w}$ [17]. One can see by dividing through by $|\mathbf{w}|$ that inclusion monotonicity relation Eq. (2.22) gives that the magnitude of the drag force on particle 1 is greater than magnitude of the drag force on particle 2. Proof of inclusion monotonicity principle Eq. (2.22) is illustrated in Fig. 2.3 and given below.

Let $\mathbf{u}_1$ be the Stokes flow around the larger particle 1 and $\mathbf{u}_2$ be the Stokes flow around the smaller particle 2. The energy dissipation rate per unit time in the fluid is proportional to the drag [20]

$$\epsilon^{\mathbf{u}_1} = \mathbf{f}_1 \cdot \mathbf{w} \tag{2.23}$$

$$\epsilon^{\mathbf{u}_2} = \mathbf{f}_2 \cdot \mathbf{w} \tag{2.24}$$

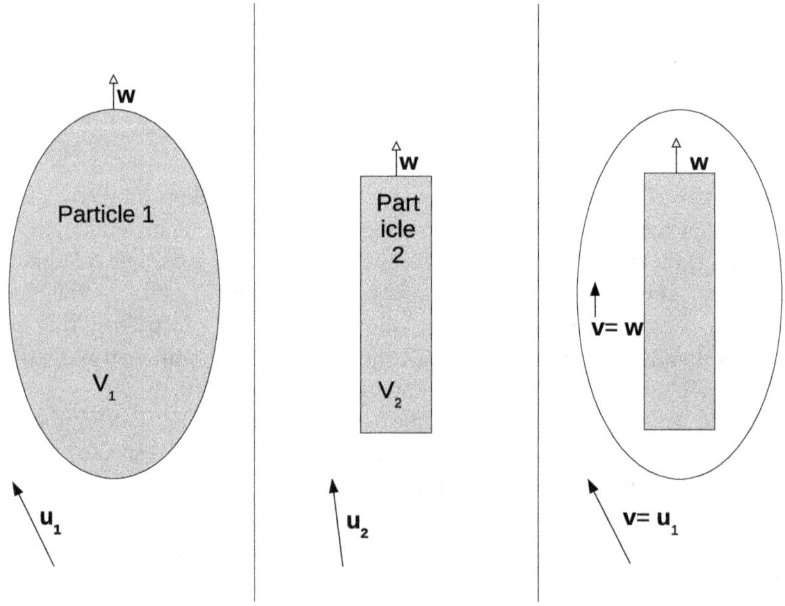

Fig. 2.3 Proof of inclusion monotonicity principle, illustrated. Three panels are drawn with particles in grey and fluid in white. In panel 1 and 2 particle 1 and particle 2 displace volumes such that $V_2 \subset V_1$. The particles are moving with the same velocity $\mathbf{w}$—shown with white tipped arrows—creating fluid velocity fields $\mathbf{u}_1$ and $\mathbf{u}_2$ shown with black tipped arrows. The last panel depicts a non-physical velocity field $\mathbf{v}$ which is equal to $\mathbf{u}_1$ outside of V_1 and $\mathbf{w}$ in $V_1 - V_2$

From the above equations it is easily seen that Eq. (2.22) is equivalent to $\epsilon^{\mathbf{u}_2} \leq \epsilon^{\mathbf{u}_1}$. However because these are Stokes flows for different geometries, the energy dissipation rates cannot be directly compared. Therefore, we will construct a (non-physical) vector field $\mathbf{v}$ in order to compare the energy dissipated by the motion of the two particles. Define $\mathbf{v}$ piecewise to be $\mathbf{u}_1$ outside of V_1 and $\mathbf{v} = \mathbf{w}$, the translational velocity, inside of $V_1 - V_2$. The vector field $\mathbf{v}$ is continuous because of the no-slip boundary condition. Now we will compare the energy dissipation rates of the different vector fields $\mathbf{u}_2$, $\mathbf{u}_1$ and $\mathbf{v}$.

We start by comparing $\mathbf{u}_1$ and $\mathbf{v}$. Because $\mathbf{v}$ is rigid body motion on $V_1 - V_2$, $\mathbf{v}$ does not dissipate any energy there. Outside that set, $\mathbf{v} = \mathbf{u}_1$. Therefore

$$\epsilon^{\mathbf{v}} = \epsilon^{\mathbf{u}_1} \tag{2.25}$$

We now move on to the comparison between $\mathbf{u}_2$ and $\mathbf{v}$. By definition, outside of V_1, $\mathbf{v} = \mathbf{u}_1$ which is a divergence-less vector field. On $V_1 - V_2$, $\mathbf{v}$ is constant, so it is automatically divergence free there. Therefore $\mathbf{v}$ is a divergence-free vector field defined on the same volume of fluid as $\mathbf{u}_2$. Therefore, by minimum energy dissipation theorem we have that $\mathbf{v}$ cannot dissipate less energy than $\mathbf{u}_2$, i.e.

$$\epsilon^{\mathbf{u}_2} \leq \epsilon^{\mathbf{v}} \tag{2.26}$$

Substituting the formulas for the energy dissipation rates Eqs. (2.23) and (2.24) into the above gives inclusion monotonicity Eq. (2.22).

2.5 Limits of the Stokes Approximation

2.5.1 Example of a System Where the Stokes Approximation Does Not Work

The examination of the validity of the Stokes approximation is very revealing of the logical structure of the features of Stokes flow (negligibility of inertial forces, reversibility and the minimum energy dissipation theorem). The most dramatic setting to consider is the famous "Stokes paradox". This paradox arises in the uniform Stokes flow past an infinite rigid cylinder. Suppose that such a cylinder is translating through a fluid with constant non-zero velocity $\mathbf{u}_0$ and has "no slip" on its surface. We suppose that very far from the cylinder, the fluid is at rest: $\mathbf{u}(\mathbf{x}) \to 0$ as $\mathbf{x} \to \infty$. Unfortunately, there is no solution to Eqs. (2.6) and (2.7) consistent with these boundary conditions [16, 19]. In a more general context, Stokes paradox occurs when a non-trivial two-dimensional solution of the Stokes equations (2.6) and (2.7) has no-slip boundary conditions on an object whose surface is a simple closed curve. The velocity is then necessarily logarithmically unbounded as one gets far from the object [18, 40]. More physically, Stokes paradox occurs because the energy dissipated by the cylinder does not decline far from the particle—in other words, due to the minimum dissipation principle.

Other Linear Flow Equations

Because the Stokes approximation is not always justified and Navier–Stokes equations (2.1)–(2.2) are mathematically complicated, it is desirable to have other linear equation systems for fluid flow. We will very briefly give two such example systems in which Stokes paradox demonstrably does not occur but are still tractable: the Oseen and Brinkman equations.

We start with the well-known Oseen equations [19]. Let there be some constant background flow $\mathbf{u}_\infty$ imposed on the fluid. As mentioned before, the Navier–Stokes equations give that the inertial force have the nonlinear form $\rho(\mathbf{u} \cdot \nabla)\mathbf{u}$. If the characteristic velocity of the flow is much less than $|\mathbf{u}_\infty|$, then the main component of inertia is the resistance of the fluid flow against the background flow. We can decompose the local flow as $\mathbf{u} = \mathbf{u}_\infty + \mathbf{u}^O$ and call $\mathbf{u}^O$ the Oseen flow. The inertial force has therefore $\rho\left[(\mathbf{u}_\infty + \mathbf{u}^O) \cdot \nabla\right](\mathbf{u}^O + \mathbf{u}_\infty) = \rho(\mathbf{u}_\infty \cdot \nabla)\mathbf{u}^O + \rho(\mathbf{u}_\infty \cdot \nabla)\mathbf{u}_\infty + \rho(\mathbf{u}^O \cdot \nabla)\mathbf{u}_\infty + \rho(\mathbf{u}^O \cdot \nabla)\mathbf{u}^O$. Because $\mathbf{u}_\infty$ is constant, the middle terms are zero. Furthermore, we are looking for a linear equation, we assume that $|\mathbf{u}^O| \ll |\mathbf{u}_\infty|$. Therefore, we can neglect the nonlinear term. We use the term $\rho(\mathbf{u}_\infty \cdot \nabla)\mathbf{u}^O$ to incorporate inertial forces into linear equations. The equations resulting from the addition of this term to Eq. (2.6) are termed the Oseen equations [41, 42]. The Oseen equations for a steady, incompressible fluid have the form

$$\begin{aligned} \rho(\mathbf{u}_\infty \cdot \nabla)\mathbf{u}^O &= \mu\nabla^2\mathbf{u}^O - \nabla p^O \\ \nabla \cdot \mathbf{u}^O &= 0 \end{aligned} \tag{2.27}$$

where p^O is the pressure associated with such a flow.

There are considerations other than inertial forces that one can take into account for fluid motion in systems described by linear equations. For example, fluid flows in porous media can be described by linear equations. The solid skeleton causes an additional hydrodynamic resistance, which in the Brinkman model of porous media is introduced as a new term. This results in the following equations for fluid velocity $\mathbf{u}^B$ and fluid pressure p^B:

$$\begin{aligned} \mu\nabla^2\mathbf{u}^B - \nabla p^B &= c\mathbf{u}^B \\ \nabla \cdot \mathbf{u}^B &= 0 \end{aligned} \tag{2.28}$$

where c is the ratio of the fluid dynamic viscosity and the permeability of the porous media.

2.5.2 Departures from Reversibility Caused by Inertia

The Stokes approximation—which involves the deliberate neglecting of inertia—cannot be applied to systems in which inertial forces materially contribute to motion. This can be seen in flow visualisation. In symmetric environments, reversibility implies that the flow will also be symmetric [33]. For non-Stokes flows (i.e. $Re \gg 0$), the symmetry in the flow lines breaks down [43]. This departure from reversibility grows with the Reynolds number [44].

Reversible flow was shown in Sect. 2.3.1 to have the interesting property that a spherical particle under an external force parallel to a wall would not experience any lateral motion. In an inertial flow, however, a spherical particle tends to drift away from walls, breaking the reversibility, and causing the "tubular pinch effect" [44], with a different pattern of fluid streamlines.

In the analysis given in Sect. 2.3.1, a sphere rotating in a non-inertial fluid was considered. This leads to a reversible, time symmetric fluid flow [33]. However, a sphere (or a cylinder) which experiences inertial effects while rotating will create an irreversible flow. The inertial forces will cause the cylinder to irreversibly create vortices, which then interact with the rotation of the cylinder in a complex, non-time symmetric way, as shown in Ref. [45].

2.5.3 Accelerating Fluid Example

Even when the Stokes approximation is mathematically coherent, one should think with care how to interpret their results. As an illustrative example, consider a fluid contained within an infinite, impenetrable cylinder with radius R rotating with angular velocity Ω and with no-slip boundary conditions at its surface. The explicit solution of the Stokes equations has the form,

$$\mathbf{u}^S = \Omega r \hat{\boldsymbol{\theta}} \tag{2.29}$$

$$p^S = c \tag{2.30}$$

where c is a constant and $\hat{\boldsymbol{\theta}}$ is the unit vector in the azimuthal direction of the corresponding cylindrical coordinates. The flow velocity is, effectively, rigid body rotation. Pressure is constant in space and therefore there are clearly no centrifugal forces in the radial direction.

Moving on to consider the Navier–Stokes equations, we find that the solution becomes

$$\mathbf{u}^{NS} = \Omega r \hat{\boldsymbol{\theta}} \tag{2.31}$$

$$p^{NS} = \frac{1}{2}\rho\, \Omega^2 r^2 + c \tag{2.32}$$

Like the Stokes case, the fluid is undergoing rigid body rotation. But now a centrifugal force appears in the form of a pressure gradient in the radial direction. The Stokes solution ($\mathbf{u}^S$, p^S) has no forces in the radial direction, but in practice we would expect a centrifugal force in the presence of rotation. In the steady Navier–Stokes case, the centrifugal force per unit fluid volume is pressure gradient. Therefore, the centrifugal term p^{NS} is much more realistic than the constant term p^S.

2.6 Conclusions

Summarising, various properties essential to the understanding of Stokes flow, have been discussed, including negligibility of inertial forces, reversibility and the minimum energy dissipation theorem. Illustrative examples related to these properties have been provided: irreversible trajectories in Stokes flow, inertial terms for the fluid flow generated by a rotating cylinder, force on a rotating sphere close to a solid plane wall, Stokes paradox, energy dissipation for particles of different shapes. The meaning and the limits of the Stokes approximation have been discussed in the context of more general equations.

We will conclude with some analysis of the logical relationship between the assumption of the negligibility of inertial forces, the assumption of reversibility and the minimum energy dissipation theorem.

The assumption that a flow minimises the energy dissipation rate entails that the flow satisfies the Stokes equations. This means that minimum energy dissipation implies both reversibility and the negligibility of inertial forces. Stated contrapositively, irreversible flows or flows with inertial forces dissipate more energy than Stokes flows.

Furthermore, reversibility implies the negligibility of inertial forces. This is equivalent to saying that the presence of inertial forces implies irreversibility. Any term proportional to $\rho(\mathbf{u}\cdot\nabla)\mathbf{u}$, the inertial force term in the Navier–Stokes equation, will make a flow irreversible.

However, neither negligibility of inertial forces nor reversibility does not imply the minimum energy dissipation theorem. Like the Stokes equations (2.6) and (2.7), the Brinkman equations (2.28) and (2.29) are reversible and do not contain inertial terms. But one can now simply apply the proof in Sect. 2.4.1 by substituting $\mathbf{u}^B$ for the general solenoidal vector field $\mathbf{v}$ to find that the Brinkman flow dissipates more energy than the Stokes one. This also shows that, counterintuitively, reversibility is not sufficient to achieve the minimum energy dissipation achieved by Stokes flows.

Acknowledgements This work was supported in part by Narodowe Centrum Nauki under grant No. 2014/15/B/ST8/04359. We acknowledge scientific benefits from COST Action MP1305.

References

1. H.A. Stone, A. Stroock, A. Ajdari, Engineering flows in small devices: microfluidics toward a lab-on-a-chip. Annu. Rev. Fluid Mech. **36**, 381 (2004)
2. B.E. Rabinow, Nanosuspensions in drug delivery. Nat. Rev. Drug Discov. **3**(9), 785 (2004)
3. M. Peltomäki, G. Gompper, Sedimentation of single red blood cells. Soft Matter **9**(34), 8346 (2013)
4. J.K. Nunes, K. Sadlej, J.I. Tam, H.A. Stone, Control of the length of microfibers. Lab Chip **12**(13), 2301 (2012)
5. H.M. Shewan, J.R. Stokes, Review of techniques to manufacture micro-hydrogel particles for the food industry and their applications. J. Food Eng. **119**(4), 781 (2013)
6. A. Perazzo, J.K. Nunes, S. Guido, H.A. Stone, Flow-induced gelation of microfiber suspensions. Proc. Natl. Acad. USA **114**(41), E8557 (2017)
7. J.K. Oh, R. Drumright, D.J. Siegwart, K. Matyjaszewski, The development of microgels/nanogels for drug delivery applications. Prog. Polym. Sci. **33**(4), 448 (2008)
8. E. Kjeang, N. Djilali, D. Sinton, Microfluidic fuel cells: a review. J. Power Sources **186**(2), 353 (2009)
9. Y. Qiu, K. Park, Environment-sensitive hydrogels for drug delivery. Adv. Drug Deliv. Rev. **53**(3), 321 (2001)
10. N.A. Hill, M.A. Bees, Taylor dispersion of gyrotactic swimming micro-organisms in a linear flow. Phys. Fluids **14**(8), 2598 (2002)
11. E.M. Purcell, Life at low Reynolds number. Am. J. Phys. **45**(1), 3 (1977)
12. M. Garcia, S. Berti, P. Peyla, S. Rafaï, Random walk of a swimmer in a low-Reynolds-number medium. Phys. Rev. E **83**(3), 35301 (2011)
13. V. Kantsler, R.E. Goldstein, Fluctuations, dynamics, and the stretch-coil transition of single actin filaments in extensional flows. Phys. Rev. Lett. **108**(3), 38103 (2012)
14. A. Farutin, S. Rafaï, D.K. Dysthe, A. Duperray, P. Peyla, C. Misbah, Amoeboid swimming: a generic self-propulsion of cells in fluids by means of membrane deformations. Phys. Rev. Lett. **111**(22), 228102 (2013)
15. M. Harasim, B. Wunderlich, O. Peleg, M. Kröger, A.R. Bausch, Direct observation of the dynamics of semiflexible polymers in shear flow. Phys. Rev. Lett. **110**(10), 108302 (2013)
16. H. Lamb, *Hydrodynamics* (Cambridge University Press, Cambridge, 1932)
17. L.D. Landau, E.M. Lifshitz, *Fluid Mechanics: Course of Theoretical Physics*, vol. 6, 2nd edn. (Pergamon Press, Oxford, 1959)
18. G. Batchelor, *An Introduction to Fluid Dynamics* (Cambridge University Press, Cambridge, 2000)
19. J. Happel, H. Brenner, *Low Reynolds Number Hydrodynamics: With Special Applications to Particulate Media*, vol. 1 (Springer Science & Business Media, Dordrecht, 2012)
20. S. Kim, S.J. Karrila, *Microhydrodynamics: Principles and Selected Applications* (Dover, Downers Grove, 2005)
21. É. Guazzelli, J.F. Morris, *A Physical Introduction to Suspension Dynamics*, vol. 45 (Cambridge University Press, Cambridge, 2011)
22. M.D. Graham, *Microhydrodynamics, Brownian Motion, and Complex Fluids*, vol. 58 (Cambridge University Press, Cambridge, 2018)
23. G.I. Taylor, Stability of a viscous liquid contained between two rotating cylinders. Philos. Trans. R. Soc. London A Math. Phys. Eng. Sci. **223**(605–615), 289 (1923)
24. J.F. Brady, G. Bossis, Stokesian dynamics. Annu. Rev. Fluid Mech. **20**, 111 (1988)
25. B.U. Felderhof, Many-body hydrodynamic interactions in suspensions. Physica A **151**(1), 1 (1988)
26. A.J.C. Ladd, Hydrodynamic transport coefficients of random dispersions of hard spheres. J. Chem. Phys. **93**(5), 3484 (1990)
27. A.S. Sangani, G. Mo, Inclusion of lubrication forces in dynamic simulations. Phys. Fluids **6**(5), 1653 (1994)

28. B. Cichocki, B.U. Felderhof, K. Hinsen, E. Wajnryb, J. Bławzdziewicz, Friction and mobility of many spheres in Stokes flow. J. Chem. Phys. **100**, 3780 (1994)
29. B. Cichocki, M.L. Ekiel-Jeżewska, E. Wajnryb, Lubrication corrections for three-particle contribution to short-time self-diffusion coefficients in colloidal dispersions. J. Chem. Phys. **111**(7), 3265 (1999)
30. M.L. Ekiel-Jeżewska, E. Wajnryb, in *Theoretical Methods Micro Scale Viscous Flows*, ed. by F. Feuillebois, A. Sellier (Transworld Research Network, 2009), pp. 127–172
31. G.I. Taylor, *Low Reynolds Number Flows* (Educational Services Incorporated. Distributor: Encyclopaedia Britannica Educational Corporation, Chicago, Illinois, 1967)
32. G.I. Taylor, *Film Notes for "Low-Reynolds-Number Flows"* (National Committee for Fluid Mechanics Films, Education Development Center, Educational Services Incorporated. Distributor: Encyclopaedia Britannica Educational Corporation, Chicago, Illinois, 1967)
33. D.J. Jeffrey, in *Sedimentation of Small Particles in a Viscous Fluid*, ed. by E. Tory (Computational Mechanics Publications, Southampton, 1996)
34. D.J. Jeffrey, Y. Onishi, The slow motion of a cylinder next to a plane wall. Q. J. Mech. Appl. Math. **34**(2), 129 (1981)
35. K. Malysa, T. Dabros, T.G.M. de Ven, The sedimentation of one sphere past a second attached to a wall. J. Fluid Mech. **162**, 157 (1986)
36. M.L. Ekiel-Jeżewska, N. Lecoq, R. Anthore, F. Bostel, F. Feuillebois, in *Tubes, Sheets and Singularities in Fluid Dynamics*, ed. by K. Bajer, H.K. Moffatt (Kluwer Academic, Dordrecht, 2002), pp. 343–348
37. M.L. Ekiel-Jeżewska, F. Feuillebois, N. Lecoq, K. Masmoudi, R. Anthore, F. Bostel, E. Wajnryb, in *Third Conference on Multiphase Flow, ICMF 98, Lyon, France* (1998), pp. 1–8
38. K. Masmoudi, Etude des interactions hydrodynamiques particule-particules, particules-parois par interferometrie laser. Phd Thesis, Univ. Rouen, Rouen (1998)
39. M.L. Ekiel-Jeżewska, F. Feuillebois, N. Lecoq, K. Masmoudi, R. Anthore, F. Bostel, E. Wajnryb, Hydrodynamic interactions between two spheres at contact. Phys. Rev. E **59**(3), 3182 (1999)
40. I. Chang, R. Finn, On the solutions of a class of equations occurring in continuum mechanics, with application to the Stokes paradox. Arch. Ration. Mech. Anal. **7**(1), 388 (1961)
41. C.W. Oseen, Uber die Stokes' sche Formel und Uber eine verwandte Aufgabe in der Hydrodynamik. Ark. Mat., Astron. och Fys. **6**, 1 (1910)
42. H. Lamb, On the uniform motion of a sphere through a viscous fluid. Lond. Edinb. Dublin Philos. Mag. J. Sci. **21**(121), 112 (1911)
43. M.V. Dyke, *An Album of Fluid Motion* (Parabolic Press, Stanford, 1982)
44. E.G. Flekkøy, T. Rage, U. Oxaal, J. Feder, Hydrodynamic irreversibility in creeping flow. Phys. Rev. Lett. **77**(20), 4170 (1996)
45. A. Rao, B.E. Stewart, M.C. Thompson, T. Leweke, K. Hourigan, Flows past rotating cylinders next to a wall. J. Fluids Struct. **27**(5–6), 668 (2011)

3

Characterization of Nematic Fluids at the Mesoscopic Scale

Žiga Kos, Jure Aplinc, Urban Mur, and Miha Ravnik

3.1 Introduction to Nematic Fluids

Nematic liquid crystalline fluids are complex anisotropic fluids characterised by internal orientational order of its constituent building blocks [1, 2], which ranges in scales from molecules, macromolecules like DNA, to colloidal rods or platelets. Typically, the orientational order emerges at some temperature or concentration range of building blocks as a result of the geometrical shapes of prolate or oblate building blocks. More recently, nematic order emerged also as an important characteristic of various active fluids, i.e. fluids that can self-propel. Nematic fluids are inherently soft materials, with the orientational order responding as an effective elastic medium to external perturbations, like surfaces or electromagnetic fields. And it is this soft and—optically or structure wise—strong response to external fields which makes nematic fluids potent materials in various applications, including in the fields of optics, photonics, and sensors. The broadest range of applications and experiments with nematic fluids is as at the scales of multiple building elements (which, for example, for molecular nematics, is in the micrometre regime), where mesoscopic approaches prove to be the strongest to describe the systems, as compared to molecular and effective molecular approaches [3, 4], which are used at smaller scales. In view of this, this chapter will present mesoscopic approach to nematic fluids, based on continuum description of the nematic mechanisms and phenomena.

Ž. Kos · J. Aplinc · U. Mur
Faculty of Mathematics and Physics, University of Ljubljana, Ljubljana, Slovenia

M. Ravnik (✉)
Faculty of Mathematics and Physics, University of Ljubljana, Ljubljana, Slovenia

J. Stefan Institute, Ljubljana, Slovenia
e-mail: miha.ravnik@fmf.uni-lj.si

Nematic orientational order is characterised at the mesoscopic scale primarily with the average orientation of the building blocks (also called nematogens), called the director **n** with apolar $\mathbf{n} \rightarrow -\mathbf{n}$ symmetry. This symmetry can be seen from the basic model system of nematic fluids, i.e. fluid of rods, where opposite orientations of a rod are equivalent. The fluctuations and possible asymmetry in the fluctuations of the individual building blocks open additional degrees of freedom given by the nematic degree of order S and biaxiality which then are embedded in the nematic tensor order parameter Q, which further will be defined and explained in this chapter. Overall, the full anisotropic configuration of nematic ordering can vary in three spatial dimensions and can be also time dependent [1]. Equilibrium nematic configurations correspond to a minimum of the free energy. Uniform nematic ordering can be broken by electromagnetic fields or interaction with the boundaries, such as cell or particle surfaces. To satisfy these constraints, often regions of a frustrated orientational order—topological defects—emerge. In a defect, the singularity in the director field is accompanied by a reduction of S, thus effectively melting the nematic into the isotropic state. The shape of topological defects ranges from point defects to line defects and defect walls [2]. Although the topology of a local nematic field is given by the constraints at the boundary, the shape and the metastability of the nematic structures are dictated by the free energy.

Out of the equilibrium nematic alignment is strongly coupled to the velocity field. There are three main effects of the nematic ordering to the velocity field: (1) the rotation of nematic molecules induces material flow, which is known as backflow, (2) even at a fixed nematic configuration (i.e. fixed Q) the fluid viscosity is anisotropic, (3) in active nematics active force dipoles drive the fluid flow. The nematic ordering is affected by the fluid flow through the advection process and the tumbling/aligning dynamics. The coupling between fluid flow and orientational order shows remarkable complexity and provides another aspect to the low Reynolds number fluid mechanics. Out of the equilibrium dynamics can be understood as a competition between effective nematic elasticity, which drives the system towards the equilibrium, and the velocity field and other time modulating fields that promote further deformations of the orientational order. The solutions of such competition range from low Reynolds number turbulence to complex mutually interacting structures in both velocity and Q-tensor fields. Several approaches to the coupled behaviour between fluid flow and nematic ordering were developed, among the most commonly used are Beris–Edwards [5] and Qian–Sheng [6] model of nematodynamics. Velocity effects on the Q-tensor are introduced through the generalised advection terms that compete with the molecular forces that promote the relaxation to the minimum of the free energy. The effect of the orientational order on the fluid flow is given by the stress tensor that is included in the Navier–Stokes equation. The above-mentioned models are based on full Q-tensor, but rather extensively also models based on only director dynamics are used, such as Ericksen–Leslie–Parodi model [1].

Orientational order and material flow of nematic fluids can be shaped into complex microfluidic structures, as a result of combined soft response to the external electromagnetic fields, confinement, and pressure boundary conditions, coupled with the internal effective elasticity and possibly even intrinsic activity [1, 2]. Fascinating field structures in nematic fluids are revealed by theory and experiments, for example, in the context of assisted assembly of colloidal crystals [7], study of complex topological states [8, 9], and sensing applications [10]. Studied systems out of the equilibrium include quench transitions [11], back flow effects [12, 13], and nematic flow in various Poiseuille geometries [14, 15]. Dependence of the nematic viscosity on the director orientation can be used in microfluidic circuits to control the direction of flow and transport of material with selected recent works including defect line assisted transport of colloidal cargo [16] and nematic fluid resistance tunable by electric field [17]. Optical sensors are developed in Ref. [18]. A lot of attention has been also given to the subject of active nematics, both from the experimental and theoretical point of view [19]. The rich variety of structures in mutually coupled fields, each with its own intrinsic symmetry, calls for a deeper understanding of their interactions and potential applications.

Defects in liquid crystals can form regular or irregular structures, depending on the type of confinement, inserted colloidal particles, external fields, and flow [20–23]. Confinement and the surface anchoring can impose and affect defects in liquid crystal. If the confinement has a regular structure, the transition from regular to irregular structures is controllable [20] and the created system can even have a memory effect. Colloidal particles inserted in the liquid crystal introduce the topological defects on their surface. These defects attract each other, so if the colloids have an appropriate design—such as geometry and surface imposed anchoring—this can cause self-assembly of colloidal crystals [7]. Even the passive or active flow itself can cause the reorientation of the director via backflow mechanism, at certain circumstances this gives birth to the structures in the director field [24]. A recently developed system for studying topological defects in passive materials uses confined nematics and nematic colloids [11, 20, 25, 26] where defects of high complexity can be realised ranging from topological defect knots [27], handlebody topological colloids [28], chiral nematic solitons of torons [22] and hopfions [29], quasicrystalline colloidal tilings [30], and droplets with holes [31]. The joint feature of these passive complex defect structures is that defects in 3D generally become delocalised and emerge in the form of topological loops or even networks, called nematic braids. In parallel to their observation, there was also a major development of experimental and theoretical, especially topological, methods for characterisation and control of these advanced defects [26, 32–34].

Complex materials based on nematic fluids are recently attracting a lot of attention, including because of novel ways to control birefringence and the possibility to form complex topological structures. Notably, these specific systems include chiral nematics [22, 27, 35–38] and lyotropic liquid crystals [39–42] which will not be explained in this chapter due to space limitations. Additionally, active matter is a major growing field of science, where nematicity is emerging as an important characteristic of multiple systems. Finally, the goal of this chapter is to give basic introduction to mesoscopic approach to nematic fluids and show some selected exciting fields of development in nematic fluids, such as nematic colloids, topology, and microfluidics.

3.2 Nematic Order Parameters

Nematic liquid crystal fluids consist typically of rod-like or disk-like molecules (building blocks) with no long-range positional order. Due to their anisotropic shape, molecules exhibit orientational order and tend to align along some common direction, usually referred as director $\mathbf{n}$ with both directions $\mathbf{n}$ and $-\mathbf{n}$ being equivalent. The director is a vectorial-like order parameter, which corresponds to the time or ensemble average of molecular orientations $\mathbf{u}$ (see Fig. 3.1).

The director $\mathbf{n}$ bears no information about the degree of orientational order, i.e. the degree of fluctuations of molecular orientations $\mathbf{u}$; therefore, nematic degree of order (scalar order parameter) S is introduced. Nematic molecules in thermodynamic equilibrium assume some direction according to some probability distribution $\rho(\mathbf{u})$. We want to characterise the alignment by one parameter, not the full distribution function $\rho(\mathbf{u})$, which can be quite general. Without loss of generality one can choose z axis along $\mathbf{n}$ and characterise spatial directions of $\mathbf{u}$ by azimuthal angles ϕ and polar angles θ. The first idea would be to use the average $\langle \mathbf{a} \cdot \mathbf{n} \rangle = \langle \cos\theta \rangle$, but this vanishes because the molecules have no distinction between head and tail. The first non-trivial moment is the quadrupole, which is then used to determine the nematic degree of order S as:

$$S = \big\langle P_2(\cos\theta)\big\rangle = 2\pi \int_0^{\pi} P_2(\cos\theta)\rho(\theta)\sin(\theta)\mathrm{d}\theta, \tag{3.1}$$

where P_2 is the associated Legendre polynomial of the second order and $\langle\cdot\rangle$ average over all molecular orientations. The values of S lie on the interval $\left[-\frac{1}{2}, 1\right]$. When all the molecules are perfectly aligned with the director, the nematic degree of order is $S = 1$, whereas $S = 0$ corresponds to the isotropic phase, in which the molecules are oriented randomly, and $S = -\frac{1}{2}$ represents the state where all the molecules are aligned perpendicular to the director.

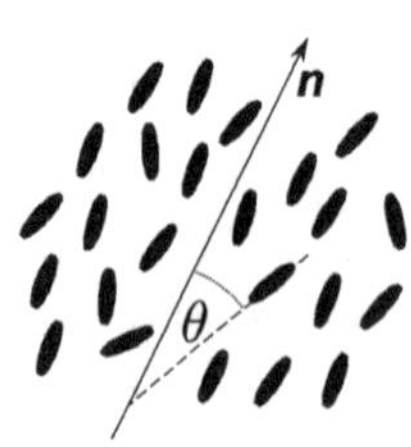

Fig. 3.1 Nematic ordering of rod-like molecules along the director $\mathbf{n}$. θ is the tilt angle of individual molecules with respect to director

The full orientational order in liquid crystals is described by the tensor order parameter Q_{ij} that contains both degree of order S and director $\mathbf{n}$ and also possible biaxiality P. Q_{ij} reads

$$Q_{ij} = \frac{S}{2}\left(3n_i n_j - \delta_{ij}\right) + \frac{P}{2}\left(e_i^{(1)}e_j^{(1)} - e_i^{(2)}e_j^{(2)}\right), \tag{3.2}$$

where $\mathbf{e}^{(1)}$ is the secondary director (perpendicular to $\mathbf{n}$) that characterises the biaxial ordering, and $\mathbf{e}^{(2)} = \mathbf{n} \times \mathbf{e}^{(1)}$. Values of P are in the interval $\left[-\frac{3}{2}, \frac{3}{2}\right]$, where $P = 0$ characterises uniaxial ordering and $|P| = \frac{3}{2}$ corresponds to the perfect ordering along the secondary director $\mathbf{e}^{(1)}$. Order parameter tensor Q_{ij} is a real, symmetric, and traceless tensor. It has five degrees of freedom: nematic degree of order S, biaxiality P, orientation of the director $\mathbf{n}$ (two angles), and orientation of the secondary director $\mathbf{e}^{(1)}$ relative to the director (one angle). Order parameter tensor has three eigenvalues, namely S, $-\frac{1}{2}(S+P)$ and $-\frac{1}{2}(S-P)$, with the corresponding eigenvectors $\mathbf{n}$, $\mathbf{e}^{(1)}$, and $\mathbf{e}^{(2)}$.

3.3 Landau–de Gennes Free Energy Approach

A strong approach at the mesoscopic level to characterise the equilibrium properties of nematic fluids is to use the Landau–de Gennes free energy volume density f, which can be written as:

$$f = f_{\mathrm{NI}} + f_{\mathrm{E}}. \tag{3.3}$$

The model consists of two main bulk free energy density contributions: first contribution f_{NI} describes the phase transition from nematic to isotropic mesophase and second contribution f_E accounts for the spatial elastic deformations of tensor order parameter Q_{ij}.

3.3.1 Landau Theory of Nematic Phase Transition

The stability of nematic mesophases in most nematic fluids depends either on temperature or concentration of the building blocks. In thermotropic nematic liquid crystals, temperature drives the phase transition between the nematic phase with orientational order and isotropic phase with no long-range orientational order, with consequently the order parameter Q_{ij} abruptly dropping to zero. Phenomenological Landau theory is a mean field theory and a well-established approach to model phase transitions. The Gibbs free energy is expanded in vicinity of the transition with respect to the scalar invariants of the order parameter tensor Q_{ij} up to the fourth order [1]. Expansion reads

$$f_{\mathrm{NI}} = \frac{1}{2}A(T)Q_{ij}Q_{ji} + \frac{1}{3}BQ_{ij}Q_{jk}Q_{ki} + \frac{1}{4}C(Q_{ij}Q_{ji})^2, \tag{3.4}$$

where A, B, and C are material parameters and summation over repeated indices is assumed. Parameter $A = a(T - T^*)$ contains temperature dependence which governs the nematic to isotropic transition [2]. In nematic phase, below the temperature T^*, both A and B are negative, but C must be positive to ensure that the free energy density functional is bounded from below. Typical values for molecular nematic liquid crystal are $\approx 10^6\ \mathrm{J/m^3}$. Free energy functional can be rewritten only with S, assuming homogeneous $\mathbf{n}$ and no biaxiality, as:

$$f_{\mathrm{NI}}^{SI} = \frac{3}{4}a(T - T^*)S^2 + \frac{1}{4}BS^3 + \frac{9}{16}CS^4, \tag{3.5}$$

where now the free energy functional exhibits dependence shown in Fig. 3.2. First term drives the transition, second breaks the symmetry of S, and the third bounds the f_{NI} from below. The minimisation of the free energy gives the equilibrium nematic degree of order S_{eq}

$$S_{\mathrm{eq}} = \frac{1}{2}\left(-\frac{B}{3C} + \sqrt{\left(\frac{B}{3C}\right)^2 - \frac{8A(T)}{3C}}\right) \tag{3.6}$$

that holds for $T < T_{\mathrm{c}}$ and homogeneous nematic under no external field.

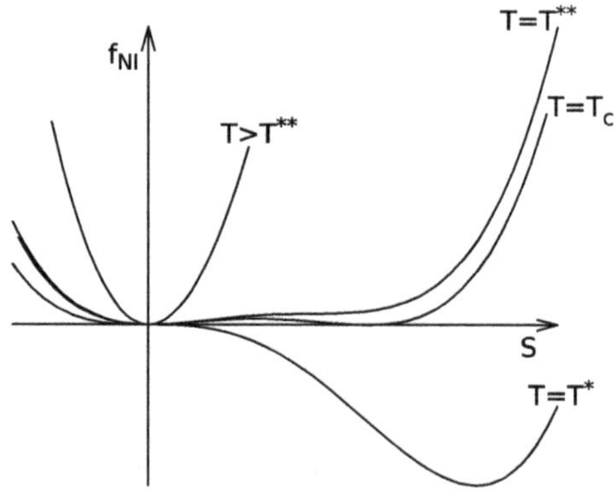

Fig. 3.2 Free energy density f_{NI}^{SI} as a function of the nematic degree of order S for some typical temperatures. Minimum of the free energy density changes with variation of temperature. For $T > T_{\mathrm{c}}$ free energy has stable minimum at $S = 0$ (isotropic phase), but for $T < T_{\mathrm{c}}$, the minimum corresponds to $S \neq 0$ (nematic phase). At temperature T_c, we have coexistence of both phases and transition is of first order. Below the super-cooling temperature T^* the isotropic phase is unstable. T^{**} represents super-heating temperature of nematic phase

3.3.2 Elastic Free Energy

Organisation of nematic into a uniform and homogeneous pattern is energetically preferred. However, such organisation is usually not compatible with boundary conditions and external fields. While subjected to spatial variations of orientational ordering, nematic material effectively acts as an elastic medium, where the elastic deformation can be decomposed into three basic deformation modes, splay, twist, and bend, presented in Fig. 3.3. The elastic free energy is an expansion in small gradients of the order parameter tensor **Q** that penalise nematic distortions from the uniform configuration. The terms are second order in derivatives because first-order terms are disallowed by the symmetry of an achiral nematic [2]. Free energy density reads

$$f_{\mathrm{E}} = \frac{1}{2}L_1\frac{\partial Q_{ij}}{\partial x_k}\frac{\partial Q_{ij}}{\partial x_k} + \frac{1}{2}L_2\frac{\partial Q_{ij}}{\partial x_j}\frac{\partial Q_{ik}}{\partial x_k} + \frac{1}{2}L_3 Q_{ij}\frac{\partial Q_{kl}}{\partial x_i}\frac{\partial Q_{kl}}{\partial x_j}, \tag{3.7}$$

where L_1, L_2, and L_3 are tensorial elastic constants, x_i are Cartesian coordinates, and summation over repeated indices is assumed. Three elastic constants are introduced to quantify all three basic elastic modes. More third-order terms in **Q** are possible in Eq. (3.7), but the choice of three terms is sufficient for matching into three Frank elastic constants. If one assumes uniaxial approximation of the order parameter tensor **Q** ($S =$ const. and $P = 0$), the free energy can be rewritten into the Frank–Oseen free energy density, which is expressed in terms of director **n** and its derivatives [43, 44]:

$$f_{\mathrm{E}}^{\mathrm{FO}} = \frac{1}{2}K_1(\nabla\cdot\mathbf{n})^2 + \frac{1}{2}K_2(\mathbf{n}\cdot(\nabla\times\mathbf{n}))^2 + \frac{1}{2}K_3(\mathbf{n}\times(\nabla\times\mathbf{n}))^2, \tag{3.8}$$

where the terms directly account for splay, twist, and bend deformation modes of the nematic. By comparing both expansions [Eqs. (3.7) and (3.8)] one can get the mapping between tensorial constants L_i and Frank elastic constants K_i, which are usually measured in experiments [45]:

$$K_1 = \frac{9S^2}{4}(2L_1 + L_2 - L_3 S), \tag{3.9}$$

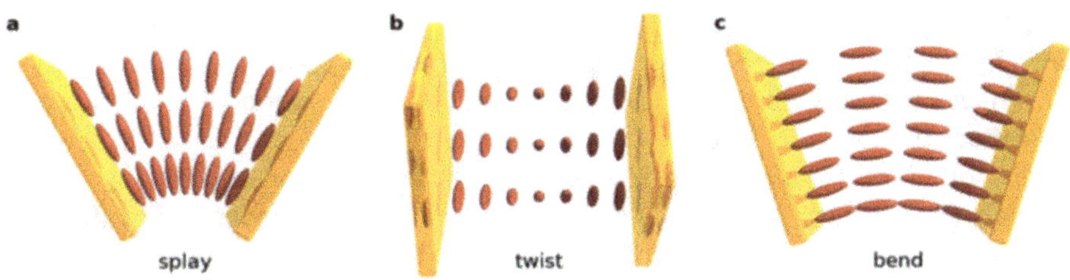

Fig. 3.3 Basic nematic elastic deformation modes: (**a**) splay, (**b**) twist, and (**c**) bend

$$K_2 = \frac{9S^2}{4}(2L_1 - L_3S), \tag{3.10}$$

$$K_3 = \frac{9S^2}{4}(2L_1 + L_2 + 2L_3S). \tag{3.11}$$

Usually, a single elastic constant approximation is used, which sets $L_1 = L$, $L_2 = L_3 = 0$, and $K_1 = K_2 = K_3 = K$. The elastic free energy densities than reduce to

$$f_{\mathrm{E}} = \frac{1}{2}L\frac{\partial Q_{ij}}{\partial x_k}\frac{\partial Q_{ij}}{\partial x_k}, \tag{3.12}$$

$$f_{\mathrm{E}}^{\mathrm{FO}} = \frac{1}{2}K\left[(\nabla\cdot\mathbf{n})^2 + (\nabla\times\mathbf{n})^2\right]. \tag{3.13}$$

The eligibility of one-constant approximation strongly depends on the choice of the material. For a nematic liquid crystal such as 5CB, the values of elastic constants are in the range 10^{-12}–10^{-11} N and differ for around 40% [46]. Elastic constants K_i can also be strongly temperature dependent, but usually not with the same rate; therefore, ratios K_3/K_1 and K_2/K_1 typically also vary [47].

The effective ratio between the ordering free energy f_{NI} and elastic f_{E} contribution to the Landau–de Gennes free energy determines a characteristic length scale of nematics—the nematic correlation length ξ_{N}. Within single elastic constant approximation and uniaxial order parameter tensor, ξ_{N} equals [48]

$$\xi_{\mathrm{N}} = \sqrt{\frac{L}{A + BS_{\mathrm{eq}} + \frac{9}{2}CS_{\mathrm{eq}}^2}}. \tag{3.14}$$

Nematic correlation length determines spatial length scale for the variation of nematic degree of order; therefore, it roughly sets the defects size. For example, in molecular nematics, ξ_{N} is of the order of few nm.

3.3.3 *Surface Anchoring*

The Landau–de Gennes free energy density (Eq. (3.3)) determines the distortion energies in the bulk of the nematic fluids, and needs to be extended in the presence of surfaces with also surface free energy terms. Surfaces surrounding the nematic can affect the nematic ordering by imposing both preferred orientation and the degree of order. Surfaces can in principle impose arbitrary direction in space, with planar (tangential) anchoring and homeotropic (normal) anchoring being most common [49].

Uniform surface anchoring (homeotropic or other fixed direction) can be well described by using Rapini–Papoular like surface free energy density functional [50]:

$$f_{\mathrm{H}} = \frac{1}{2} W_{\mathrm{H}} (Q_{ij} - Q_{ij}^{0})^2, \tag{3.15}$$

which quadratically penalises all deviations from the surface-preferred order parameter tensor Q_{ij}^0 with the strength W_{H}. Besides the preferred direction at the surface, tensor Q_{ij}^0 imposes also surface degree of order and biaxiality. In case of homeotropic anchoring, the preferred tensor is constructed using the surface normal $\boldsymbol{\nu}$, so that $Q_{ij}^0 = \frac{S_{\mathrm{eq}}}{2}(3\nu_i\nu_j - \delta_{ij})$. Typically values of the anchoring strength W_{H} range from 10^{-3} J/m^2 (strong anchoring) to 10^{-7} J/m^2 (weak anchoring) [51].

Some surfaces favour planar degenerate anchoring, where molecules have tendency to align along any direction within a plane, so all azimuthal angles are equally possible. Such anchoring can be described by introducing surface free energy potential [52]:

$$f_{\mathrm{PD}} = W_{\mathrm{PD}} \left(\tilde{Q}_{ij} - \tilde{Q}_{ij}^{\perp} \right)^2, \tag{3.16}$$

where W_{PD} is a constant measuring surface anchoring strength. Model penalises any deviations of $\tilde{Q}_{ij} = Q_{ij} + \frac{S_{\mathrm{eq}}}{2}\delta_{ij}$ from its projection to the surface $\tilde{Q}_{ij}^{\perp} = P_{ik}\tilde{Q}_{kl}P_{lj}$. The projection matrix is defined using the surface normal ν_i as $P_{ij} = \delta_{ij} - \nu_i\nu_j$. The term quadratically penalises deviations of $\tilde{Q}_{ij}$ from its projection. Anchoring strength is frequently characterised with the Kleman–de Gennes extrapolation length ξ_{S}, which is defined as follows [1]:

$$\xi_{\mathrm{S}} = K/W. \tag{3.17}$$

It effectively measures relative strength between nematic elasticity and surface anchoring. Typically, extrapolation length is of the order of 10 nm for surfaces with strong anchoring and ranges to $\xi_{\mathrm{S}} \sim 10\,\mu$m for surfaces with weak anchoring.

3.3.4 Electric Field Effects

Due to their polarisability, nematics are highly responsive to external electric fields. In the Landau–de Gennes framework, the dielectric coupling between nematic and external electric field can be introduced as an additional free energy contribution f_{D} as

$$f_{\mathrm{D}} = -\frac{1}{2}\epsilon_0 \left(\bar{\epsilon}\delta_{ij} + \frac{2}{3}\epsilon_a^{\mathrm{mol}} Q_{ij} \right) E_i E_j, \tag{3.18}$$

where E_i is the external electric field, ϵ_0 the dielectric vacuum permittivity constant, $\bar{\epsilon} = (2\epsilon_\perp + \epsilon_\parallel)/3$ the average liquid crystal permittivity, and $\epsilon_{\mathrm{a}}^{\mathrm{mol}} = \epsilon_\parallel^{\mathrm{mol}} - \epsilon_\perp^{\mathrm{mol}}$ the molecular dielectric anisotropy, which is connected to macroscopic dielectric anisotropy $\epsilon_\mathrm{a} = S\epsilon_\mathrm{a}^\mathrm{mol}$. $\epsilon_\perp^\mathrm{mol}$ and $\epsilon_\parallel^\mathrm{mol}$ are eigenvalues of dielectric permittivity tensor and correspond to eigenvectors perpendicular and parallel to the director. Typical values for 5CB at room temperature are $\epsilon_\mathrm{a} = 11$ and $S = 0.525$, giving $\epsilon_\mathrm{a}^\mathrm{mol} = 21$ [53].

The strength of the electric field can be characterised by introducing another length scale—the electric coherence length ξ_E. The comparison between free energy due to the electric field [Eq. (3.18)] and elasticity (Eq. (3.7)) gives [1]

$$\xi_\mathrm{E} = \frac{1}{E}\sqrt{\frac{L_1}{\epsilon_a \epsilon_0}}, \tag{3.19}$$

where E is a typical electric field in the sample. The effects of electric field are perceptible when the ξ_E is small compared to the system size. In a typical nematic ($L_1 \approx 10^{-11}$ N, $\epsilon_a \approx 10$) and electric field $E = 1$ V/μm the electric coherence length is $\xi_\mathrm{E} \approx 0.3$ μm.

The competition between elasticity and electric field can be more comprehensively studied in the Fréedericksz cell [54]. It consists of nematic liquid crystal being oriented between two solid plates with strong anchoring. The preferential direction imposed by the surfaces may be parallel or perpendicular to the plates, whereas the electric field is always applied perpendicular to the orientational axis imposed by the surface anchoring. There are three different cell setups and each exactly refers to one of the three basic elastic deformation modes (Fig. 3.3), namely splay, twist, and bend [55]. If the electric field exceeds the threshold E_c, the director field deforms. Critical electric field E_c is given as:

$$E_\mathrm{c} = \frac{\pi}{d}\sqrt{\frac{K_i}{\epsilon_0 |\epsilon_\mathrm{a}|}}, \tag{3.20}$$

where ϵ_0 is the vacuum permittivity, ϵ_a the dielectric anisotropy, d the thickness of the cell, and K_i the elastic constant for splay, twist, and bend, respectively. In typical liquid crystals ($K_i \approx 10^{-11}$ N and $\epsilon_\mathrm{a} \approx 10$) in a Freedericksz cell with separation $d = 20$ μm the critical electric field is $E_\mathrm{c} \approx 0.05$ V/μm.

3.3.5 Magnetic Field Effects

Many nematic fluids are diamagnetic. For example, the diamagnetism is especially enhanced when the molecule is aromatic, because the benzene ring effectively acts as a coil. Free energy contribution describing the diamagnetic coupling f_H can be introduced in analogous way as for the electric field, namely

$$f_\mathrm{H} = -\frac{1}{2}\mu_0 \left(\bar{\chi}\delta_{ij} + \frac{2}{3}\chi_\mathrm{a}^\mathrm{mol} Q_{ij} \right) H_i H_j, \tag{3.21}$$

where H_i is the external magnetic field, μ_0 the vacuum permeability constant, $\bar{\chi} = (2\chi_\perp + \chi_\parallel)/3$ the average liquid crystal magnetic susceptibility, and $\chi_a^{\mathrm{mol}} = \chi_\parallel^{\mathrm{mol}} - \chi_\perp^{\mathrm{mol}}$ the molecular dielectric anisotropy, which is connected to macroscopic magnetic susceptibility $\chi_a = S\chi_a^{\mathrm{mol}}$. $\chi_\perp^{\mathrm{mol}}$ and $\chi_\parallel^{\mathrm{mol}}$ are eigenvalues of magnetic susceptibility tensor and correspond to eigenvectors perpendicular and parallel to the director. Values for a typical representative of thermotropic molecular nematic fluids MBBA at room temperature are $\chi_a = 1.23 \times 10^{-7}$ and $S = 0.525$, giving $\chi_a^{\mathrm{mol}} = 2.34 \times 10^{-7}$ [1].

The strength of the magnetic field can be characterised by introducing the magnetic coherence length ξ_B which determines the relative comparison between the free energy contributions due to the magnetic field [Eq. (3.21)] and the nematic elasticity (Eq. (3.7)) gives [1]

$$\xi_B = \frac{1}{B}\sqrt{\frac{L_1}{\chi_a \mu_0}}, \tag{3.22}$$

where B is typical magnetic field in the sample.

3.4 Topological Defects

Frustration of nematic ordering by opposing surfaces or external fields leads to formation of defect regions, where molecular orientation is frustrated and has no preferential orientation. Defect regions are characterised by severe drop of nematic degree of order S (to $S = 0$) and strong spatial distortions of the nematic director $\mathbf{n}$. Defects in nematic liquid crystal can be either points or lines [56] and are usually characterised by topological charges and winding numbers [2, 57–59].

Singular point defects form either in the bulk or on the surfaces. Frequently, point defects in the bulk are named "hedgehogs", whereas those on the surfaces are called "boojums". The topological charge q of point defects can be introduced as an integral over a closed defect-free surface Ω surrounding the defect [2]

$$q = \frac{1}{8\pi} \oint_\Omega \epsilon_{ijk} \mathbf{n} \cdot \left(\frac{\partial \mathbf{n}}{\partial x_j} \times \frac{\partial \mathbf{n}}{\partial x_k} \right) \mathrm{d}S_i, \tag{3.23}$$

where ϵ_{ijk} is the Levi-Civita totally antisymmetric tensor and x_i are Cartesian coordinates. Notice that q is odd in $\mathbf{n}$, which causes that topological charge in nematics is not uniquely defined due to the $\mathbf{n} \to -\mathbf{n}$ symmetry. Three typically observed configurations of point defects with charge magnitude $|q| = 1$ are: radial $\mathbf{n} = (x, y, z)/\sqrt{x^2 + y^2 + z^2}$, circular $\mathbf{n} = (y, -x, z)/\sqrt{x^2 + y^2 + z^2}$, and hyperbolic $\mathbf{n} = (-x, -y, z)/\sqrt{x^2 + y^2 + z^2}$ hedgehog.

The $\pm$ sign differentiates two vector fields with opposite topological charge ($\pm$), but represents the same physical director field. This sign ambiguity is always present in the nematic systems. Frequently, a convention of assigning $+1$ to the radial charge and -1 to the hyperbolic is used, but in general, the vectors in the entire sample must be oriented consistently and the topological charges assigned accordingly [33]. The elastic free energy of isolated singular point defects scales as KR, where R is the size of liquid crystal volume and K some value of Frank elastic constants.

Line defects, named also disclinations, are locally quantified with winding number (strength) m which characterises the symmetry of surrounding director field at some cross-section. For simplification let the disclination line be aligned parallel with the z-axis and the director field is observed in plane perpendicular to it. The in-plane director field can be parameterised with the director azimuthal angle α and the integral over closed loop Γ gives local winding number

$$m = \frac{1}{2\pi} \oint_{\Gamma} \mathrm{d}\alpha. \tag{3.24}$$

Winding number m can be integer and also half-integer, since the states $\mathbf{n}$ and $-\mathbf{n}$ are physically indistinguishable (Fig. 3.4). Note that the definition of winding number assumes that the director is confined to the 2D plane, perpendicular to the disclination. The in-plane director field of a disclination at the coordinate origin can be written as:

$$\mathbf{n} = (\cos\alpha, \sin\alpha, 0) = (\cos(m\phi + c), \sin(m\phi + c), 0), \tag{3.25}$$

with c being typically a constant, which sets the shape and relative orientation of the director field regarding the coordinate frame.

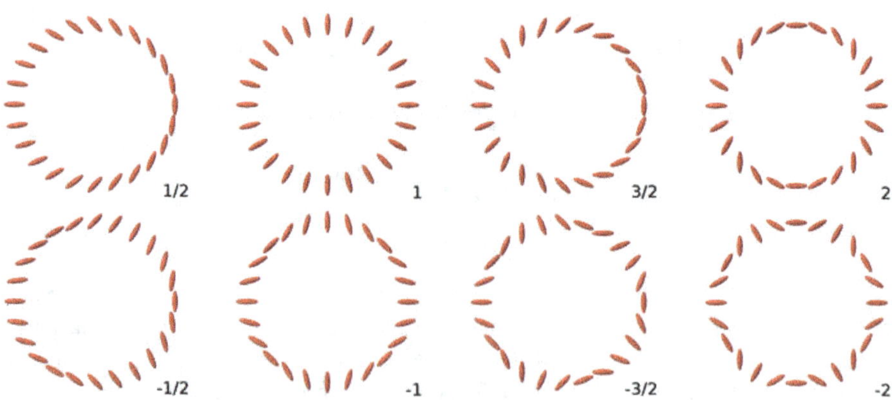

Fig. 3.4 Schematic representation of director field surrounding disclination lines with various winding numbers m. $c = 0$ was used

By surface integrating the Frank–Oseen free energy density (Eq. (3.13)) over the in-plane director field (using single elastic approximation), one can calculate free energy per unit length of the disclination line [2]

$$W_{\text{def}}(m) = \pi K m^2 \ln\left(\frac{R}{r_{\text{core}}}\right) + W_{\text{core}}, \tag{3.26}$$

where r_{core} is the core radius with energy per unit length $W_{\text{core}} \sim \pi m^2 K$, K is some function of Frank elastic constants, and R is the system size. Frank–Oseen approach does not apply for large gradients of the director; hence, the core is introduced to avoid the discontinuity of the director field in the centre of disclination. The proportionality $W \propto m^2$ implies that one disclination of strength $m = \pm 1$ bears two times more energy than two disclinations of strength $m = \pm 1/2$. As a result in 2D systems only $\pm 1/2$ disclinations are stable. Note that total energy of the disclination line is linearly proportional with its length. Line defects can be also closed to loops.

3.4.1 Umbilic Defects

In contrast to the standard defects in liquid crystals, with a discontinuity of the director field at the centre, umbilic defects are continuous everywhere and have no melted (isotropic) core. However, the discontinuity emerges only in the *projection* of the director field to a distinct plane (xy) perpendicular to the far-field orientation (z-axis) [60]. Therefore, it is necessary to bear in mind that umbilics are not fundamentally topological, as they can be continuously transformed into a homogeneous field. Umbilic defects are most commonly created by using electric fields in Hele-Shaw cells [61] with strong homeotropic anchoring, containing a nematic monocrystal with a negative dielectric anisotropy. The external field (above the critical value E_c (Eq. (3.20))), applied perpendicular to the cell surfaces, induces bend distortions to the homogeneous alignment of the director field [55]. The situation is reminiscent to the Fréedericksz transition, the molecules tend to lie parallel to the surface in order to minimise free energy, but importantly no particular direction is preferred in this plane. This degeneration of the tilt direction leads to formation of umbilic defects (Fig. 3.5).

The distorted director field of umbilic defects can be written as [60–62]:

$$\mathbf{n} = (\cos(m\phi + \phi_0)\sin\theta, \sin(m\phi + \phi_0)\sin\theta, \cos\theta), \tag{3.27}$$

where ϕ is the azimuthal angle in xy plane, ϕ_0 some arbitrary constant, and θ the tilt angle. In Hele-Shaw cell of thickness d (plates being at $-d/2$ and $d/2$) the tilt angle θ depends on the intensity of the electric field E and can be written as:

$$\theta(z) = \theta_0 \cos\left(\frac{\pi z}{d}\right), \tag{3.28}$$

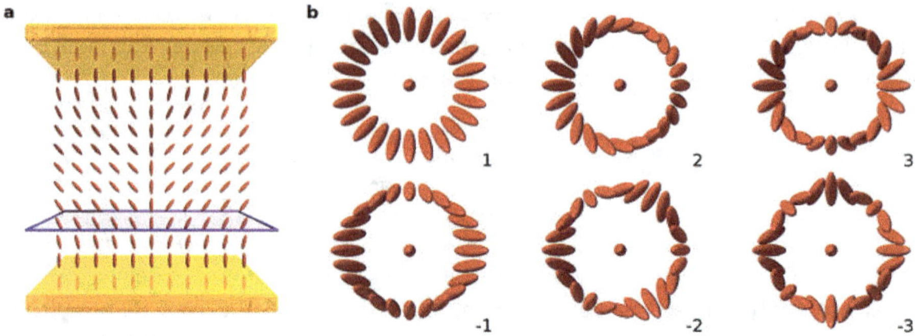

Fig. 3.5 Schematic representation of umbilic defects with $\phi_0 = 0$. (**a**) Hele-Shaw cell with the side view of the umbilic defect $m = +1$. The discontinuity emerges only in the projection of the director field to the plane parallel with the plates (purple). (**b**) Umbilic defects of various umbilic charge (top view)

where

$$\theta_0 \approx 2\sqrt{\frac{E - E_c}{E_c}}. \tag{3.29}$$

Direction of the tilt may be described with the two-dimensional unit vector **c**, which resides in the xy plane, and gives the projected structure of the defect. The tilt directions **c** and −**c** are not equivalent, because **c** is an oriented vector, as a result umbilic charge m must be an integer. The core of umbilics is continuous, which allows for a full calculation of the elastic free energy. For example, for umbilics of strength $m = +1$, the free energy per unit length is proportional to $W \propto K_1 \cos^2 \phi_0 + K_2 \sin^2 \phi_0$, whereas for the umbilics of strength $m = -1$ it is $W \propto \frac{K_1+K_2}{2}$ [60, 63].

3.4.2 *Basics of Topological Theory of Defects*

A comprehensive topological description of defects in liquid crystals requires involvement of the theory of homotopy [64]. The order parameter, namely the director, is a map from a real space, excluding the singularities, to the ground state manifold, which is the topological space of all possible states the director can occupy. The ground state manifold of nematic is real projective plane $\mathbb{R}P^2$, the top half of a unit sphere with opposite points on the equator identified [58]. The main consideration are mappings of i-dimensional spheres enclosing the defects in real space. A line defect can be enclosed by a linear contour ($i = 1$), whereas the point defect by a sphere ($i = 2$); therefore, they are mapped to $\mathbb{R}P^2$ with different homotopy groups $\mathcal{H}_i$. Each element of homotopy group corresponds to a class of topologically stable defects, which can be continuously deformed one to the other [2]. They are topological invariants, previously referred to as the topological charges of the defects [59]. A defect-free state, where director field **n** is equivalent to a constant, corresponds to an identity element of the homotopy group and zero topological charge.

Defect loops and point defects can both be enclosed in a sphere S^2 and thus have a topological index in the second homotopy group. We say they are equivalent in sense of topology as they can be continuously deformed one into another. The winding number m defines the local symmetry of the director field surrounding the defect loop. However, it is the topological charge of a loop q (Eq. (3.23)), as in the case of point defects, that defines its global topological properties.

Topological charge needs to be preserved, also in the process of annihilation or creation of defects. As a result defects in form of points or lines can be created and annihilated in pairs of opposite sign. However, if the confining surfaces impose some preferential direction, this may, in addition with the genus of the surface, determine a non-zero net total topological charge. For example, in the case of droplet with homeotropic anchoring at the surface, the total net topological charge of the nematic within the droplet is $q = 1$. Similarly, the colloidal particles with certain surface anchoring effectively behave as point defects of certain topological charge q. This is compensated by the surrounding nematic with introduction of the defect with the opposite sign $-q$ in order to preserve the total topological charge.

3.5 Nematodynamics

In this section we shall discuss hydrodynamics of nematic liquid crystals, considering the coupling between the nematic orientational ordering and the material flow. The flow field of nematic is given by the generalised Navier–Stokes equation

$$\rho \left[\frac{\partial v_i}{\partial t} + \left(v_j \partial_j \right) v_i \right] = \partial_j \sigma_{ij}, \tag{3.30}$$

where ρ is the density, **v** the velocity, and $\boldsymbol{\sigma}$ the stress tensor which includes beside the standard pressure also the dependence on the anisotropic nematic order in the system. The stress tensor can be written as a sum of the Ericksen stress tensor $\boldsymbol{\sigma}^{\mathrm{Er}}$ which includes the elasticity effects, and viscous stress tensor $\boldsymbol{\sigma}^{\mathrm{viscous}}$. The incompressibility condition

$$\partial_j v_j = 0 \tag{3.31}$$

is assumed. Equations (3.30) and (3.31) have to be complemented by the equation for the evolution of the nematic order parameter, written either in the director or in the Q-tensor form. In the presented formulation, we use typical assumptions from the literature when considering dynamics of liquid crystals (in comparison to statics): two elastic constants and negligible moment of inertia of nematic molecules. Also, we omit the contributions of the electric and magnetic fields to the nematodynamics, which are sufficiently discussed elsewhere [1, 6].

3.5.1 Ericksen Stress Tensor

Elasticity of the nematic internal structure can transmit stresses through the bulk. Consider, for instance, a pair of colloidal particles in a nematic held by an external force at distance d apart. The distortion of the director field and the nematic free energy depend on d. The force between colloidal particles is mediated by the director field and can be calculated through the Ericksen stress tensor. Ericksen stress tensor is derived by considering changes in the free energy due to displacement of nematic molecules [1]. In the director formulation, Ericksen stress is given by

$$\sigma_{ij}^{\text{Er}} = -\frac{\delta\mathcal{F}}{\delta\partial_j n_k}\partial_i n_k - \left(p_0 - f\right)\delta_{ij}. \tag{3.32}$$

In the tensorial formulation, a similar expression holds:

$$\sigma_{ij}^{\text{Er}} = -\frac{\delta\mathcal{F}}{\delta\partial_j Q_{kl}}\partial_i Q_{kl} - \left(p_0 - f\right)\delta_{ij}, \tag{3.33}$$

where p_0 is the external pressure, $\mathcal{F}$ the total free energy of the nematic, and f the bulk free energy density (Eq. (3.3)). In equilibrium, nematic may exert stress on the confining boundaries; however, equilibrium bulk forces, calculated from the divergence of the stress tensor, are exactly zero.

3.5.2 Ericksen–Leslie–Parodi Approach

Ericksen–Leslie–Parodi (ELP) approach formulates the description of nematic hydrodynamics, which determines the coupling between the nematic director field and the velocity field. The model is written in terms of stress tensor σ_{ij}, molecular field $h_i = -\frac{\delta\mathcal{F}}{\delta n_i}$, rotation rate of the director with respect to the background fluid $N_i = \dot{n}_i - \left((\nabla\times\mathbf{v})\times\mathbf{n}\right)_i/2$, and symmetric velocity gradient tensor $A_{ij} = \left(\partial_i v_j + \partial_j v_i\right)/2$. ELP approach relies on considering the processes that contribute to the entropy production as expressions of *thermodynamic forces* (σ_{ij} and h_i) and *thermodynamics fluxes* (A_{ij} and N_i). Phenomenological relations between forces and fluxes must reflect the nematic symmetry $\mathbf{n}\to-\mathbf{n}$ and obey the Onsager reciprocal relations. The resulting expression for the molecular field is [1]

$$h_i = \gamma_1 N_i + \gamma_2 A_{ij} n_j, \tag{3.34}$$

where γ_1 and γ_2 are the viscosity coefficients discussed below. The equation for the time derivative of **n**, derived from Eq. (3.34) has to include a Lagrange multiplier Λ to preserve the unit length of the director:

$$\dot{n}_i = \frac{1}{2}\left((\nabla \times \mathbf{v}) \times \mathbf{n}\right)_i + \frac{1}{\gamma_1} h_i - \frac{\gamma_2}{\gamma_1} A_{ij} n_j - \Lambda n_i. \tag{3.35}$$

Stress tensor is written within the ELP theory as:

$$\sigma_{ij}^{\text{viscous}} = \alpha_1 n_i n_j n_k n_l A_{kl} + \alpha_2 n_j N_i + \alpha_3 n_i N_j + \alpha_4 A_{ij} + \alpha_5 n_j n_k A_{ik} + \alpha_6 n_i n_k A_{jk}. \tag{3.36}$$

The viscosity coefficients γ_1 and γ_2 are functions of the Leslie viscosities α_i:

$$\gamma_1 = \alpha_3 - \alpha_2, \tag{3.37}$$

$$\gamma_2 = \alpha_6 - \alpha_5 = \alpha_2 + \alpha_3. \tag{3.38}$$

Six Leslie viscosities α_i are constrained by Eq. (3.38), meaning that there are five independent parameters within the ELP approach. For typical thermotropic liquid crystals, such as 5CB or MBBA, they are of the order of magnitude 0.001–0.1 Pa s [1] and can be measured by a variety of experimental techniques, such as observing liquid crystals under laminar flow, sound attenuation, time-dependent variation of orienting external fields, or scattering of light [1].

In order to better quantify nematic flow, one can construct and use relevant dimensionless numbers, as is also extensively used in general fluid dynamics. In experiments and simulation involving nematic flow, Reynolds number is typically smaller than 1. Also, Reynolds number does not include effects of orientational order. Better insight into nematic nature of flow is given by comparing elastic forces to the viscous forces in Eq. (3.34), which gives the Ericksen number

$$\text{Er} = \frac{\gamma_1 v/l}{K/l^2} = \frac{\gamma_1 v l}{K}, \tag{3.39}$$

where v is a typical velocity of the problem and l a typical length scale. At small Ericksen numbers, the director dynamics is governed by the elastic terms. At large Ericksen numbers the dynamics is dictated by the velocity profile. Typical values for Ericksen number are Er $\sim$ 1 when considering annihilation of defect pairs [12, 13] or moderately slow flow in microchannels [14] and Er $\sim$ 20 for strong flow in microchannels [14].

Looking at the governing equations of the ELP approach, one can identify the basic mechanisms of nematic hydrodynamics. Below we show few selected basic examples of nematic flows. In Sect. 3.3 we have discussed how the nematic equilibrium orientation profile is defined by a minimum of the free energy. However, out of the equilibrium, nematic orientation field is deformed also by the velocity effects, as described, for example, by Eq. (3.35). Within ELP approach, the director field is distorted by velocity gradients that impose a hydrodynamic torque $\mathbf{\Gamma} = \mathbf{n} \times \mathbf{h}$ upon nematic molecules. At strong flows (i.e. strong Ericksen numbers) the director tends to align in the direction where the hydrodynamic torque $\mathbf{\Gamma}$ vanishes. Director tilt angle, at which this condition is satisfied, is in simple geometries given by the Leslie angle $\theta_{\rm L} = \frac{1}{2} \arccos \frac{1}{\lambda}$, where λ is the alignment parameter, calculated from Leslie viscosities $\lambda = -\frac{\gamma_2}{\gamma_1}$. Figure 3.6 shows director structure for (a) Couette and (b) Poiseuille flow of a nematic fluid at Er $\sim$ 80 and $\lambda = 1.1$. The cell surface imposes strong homeotropic anchoring and the director deforms in the bulk due to hydrodynamic torques. For a Couette flow the director tilt angle in the middle of the sample is close to Leslie angle; however, next to the surfaces, it is continuously deformed to satisfy the boundary condition. Similar situation takes place in the Poiseuille geometry, only that the director tilts in the opposite direction when the shear is reversed. Alignment parameter λ typically reflects the shape of nematic molecules. For $|\lambda| < 1$, hydrodynamic torque does not vanish and the director field continuously deforms in time. An example of such tumbling motion is discussed in Sect. 3.6. λ can also have negative values, in which case it is associated with discotic molecules [65].

Viscous stress tensor (Eq. (3.36)) includes six terms that couple local viscous losses to the director and its time derivative. The meaning of the anisotropy in the stress tensor is clearly seen in a simple geometry, as first considered by Miesowicz. In Miesowicz geometry a nematic is confined between parallel plates and subjected to shear flow. Nematic director is fixed by a strong magnetic or electric field in the direction (Fig. 3.7a) perpendicular to the flow and to the shear, (Fig. 3.7b) along the flow, or (Fig. 3.7c) along the shear. In each of the three cases, stress tensor is substantially simplified and an effective viscosity can be determined. For MBBA, values for the Miesowicz viscosities are $\eta_a \approx 0.042$ Pa s, $\eta_b \approx 0.024$ Pa s, and $\eta_c \approx 0.104$ Pa s [1]. Measurement of Miesowicz viscosities is an important contribution when determining a full set of Leslie viscosities α_i in nematic fluids [1]. For MBBA, the lowest effective viscosity is in the case of the director pointing

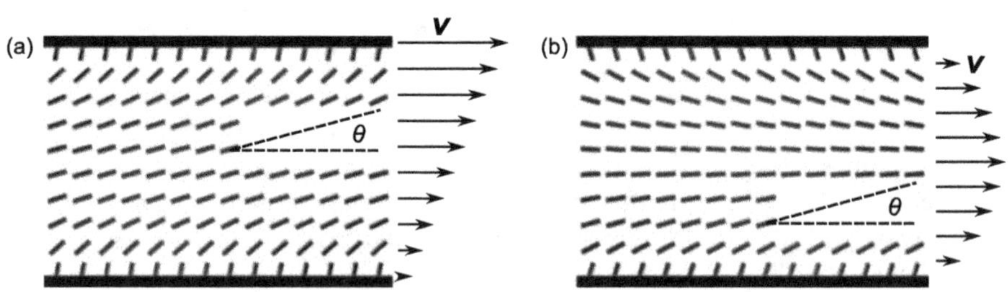

Fig. 3.6 Director distortion in (**a**) Couette and (**b**) Poiseuille flow at Er $\sim$ 80. Hydrodynamic torque due to velocity gradient tends to align the director tilt angle θ towards Leslie angle $\theta_{\rm L}$

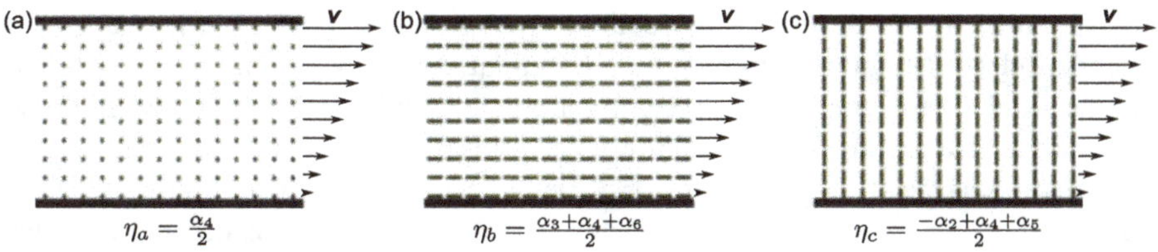

Fig. 3.7 Miesowicz geometry of shear flow in nematic microfluidics. Homogeneous director field is fixed by a strong external field (**a**) perpendicular to the flow and the shear, (**b**) along the flow, or (**c**) along the shear. An effective viscosity η_a, η_b, or η_c can be determined from the viscous stress tensor

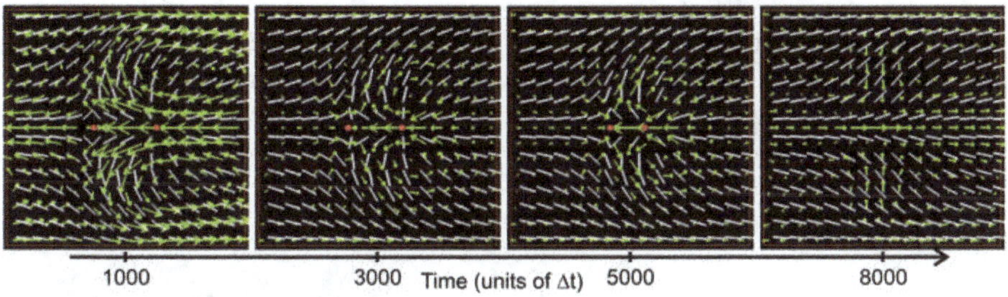

Fig. 3.8 Director and flow field during the annihilation process of two opposite-charged umbilic defects. Director field is shown with white rods and flow field with green arrows. The position of approaching −1 and +1 umbilic is indicated by red dots. Reprinted figure with permission from [I. Dierking, M. Ravnik, E. Lark, J. Healey, G.P. Alexander, J.M. Yeomans, Phys. Rev. E **85**, 21703 (2012)].

along the velocity field. For example, this characteristic can be used in flow-guiding through microfluidic junctions, as discussed in Sect. 3.6.

Viscous stress tensor (Eq. (3.36)) is dependent not only on the direction of the director, but also on its time derivative, which means that time-variation of the director field may induce flows in a nematic, as, for example, in the process of relaxation to a free energy minimum or adaptation to time-varying external fields. Such example is the annihilation of a defect pair, which induces a flow field due to the relaxation of the orientational structure. Interestingly, the speed of the defects is substantially altered by the presence of flow. In annihilation of a $\pm\frac{1}{2}$ defect pair, in particular $+\frac{1}{2}$ defect is advected by the flow, increasing the rate at which the defects are approaching each other [12, 13]. Figure 3.8 shows an example of the annihilation of two opposite-charged umbilic defects, which also approach each other and annihilate [61].

ELP approach gives the nematic contribution to the material flow and velocity contribution to the orientational dynamics in a formulation, where individual mechanisms are easily recognisable. It is particularly useful when considering analytical solutions to the problems of nematic hydrodynamics. However, it suffers from the drawbacks of a director formulation, in particular the defect cores have to be always exempt from the calculation, which can make modelling or theoretical analysis difficult. In the next two sections, we present Beris–Edwards and Qian–Sheng models that not only recover the ELP equations at uniform degree of order S, but also include the coupling between the flow field and the degree of order. These two models are among most commonly used among many different formulations of nematodynamics within the tensorial nematic order parameter [66–69]. Since formulation of nematodynamic equations in terms of the tensor order parameter eliminates the need for the special treatment of defects, it allows to explore problems of further complexity.

3.5.3 Beris–Edwards Model

Beris and Edwards formulate their equations for nematic hydrodynamics through tensorial description of nematic order, where they utilise a generalisation of the Poisson bracket description of thermodynamics [5]. In a typical formulation, their equations are written as [70]:

$$\dot{Q}_{ij} = S_{ij} + \Gamma H_{ij}, \tag{3.40}$$

$$\begin{aligned} S_{ij} =& (\zeta A_{ik} - \Omega_{ik}) \left(Q_{kj} + \frac{\delta_{kj}}{3} \right) + \left(Q_{ik} + \frac{\delta_{ik}}{3} \right) (\zeta A_{kj} + \Omega_{kj}) \\ & - 2\zeta \left(Q_{ij} + \frac{\delta_{ij}}{3} \right) Q_{kl} \frac{\partial v_k}{\partial x_l}, \end{aligned} \tag{3.41}$$

$$\begin{aligned} \sigma_{ij}^{\text{viscous}} =& -\zeta H_{ik} \left(Q_{kj} + \frac{\delta_{kj}}{3} \right) - \zeta \left(Q_{ik} + \frac{\delta_{ik}}{3} \right) H_{kj} + 2\zeta \left(Q_{ij} + \frac{\delta_{ij}}{3} \right) Q_{kl} H_{kl} \\ & + Q_{ik} H_{kj} - H_{ik} Q_{kj} + 2\eta A_{ij}, \end{aligned} \tag{3.42}$$

where $\Omega_{ij} = \left(\partial_i v_j - \partial_j v_i \right)/2$ and $\mathbf{H}$ is the molecular field defined as:

$$H_{ij} = -\frac{1}{2} \left(\frac{\delta \mathcal{F}}{\delta Q_{ij}} + \frac{\delta \mathcal{F}}{\delta Q_{ji}} \right) + \frac{1}{3} Tr \left(\frac{\delta \mathcal{F}}{\delta Q_{kl}} \right) \delta_{ij}. \tag{3.43}$$

Beris–Edwards model as formulated above has three independent viscosity parameters Γ, ζ, and η, from which Leslie viscosities can be determined [70]:

$$\begin{aligned} \alpha_1 &= \frac{\zeta^2}{\Gamma} \frac{9S^2}{2} \left(3S^2 - 2S - 1 \right), & \alpha_4 &= \frac{\zeta^2}{\Gamma} \left(S - \frac{2}{3} \right)^2 + 2\eta, \\ \alpha_2 &= -\frac{\zeta}{\Gamma} \frac{S}{4} (3S + 4) - \frac{1}{\Gamma} \frac{9S^2}{4}, & \alpha_5 &= -\frac{\zeta^2}{\Gamma} \frac{S}{4} (3S - 8) + \frac{\zeta}{\Gamma} \frac{S}{4} (3S + 4), \\ \alpha_3 &= -\frac{\zeta}{\Gamma} \frac{S}{4} (3S + 4) + \frac{1}{\Gamma} \frac{9S^2}{4}, & \alpha_6 &= -\frac{\zeta^2}{\Gamma} \frac{S}{4} (3S - 8) - \frac{\zeta}{\Gamma} \frac{S}{4} (3S + 4). \end{aligned} \tag{3.44}$$

The parameters in the Beris–Edwards model have a clear physical meaning. Rotational diffusion constant Γ sets up the typical time scale of the dynamical processes in the nematic at a given length scale. Parameter ζ is directly related to the alignment parameter in the ELP representation $\lambda = \frac{3S+4}{9S}\zeta$, thus prescribing the Leslie angle in the shear flow or tumbling nature of the nematic. Parameter η affects the isotropic viscosity in the system.

3.5.4 Qian–Sheng Model

A different nematodynamic model based on Q-tensor was formulated by Qian and Sheng [6]. Similar to ELP approach, in their derivation they follow the formalism of thermodynamic fluxes and forces, only within the description of the tensorial nematic order. Viscous stress tensor is written in Qian–Sheng formulation as:

$$\begin{aligned}\sigma_{ij}^{\text{viscous}} &= \beta_1 Q_{ij} Q_{kl} A_{kl} + \beta_4 A_{ij} + \beta_5 A_{ik} Q_{kj} + \beta_6 Q_{ik} A_{kj} \\ &+ \frac{1}{2}\mu_2 N_{ij} - \mu_1 N_{ik} Q_{kj} + \mu_2 Q_{ik} N_{kj},\end{aligned} \tag{3.45}$$

where N_{ij} is the corotational derivative of the Q-tensor

$$N_{ij} = \dot{Q}_{ij} + \Omega_{ik} Q_{kj} - Q_{ik}\Omega_{kj}. \tag{3.46}$$

Time evolution of the Q-tensor is given by

$$\dot{Q}_{ij} = \frac{H_{ij}}{\mu_1} - \frac{\mu_2 A_{ij}}{2\mu_1} + Q_{ik}\Omega_{kj} - \Omega_{ik} Q_{kj}. \tag{3.47}$$

Note that from Eqs. (3.46) and (3.47) the corotational derivative N_{ij} in the equation for the stress tensor can be expressed in terms of the molecular field H_{ij}, which is a form more similar to Beris-Edwards expression (Eq. (3.42)).

Qian–Sheng model is formulated with six viscosity coefficients β_1, β_4, β_5, β_6, μ_1, and μ_2, linked by relation $\beta_6 - \beta_5 = \mu_2$ The number of coefficients is exactly the same as in the Ericksen–Leslie theory. At a constant degree of order, coefficients can be exactly mapped between the two theories, thus allowing for the use of all of the experimentally measured viscosity coefficients given within the ELP formalism.

3.5.5 Towards Active Nematics

Materials that exhibit inherent activity are chemically or biologically different than standard nematic liquid crystalline fluids. However, selected active materials show nematic order, as, for example, kinesin driven microtubule, bacterial colonies, or flocks of animals [19]. A possible approach to describe in particular dense suspensions of such active constituents is to adapt equations of nematic hydrodynamics, as discussed in previous sections, for example, by including an active stress tensor. Active stress arises due to the force profiles that active particles apply on the surrounding, and can be written in the form of [71]

$$\sigma_{ij}^{\text{active}} = -\alpha Q_{ij}. \tag{3.48}$$

The active stress is proportional to nematic tensor order parameter, with the proportionality constant α being the activity. For active particles that exert contractile stress $\alpha < 0$, and for extensile $\alpha > 0$. In such model, if nematic alignment is homogeneous, divergence of the active stress tensor is zero and there are no effective active forces present. However, even in homogeneous alignment, active nematics and polar gels are prone to instabilities [72]. Active forces are particularly high close to defects, where gradients of $\boldsymbol{Q}$ are high. As shown in Fig. 3.9, active forces give rise to the self-propulsion of $+1/2$ defects, which is an important mechanism in chaotic flows in active layers [73], or—at the interface with the passive nematic—active defects can even drive the distortion of the passive medium [74]. Note that in the addition to the presented there are other approaches for describing active nematic systems, such as Vicsek-like models [75], multiscale approaches [76], and minimal hydrodynamic models in terms of solely the velocity field [77].

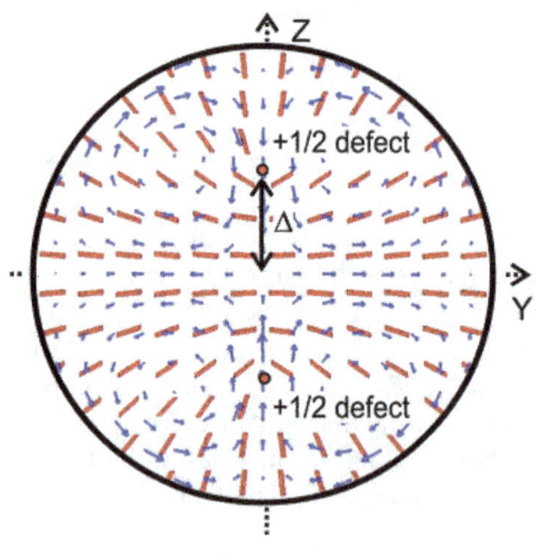

Fig. 3.9 Active flow, developed around $+1/2$ topological defect for active extensile nematic in cylindrical confinement. The director field (solid red lines) and the velocity field (blue arrows) show the mechanism of self-propulsion of a $+1/2$ defect, which competes with the elastic forces on the defect, leading to a fixed nematic structure in time. For contractile active nematics, the direction of self-propulsion of $+1/2$ defects is exactly opposite. Reprinted figure with permission from [M. Ravnik, J. Yeomans, Phys. Rev. Lett. **110**, 26001 (2013)].

3.6 Nematic Microfluidics

In this section we show selected examples of nematic flow in typically confined environment. Fluidity of nematics can have important consequences in many applications, such as in liquid crystal displays [78, 79], or it can lead to complex pattern formation, as, for example, in the process of electroconvection [80, 81]. Rheological properties have been studied for variety of liquid crystalline materials, ranging from thermotropic liquid crystals [14] to cholesterics [82] and suspensions of viruses [83].

3.6.1 Nematic Flows in Channels

In Fig. 3.6 we showed nematic orientation in Couette and Poiseuille geometry at large Ericksen number and with strong homeotropic anchoring at the walls. Similar setup where the preferred alignment of the director at the walls is planar and perpendicular to the flow was investigated by Pieranski and Guyon [84]. An undistorted configuration in such Poiseuille geometry is shown in Fig. 3.10a. In particular at moderate Ericksen numbers, where nematic orientation is at a competition between elastic and hydrodynamic effects, there are two possible director orientations in such geometries of the shear flow, as shown in Fig. 3.10b. These two conformations occur in nematics with Leslie viscosities α_2 and α_3 of the same sign due to hydrodynamic torques that act on the director as soon as it slightly fluctuates from the undistorted alignment in Fig. 3.10a [2]. Since these

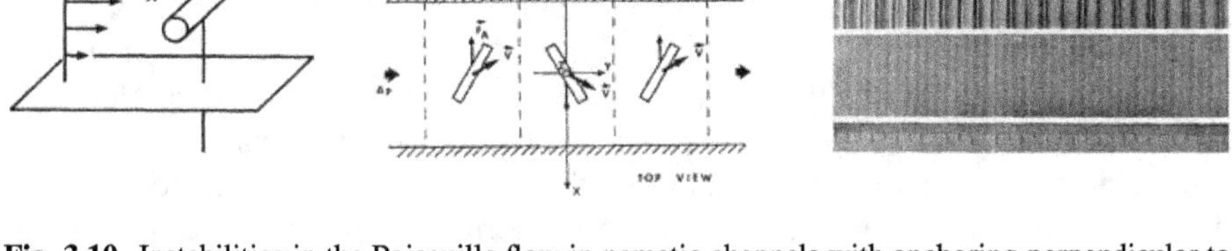

Fig. 3.10 Instabilities in the Poiseuille flow in nematic channels with anchoring perpendicular to the flow and flow gradient. (**a**) Geometry of the problem with the director field undistorted by the flow. (**b**) Two possible configurations of the director in the weak shear gradient. Left cylinder shows the alignment preferred by the nematic elasticity and surface anchoring. (**c**) Creation of the rolls in the velocity field due to the force F_A, two solutions shown in (**b**) alternate along the channel. (**d**) Experimental photograph of flow instabilities in Poiseuille geometry. Reused and adapted with permission from publisher [E. Guyon, P. Pieranski, Poiseuille flow instabilities in nematics. J. Phys. Colloq. **36**, C1 (1975)]

director distortions compete with the planar anchoring at the surfaces, the transition to the distorted state occurs after a certain finite shear threshold is exceeded [84]. Above the threshold the director configuration varies in the z direction, leading to an additional force in the x direction $F_A = \partial_z \sigma_{xz}$ that is shown in Fig. 3.10c. The force leads to the creation of rolls in the velocity profile (Fig. 3.10c), which have a well-defined wavelength. The rolls coincide with the two solutions for the director reorientation. In fact, depending on the driving pressure difference and frequency, a range of mechanisms can be found, which lead to hydrodynamic instability and creation of rolls [84, 85]. Figure 3.10d shows a photograph, revealing rolls with different wavelengths due to different hydrodynamic instabilities in nematics. This example shows the creation of instabilities in the velocity field and in the nematic orientation, once the coupling terms are introduced into the stress tensor and the equation for the director orientation.

As discussed in Sect. 3.5.2, in nematics with alignment parameter $|\lambda| > 1$, hydrodynamic torque disappears for certain angles of the director with respect to shear flow. For $|\lambda| < 1$ this is no longer the case and the hydrodynamic torque prefers to continuously rotate the director. In Ref. [86] two-dimensional nematic channels were explored in aligning and in tumbling regime. In the aligning regime, the director profile reaches a stationary orientation, which is not the case for tumbling motion. For straight channels, tumbling regime shows a series of π turns of the director across the channel. This turns are continuously generated and annihilated in time. In channels with variable width, director structure with π turns becomes unstable and pairs of opposite-charged defects are generated.

Dependence of the nematic viscosity on the director orientation can be used in microfluidic circuits to control the direction of flow and transport of material. One such example is given in Ref. [17], where electric field is used to switch nematic orientation in a channel. In Ref. [17] the preferred alignment of the director is along the channel, providing an effective Miesowicz viscosity η_b to the flow. Local electric field was used to impose director alignment along the shear, effectively increasing the viscosity for a factor of ~4 to the Miesowicz value of η_c. When flow reached a Y-junction, most of the nematic fluid flows to the branch with the lower effective viscosity. It was shown that this mechanism could be used for particle sorting by turning the electric field on and off in individual branches of a Y-junction and by doing so controlling the flow through the junction. The colloidal particles go into the channel which has a stronger flow rate. Note that switching nematic orientation in the channel is not the only mechanism to transport and guide the cargo in nematic microfluidic circuits. Sengupta and co-workers demonstrated that in channels with hybrid anchoring conditions (anchoring along the normal at three sides of the rectangular channel, and anchoring along the channel at on side), a defect line can be guided through crossings of different channels. Colloidal cargo is then pinned to the defect line and advected along it by the flow [16].

3.6.2 Nematic Microfluidic Junctions

Complex flow field profiles, as induced, for example, in junctions of nematic microchannels, can be used to create and study topological nematic defects in a controlled environment. In Ref. [87] junctions of 4, 6, and 8 microchannels are used to create nematic defects with effective charge −1, −2, and −3, respectively (Fig. 3.11a–c). The main mechanism for the creation of such defects is the fact that in the centre of a nematic microchannel at sufficiently large Ericksen numbers, the director turns along the channel, with the mechanism shown in Fig. 3.6b. Total net topological charge of −2 and −3 imposed in the junctions is realised by multiple defects with 3D topological charge of −1 as this is indeed energetically more favourable state that cannot be further divided in smaller individual charges. In observed microjunctions two topological structures are present—a defect in the orientational field of the nematic and a stagnation point in the velocity field. Figure 3.11d–g shows how the cross-talk between this two topological singularities is probed by applying a pressure pulse in one of the channels. A fast shift of the stagnation point is always followed by a slow response of the nematic defect. In a stable configuration the position of both structures coincides. If the pressure was reduced in one of the outflowing channels (Fig. 3.11e, g) nematic defect first moves

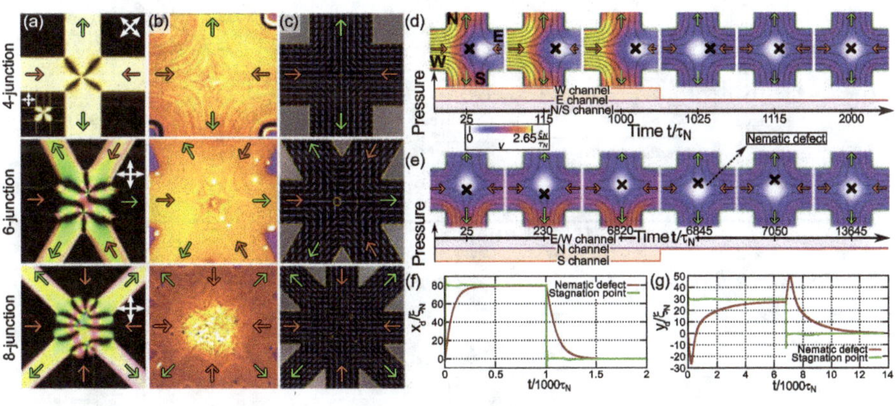

Fig. 3.11 Creation of topological defects within nematic microfluidic junctions [87]. One, two, and three defects of charge −1 are created in junctions of 4, 6, and 8 microchannels, respectively. (**a**) Polarisation micrograph of the nematic structure. (**b**) Hydrodynamic stagnation point in the centre of a junction shown by epifluorescent imaging of fluorescent tracers. Panels (**a** and **b**) are courtesy of A. Sengupta. (**c**) Details of the nematic structure revealed by numerical simulations. The effective interaction between the stagnation point and the defect in the topological nematic orientation is probed by inducing a pressure pulse in (**d**) West or (**e**) South channel. The pressure pulse quickly shifts the stagnation point, marked by a white spot in the colourmap of the velocity magnitude. In (**d**) the relocation of the stagnation point is followed by a gradual shift of the nematic defect, as shown in (**f**). In (**e** and **g**) after the stagnation point is shifted, nematic defect first moves away from the stagnation point and only then gradually moves towards it. In both cases the stagnation point and the nematic defect return to original position after the pressure is restored

downstream and then gradually approaches the stagnation point, moving against the velocity direction. During the shift of the nematic defect, the stagnation point is more or less stationary. This is an example of a cross-talk between topological structures of different fields.

Nematic liquid crystals confined to porous networks are of particular interest due to their memory effects and switching possibilities, providing a route towards new optic and photonic materials [20, 88, 89]. The nematic alignment inside porous confinement can be controlled by flow [90, 91]. In Fig. 3.12 we show flow-induced dynamics of a defect structure inside a junction of six cylindrical capillaries. In a cylindrical confinement with homeotropic anchoring and without flow, nematic director prefers the escaped alignment, in which case the director in the middle of the channel points along the channel direction. This leads to a variety of equilibrium structures, depending on the direction of the director escape in individual channels [89]. One of such structures is shown in the first snapshot of Fig. 3.12, where a -1 topological defect resides in the centre of the junction. Preferred nematic alignment in a capillary when flow is switched on is with the direction of the director escape along the flow. This leads to the flow-induced

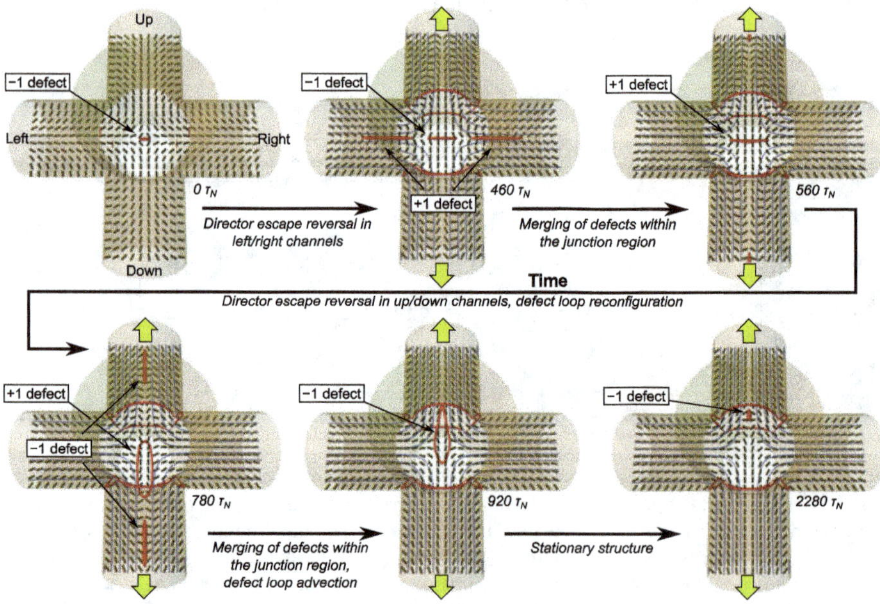

Fig. 3.12 Flow-induced nematic structures in porous microfluidic channel networks [92]. Transformational dynamics of a -1 nematic defect in a junction of six cylindrical micropores is observed with the direction of the director escape in the initial equilibrium configuration away from the junction in left and right channel, and towards the junction in up, down, front, and back channel. Transformational dynamics is characterised by flow-induced director escape reversal in individual channels and merging of multiple defects into one. Time is measured in units of nematic characteristic time scale $\tau_{\mathrm{N}} = \frac{\xi_{\mathrm{N}}^2}{\Gamma L}$

reconfiguration of the defect structure in a microjunction (Fig. 3.12). Upon the director escape reversal in left and right channel two $+1$ defects are created. They merge with the -1 defect in the junction centre, forming a defect structure with topological charge of $+1$. Similar process is repeated as a -1 defect is created in the up and in the down channel which merge with the preexisting $+1$ defect, leading to the formation of a -1 defect in the junction. The position of the defect is slightly off-centre since it is advected by the flow. Fig. 3.12 shows a transformational dynamics of a -1 nematic defect, induced by the flow through the channels. Depending on the geometry of the initial equilibrium structure and the arrangement of the flow towards and away from the junction, a variety of switching processes and flow-stabilised structures is possible [92]. This example shows how porous networks with microfluidic functionality can be turned into advanced platform for generation of various topological field states.

3.6.3 Colloidal Particles in Nematic Microfluidic Environment

In nematic colloidal dispersions, the drag force exerted upon the spherical particles is dependent on the particle velocity with respect to the director far field and the nematic structure around the particle [93]. Similar to the viscosity in the Miesowicz geometry, the effective viscosity for spherical particles is higher if they are dragged perpendicular to the director, compared to the movement along the director. The problem of drag force on spherical particles even gains in complexity, if colloids are introduced in chiral nematic liquid crystals, as, for example, in Ref. [94], where spherical particles with planar degenerate anchoring were dragged through cholesteric by a constant force at small Ericksen numbers. It was observed that for $\mathrm{Er} \ll 1$ the drag force on the particle scales linearly with the velocity. However, there is a distinct dependence on the particle radius R: for the motion along the cholesteric pitch, effective viscosity scales as $\eta(R) \sim R^{0.7}$, while for the particle motion perpendicular to the cholesteric pitch no definite scaling of η with particle radius is observed [94]. In the next section we shall discuss further implications of colloidal nematic systems, in particular in the view of self-assembly and nematic configurations due to complex-shaped microparticles.

3.7 Nematic Colloids

Nematic colloids are a soft material composed of particles, droplets, or bubbles embedded in a nematic fluid [21]. Nematic colloids attract great interest as they show effective elastic interaction between the particles, which originates from the nematic elasticity, shown in Sect. 3.7, in addition to conventional colloidal interactions such as steric, Coulomb, and van der Waals interactions. The exact profile, strength, and range of such elastic interactions are strongly affected by the surface properties of the particles, their shape, size, and topology, as well as external confinement, geometry, and possible external field. Nematic colloids are notably explored as novel materials with complex topological properties and as novel birefringent photonic materials, including for use as photonic crystals and metamaterials.

In this chapter we give review of selected nematic colloidal systems. First possible nematic director field configurations around a single spherical particle immersed in a liquid crystal are shown. Then elastic interparticle interactions are explained, which allow to organise colloids into larger structures. In the last part we introduce complex-shaped particles and particles with different topologies and their features.

3.7.1 Single Spherical Particle

A colloidal particle immersed in a nematic deforms the director field where the deformation depends strongly on the boundary conditions at the surface of the particle. The resulting nematic configuration is typically governed by an interplay between the bulk elastic and surface free energy.

In case of weak homeotropic anchoring (Fig. 3.13c), the surface terms in the total free energy are smaller than the bulk terms and the director field remains almost undistorted. No topological defects occur in the bulk. However, if the anchoring is strong, the director around the particle typically imposes frustration on the surrounding bulk orientation. Therefore, the nematic director cannot adapt to this frustration without creating orientational singularities. A point like hyperbolic -1

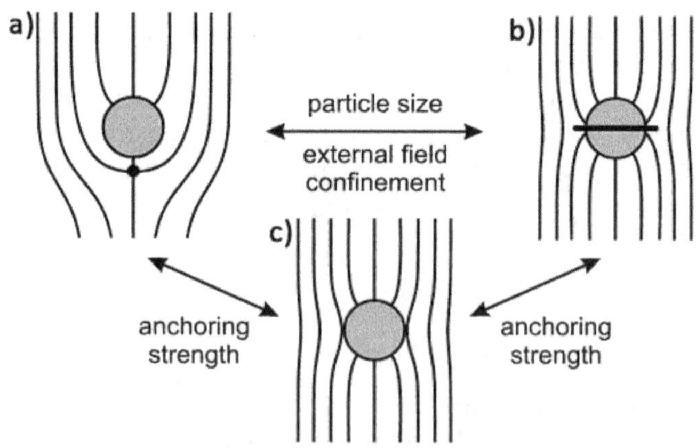

Fig. 3.13 Director field profiles around a spherical particle with homeotropic anchoring immersed in nematic fluid: (**a**) elastic dipole with hyperbolic -1 defect, (**b**) elastic quadrupole with singular Saturn ring defect, and (**c**) elastic quadrupole with no singular defect

defect (hedgehog) or a ring defect is formed around the particle with hometropic surface anchoring to achieve net zero topological charge characteristic of uniform field (Fig. 3.13a) [95]. The nematic configuration of particle accompanied with the hyperbolic -1 defect has the symmetry and profile of an elastic dipole. Namely, the defect causes the deformations of the nematic far-field director that mimic the dipolar electric field caused by electric charge distributions and have the same positional dependence in terms of multipolar expansion.

In addition to the elastic dipoles, an elastic quadrupole can form and is characterised by a $-1/2$ disclination loop encircling the particle (Fig. 3.13b) [96]. Saturn ring emerges in regimes of generally smaller anchoring strength, smaller particle size, or stronger confinement with geometry or external fields. Effectively, it can be pushed to the surface of the particle (or virtually even within the particle) if anchoring is weak enough. By opening the point defect into a ring, the head–tail symmetry is established when the ring reaches the equatorial plane and the defect structure together with the sphere represents an elastic quadrupole. Note also, that hedgehog -1 point defect is topologically equivalent to a $-1/2$ disclination loop (Saturn ring). In case of degenerate planar anchoring, two surface—boojum—defects are formed at the opposite poles of the particle, which also result in the quadrupolar nature of the structure. The nematic elasticity of the liquid crystals causes highly anisotropic interparticle interactions—i.e. with repulsive and attractive directions—and can lead to self-assembly of particles into larger structures.

3.7.2 *Interparticle Interactions*

Elastic deformations of the nematic director field, caused by colloidal particles are energetically unfavourable which leads to inter-interactions between the particles that minimise regions of such distortions. The long-range orientational order of liquid crystal is reflected also in long-range nematic interparticle interactions.

The type of long-range interactions depends on the symmetries of the distortions in the director field, induced by particles. The force between two colloidal particles in a nematic host medium can be measured experimentally and it has been shown that the interaction potential between particles with strong surface anchoring, which generates hedgehog defect, is anisotropic and proportional to the third power of inverse distance between the particles, similar as dipole electrostatic interaction [25]. The interaction potential between particles with weak anchoring or with Saturn ring defect was shown to have quadrupolar symmetry and is proportional to the fifth power of the inverse distance between the particles [97]. The binding energy of approximately micron sized colloids can reach the order of 1000 $k_B T$ for dipolar type interactions and of the order of 100 $k_B T$ for quadrupolar interactions. Landau–de Gennes free energy approach—presented above—has been used to calculate interparticle interaction, giving excellent agreement with the experiments [48, 98].

Two equally oriented elastic dipoles in a uniform nematic cell attract if they are collinear (Fig. 3.14a). However, if oriented in the opposite directions they repel in

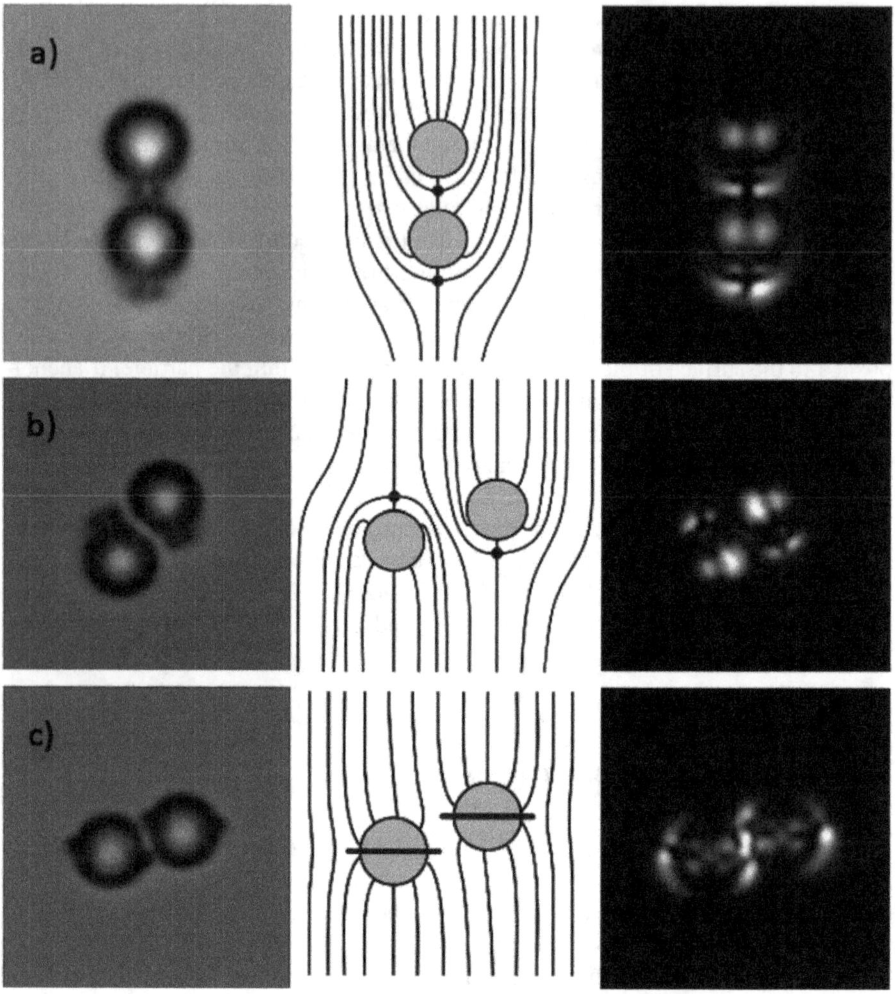

Fig. 3.14 Micrographs, director field configurations, and polarisation micrographs for stable particle pair configurations of (**a**) parallel elastic dipoles, (**b**) anti-parallel elastic dipoles, and (**c**) elastic quadrupoles. From [S. Žumer, I. Muševic, M. Ravnik, M. Škarabot, I. Poberaj, D. Babič, U. Tkalec, Nematic colloidal assemblies: towards photonic crystals and metamaterials, SPIE Proc. **6911**, 69110C (2008)]. Reprinted with permission from SPIE Publications

the direction along the far-field director, but attract sideways (Fig. 3.14b). Because there are only two attractive sites available around the sphere for dipoles oriented in the same directions, they form linear chains along the direction of the director (Fig. 3.15) [99, 100]. One should also note that the particles do not come in full contact with each other—they are separated by a small margin, which indicates the presence of short range repulsion. Typically, those short range repulsive interactions are again of elastic origin resulting from significant short-particle distance director field deformations.

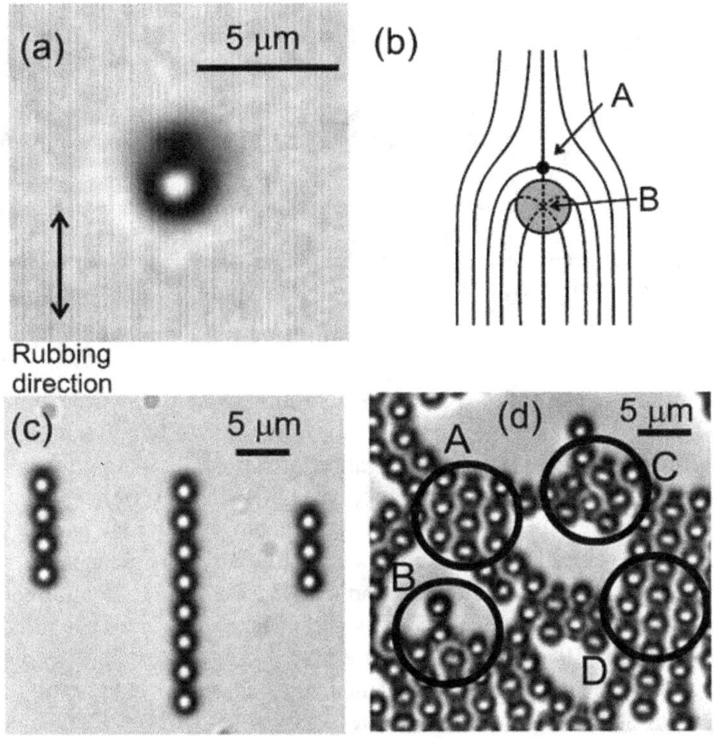

Fig. 3.15 Aggregation of dipolar nematic colloids. (**a**) Optical micrograph of a single colloidal particle with the defect structure of dipolar symmetry. (**b**) Schematic presentation of the nematic director field, where A denotes the actual hyperbolic point defect and B the virtual radial point defect inside the particle. Together they form a topological dipole. (**c**) Elastic dipoles form linear chains. (**d**) Chains bond together into 2D crystalline colloidal cluster. Letters denote different types of bonds. Reprinted figure with permission from [M. Škarabot, M. Ravnik, S. Žumer, U. Tkalec, I. Poberaj, D. Babič, N. Osterman, I. Muševič, Phys. Rev. E **76**, 51406 (2007)]

Elastic quadrupoles bind in different directions as elastic dipoles. Analogous to the electric case, quadrupoles repel if they are perfectly aligned ($\theta = 0°$) or perpendicular ($\theta = 90°$) to each other and attract for some finite angle θ (Figs. 3.14c, 3.16a, b), which depends on multiple parameters and is generally in the range of $\theta \sim 20$–$30°$. Elastic quadrupoles form zig-zag chains, generally perpendicular to the director field (Fig. 3.16c). The particles pairs can form via four different attractive sites, as shown also in 2D colloidal crystals (Figs. 3.15d, 3.16d). Typically, interactions of elastic quadrupoles are of one order of magnitude weaker than the interactions of elastic dipoles [101].

Interparticle interactions strongly depend on the configuration of the director field. In examples presented so far, colloidal particles were immersed in a nematic cell with a uniform field. However, additional types of structures can be observed

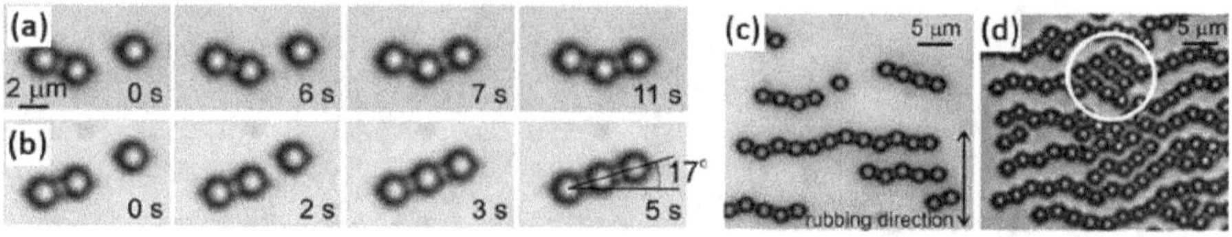

Fig. 3.16 Particles assemble along different directions and form (**a**) kinked chains, (**b**) linear chains, (**c**) longer self-assembled chain structures and (**d**) 2D quadrupolar crystal-like structures. Reprinted figure with permission from [M. Škarabot, M. Ravnik, S. Žumer, U. Tkalec, I. Poberaj, D. Babič, N. Osterman, I. Muševič, Phys. Rev. E **77**(3), 31705 (2008)].

in more complex field configurations. As a strong example, multiple emulsions of nematic liquid crystal and water droplets were used to study colloidal interactions in a radial director field [101]. Conservation of topological charge was observed experimentally and the effects of different types of anchoring were studied.

In addition to dipole and quadrupole interactions, colloidal particles can be also bound by escaped defect lines, where the director escapes in the third dimension. Two particles with homeotropic anchoring can share an escaped—i.e. non-singular—line with an effective topological charge of -2 in a so-called bubble-gum configuration [102] with the binding force being almost independent of the separation between the particles. In a nematic cell such configuration is rarely observed as the state is metastable; however, in twisted (chiral) cells such pairs form spontaneously and can also connect into larger 2D colloidal crystals [103].

Differently, two particles with Saturn ring defect can be entangled by a single escaped loop, acting as an elastic string [26]. Three different nematic configurations can be achieved by applying laser tweezers and thermally quenching (Fig. 3.17) the nematic around the particles: figure-of-eight (Fig. 3.17a), figure-of-omega (Fig. 3.17c, g), and figure-of-theta (Fig. 3.17d, h). The states are again metastable in a uniform nematic cell, but enable binding multiple particles into linear structures. Spontaneous entanglement can be realised in a twisted nematic cell, which allowed for the investigation of the knot theory on the defect lines [104].

3.7.3 Assembly and Self-assembly of Colloidal Structures

Colloidal systems attract major interest also because of their ability to interact with light [105]. Periodic structures of dielectric media with a cell size, comparable to wavelength of light, also known as photonic crystals, enable guiding and control of the light at the microscale level. Nematic colloids present an interesting platform for development of soft matter photonics due to their self- and directed-assembly, responsiveness to external stimuli, and strong binding interactions. They also show interesting potential in the development and research of topological photonic materials and metamaterials, including at nanoscale [106, 107].

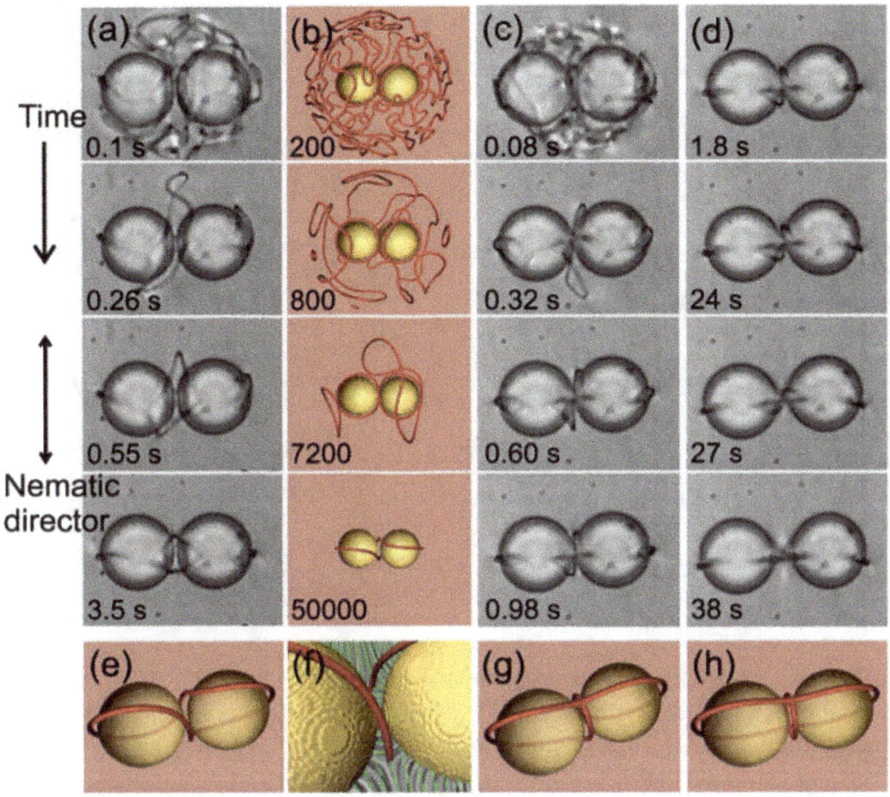

Fig. 3.17 Assembling entangled nematic colloidal pairs by thermal quench using light. (**a**) Figure-of-eight. (**b**) Numerical simulation of the time evolution of entanglement measured in the number of iteration steps. (**c**) Evolution of figure-of-omega state. (**d**) Figure-of-omega state transformed into figure-of-theta. (**e** and **f**) Numerically calculated structures. Reprinted figure with permission from [M. Ravnik, M. Škarabot, S. Žumer, U. Tkalec, I. Poberaj, D. Babič, N. Osterman, I. Muševič, Phys. Rev. Lett. **99**, 247801 (2007)].

A method to assemble larger colloidal structures is by directed assembly using laser tweezers, which can capture and guide a single colloidal particle [99]. Particles are manipulated into vicinity of each other, close enough that structural forces can bind them into stable and ordered clusters. To create 2D crystal structures, thin nematic cells with properly processed surfaces are used, so that only one layer of colloids is formed in the middle of the cell. For example, clusters and 2D colloidal crystals are assembled from elastically dipolar particles by joining two oppositely oriented linear chains, which attract due to sideways dipolar attraction, or from quadrupolar particles by joining several kinked chains, which attract due to quadrupolar interactions [108].

In addition to homogeneous crystals, a range of binary structures were realised by using combinations of elastic dipoles and quadrupoles [109] or by using different

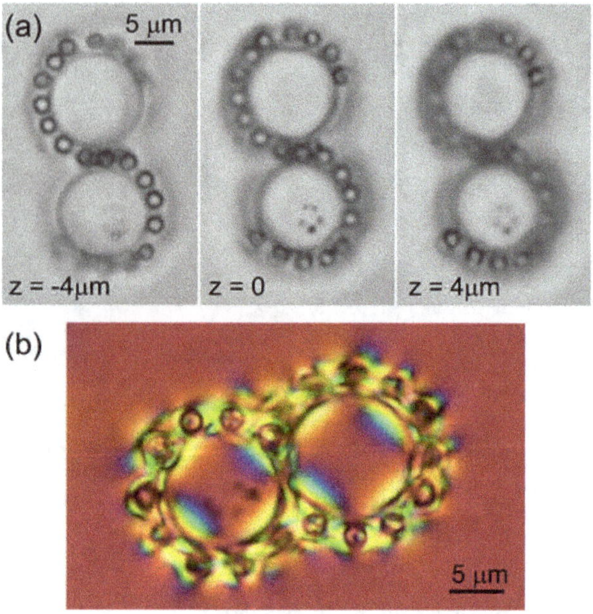

Fig. 3.18 Hierarchical multi-particle nematic colloidal structure. (**a**) Smaller colloidal particles are trapped into the topological defect loop, twisting around a larger colloidal pair. Images are taken at different height of focus. (**b**) Polarising microscope image shows the distortions of the director field around the colloids. Reprinted figure with permission from [M. Škarabot, M. Ravnik, S. Žumer, U. Tkalec, I. Poberaj, D. Babič, I. Muševič, Phys. Rev. E **77**, 61706 (2008)].

sized particles, creating hierarchical superstructures (Fig. 3.18) [110]. 3D colloidal crystal with tetragonal symmetry and interesting material properties, such as strong electrostriction and electro-rotation, has also been assembled by joining anti-parallel chains of elastic dipoles [7].

Approaches in which particles are controlled to occupy predesigned sites are developed by creating spatially variable nematic profiles. Since minimal energy configurations are different for particles with homeotropic or planar anchoring, they tend to localise in different regions, such as particles with homeotropic anchoring in regions of splay and particles with planar anchoring in regions of bend deformations [111]. Also sculpting a flat surface with a cavity that is similar to the particle in size and shape can change the sign of the interaction between the particle and the surface and lead to key-lock mechanisms for trapping the particles [112]. Topographic modulation of the surfaces can be used to select and localise particles by using convex and concave deformations [113].

The director field can also be altered by changing the geometry of the cell, containing the nematic host medium, which can lead to emergence of variety of defects, depending on the shape of the surface. For example, if a nematic medium is introduced into the cell with an array of cylindrical microposts, defects occur around them in the bulk and attract colloidal particles (Fig. 3.19a–g). Colloids assemble to mimic the defect structure in the bulk even if they are remote, i.e. on the surface of the liquid crystal layer in which microposts are submerged. In the case of high packing fraction, a triangular colloidal lattice is formed (Fig. 3.19h) [114]. In a different study, colloidal chains of elastic dipoles were found to follow the disclination lines and curved director field in the geometry of groovy cells [115].

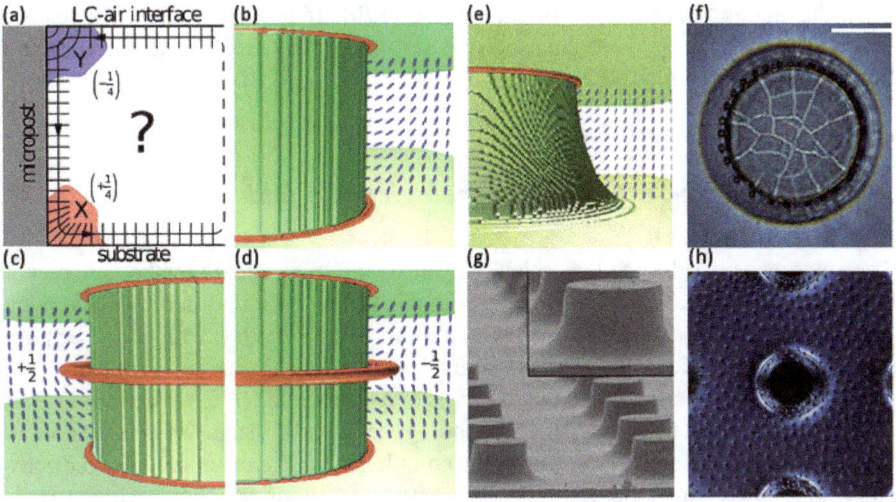

Fig. 3.19 If the geometry of the cell is altered by using microposts, the emergence of ring defects, which guide the assembly of remote colloidal particles, is enforced. (**a**) The director field in the corners of the cell with homeotropic anchoring can assume two different configurations with the opposite 2D topological charges. (**b–d**) The bulk director field corresponds to the minimum of free energy and has to satisfy the topological charge conservation. If the field in the corners has the same topological charge, the ring defect occurs to neutralise it. (**e**) By curving the edge, the configuration with positive winding number is favoured. (**f**) SEM image of experimentally realised curved microposts. (**g**) At moderate surface coverage, ordered rings assemble around the micropost due to attraction by the bulk defect and repel one another via long-range interparticle repulsion. (**h**) At higher particle density elastic interactions force them into higher order structures. Reused and adapted with permission from publishers [M. Cavallaro, M.A. Gharbi, D.A. Beller, S. Copar, Z. Shi, T. Baumgart, S. Yang, R.D. Kamien, K.J. Stebe, Exploiting imperfections in the bulk to direct assembly of surface colloids.

Nematic interparticle interactions depend strongly also on the shape of the particles. For example, polygonal particle platelets with odd number of sides exhibit dipolar symmetry and therefore dipolar interactions, while the ones with even number of sides act as elastic quadrupoles [116]. Since nonspherical particles may interact as dipoles/quadrupoles at long range, but their short range interactions depend on the geometry, they are suitable for realising 2D and 3D crystalline, quasicrystalline, and various locally ordered low-symmetry structures, which cannot be assembled from colloidal spheres [30]. Similar results were observed when using colloidal pyramidal cones and octahedrons made from thin nanofoil, which are physical analogues of mathematical surfaces with boundaries and induce no defects when flat [117]. Also switching between repulsion and attraction through re-pinning the disclinations at different edges of polygonal prism using laser tweezers has been demonstrated [118].

Among other ways of control, chemical treatment can be used to switch between different types of cell surface anchoring even at the nanoscale and control colloids by inducing defects [119] or to switch between anchoring types on colloids and therefore manipulate the type of interaction [120]. If ferromagnetic particles are used, also magnetic field can be used to control the orientation of the particles and their mutual interactions [121].

3.7.4 Complex-Shaped and Topological Colloids

The emergence of advanced chemical, physical, and biosynthetic methods in recent years enabled creation of complex-shaped anisotropic colloidal particles and even particles with different topologies. Combining such particles with liquid crystal medium lead to a wide variety of topological field states and configurations. In comparison to low temperature and magnetic systems at atomic scale, where topological phases also emerge, experiments with topology in liquid crystals can be observed at much larger optical scales. Additionally, complex structure of topological particles can lead to interesting assembly properties and interactions with light, which makes them suitable for photonic applications based on topological materials.

Topological colloidal particles with non-zero genus g (i.e. effectively, the number of holes in the particle) were demonstrated [28], which stabilise a wide variety of complex nematic profiles when immersed in nematic liquid crystal. The topology of the director field is governed by the topological charge conservation and by the Gauss–Bonnet and Poincare–Hopf theorems, which must be obeyed. By integrating the local Gaussian curvature K over the entire surface of the particle, its Euler characteristic, which is directly connected to the genus g, can be calculated as [28]:

$$2(1-g) = \chi = \frac{1}{2\pi}\oint K\,\mathrm{d}S. \tag{3.49}$$

The Euler characteristic acts as a topological invariant, which means that is preserved during the continuous transformations of the particle surface and also equals the topological charge of the surface. Net topological charge of the defects that emerge in the liquid crystal after the non-trivial particle is immersed in it is exactly determined by topological charge conservation, but the exact number of defects and their types depend on the shape and orientation of the particle with respect to the bulk field, which is determined by the minimum of the total free energy. Various possible configurations of defects in the vicinity of toroidal colloids with different genus numbers have been demonstrated [28].

Knot-shaped colloidal particles present another interesting platform to study interplay of the topologies of the particles, the nematic field, and the induced defects, which lead to knotted, linked, and other topologically non-trivial field configurations. Particle links in nematics were realised also as an example of topologically conditioned nematic colloidal material [8]. An example of configuration of linked rings, also known as Hopf link, with planar surface anchoring is shown in Fig. 3.20. Topological transitions by changing the shape and genus of the knot particles have been also studied numerically [122].

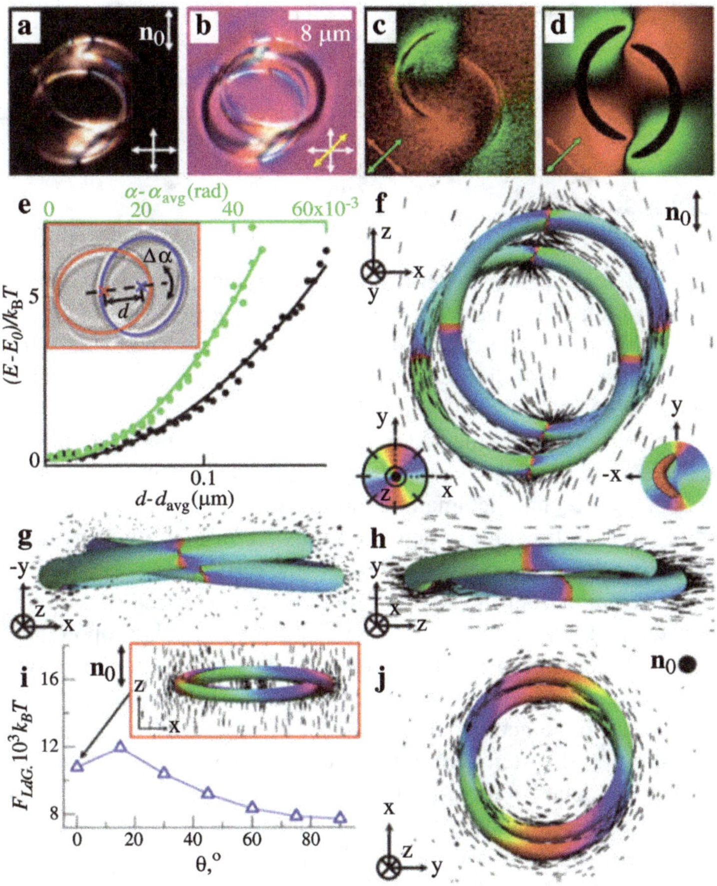

Fig. 3.20 Hopf link colloidal particle with tangential anchoring in nematic. (**a, b**) Polarising optical micrographs of a Hopf link in a nematic cell without and with a lambda plate, respectively, with far-field nematic director n_0 marked on the image. (**c, d**) Experimental and theoretical in-plane cross-section with a director field around it. (**e**) Elastic free energy dependence on the deviation from equilibrium angle and centre-to-centre distance of the two link components—rings. (**f–h**) Three perspective views from mutually orthogonal directions of a numerically calculated director field around the link. (**i**) Landau–de Gennes free energy vs. angle $\theta_{1,2}$ between bulk director orientation and the plane bisecting the angle $\theta_{1,2} = 20°$ between the rings at fixed radius R and $d = 0.22$, with the energy minimum corresponding to the configuration f. Inset in (**i, j**) Metastable configurations. Reused and adapted with permission from publishers [A. Martinez, L. Hermosillo, M. Tasinkevych, I.I. Smalyukh, Linked topological colloids in a nematic host. Proc. Natl. Acad. Sci. **112**, 4546 (2015)]

Current research in the field of nematic colloids is more and more directed towards finding new possible interactions between particles, which are governed by their topologically non-trivial or fractal shape [123] that induces tangled director fields and defects in the liquid crystal medium. Such particles can nowadays be fabricated and used in experiments and show a lot of promise in photonic applications. A strong recent direction is also in exploring motile "active" colloids in anisotropic nematic background, which can be driven by external fields or internally via self-propelled particles such as motile bacteria or molecular motors [74, 124, 125].

3.8 Conclusions

Nematic fluids cover a span of materials, from molecular fluids, colloidal dispersion to viruses, with their main material characteristic being orientational order of the building blocks. This orientational order is soft and as an effective elastic medium responsive to external stimuli, including mechanical fields, pressure, light, electric and magnetic fields. The strong susceptibility to external stimuli makes nematic fluids potent materials in systems that require controllability and tunability, which is today extensively used in display and optical applications, with strong development also towards photonics and metamaterial applications.

A major emergent direction in nematic fluids are also active nematic materials, which are inherently out-of-equilibrium systems based on motile building blocks that can show nematic ordering. Active nematic systems include systems of kinesin driven microtubules, bacterial colonies, or flocks of animals. Topological defects in active fluids are emerging as major elements that determine the active material properties. And there exists an interesting route for transfer of knowledge from passive nematic fluids to active nematic fluids. For example, in terms of structure, nematic braids realised by temperature quench in passive nematics are probably the closest passive analogues to three-dimensional active turbulence, as one of the top-level challenges in understanding of active nematics.

Finally, nematic fluids is a topic that is naturally in an interdisciplinary way reaching towards other field of science and technology, notably including complex flows and active fluids.

References

1. P.G. de Gennes, J. Prost, *Physics of Liquid Crystals* (Oxford University Press, New York, 1993)
2. M. Kleman, O.D. Lavrentovich, *Soft Matter Physics - An Introduction* (Springer, New York, 2003)
3. C. Zannoni, Molecular design and computer simulations of novel mesophases. J. Mater. Chem. **11**, 2637 (2001)
4. A.H. Lewis, I. Garlea, J. Alvarado, O.J. Dammone, P.D. Howell, A. Majumdar, B.M. Mulder, M.P. Lettinga, G.H. Koenderink, D.G.A.L. Aarts, Colloidal liquid crystals in rectangular confinement: theory and experiment. Soft Matter **10**, 7865 (2014)
5. A.N. Beris, B.J. Edwards, *Thermodynamics of Flowing Systems with Internal Microstructure* (Oxford University Press, New York, 1994)
6. T. Qian, P. Sheng, Generalized hydrodynamic equations for nematic liquid crystals. Phys. Rev. E **58**, 7475 (1998)

7. A. Nych, U. Ognysta, M. Škarabot, M. Ravnik, S. Žumer, I. Muševič, Assembly and control of 3D nematic dipolar colloidal crystals. Nat. Commun. **4**, 1489 (2013)
8. A. Martinez, L. Hermosillo, M. Tasinkevych, I.I. Smalyukh, Linked topological colloids in a nematic host. Proc. Natl. Acad. Sci. **112**, 4546 (2015)
9. D. Seč, S. Čopar, S. Žumer, Topological zoo of free-standing knots in confined chiral nematic fluids. Nat. Commun. **5**, 3057 (2014)
10. L.E. Aguirre, A. de Oliveira, D. Seč, S. Čopar, P.L. Almeida, M. Ravnik, M.H. Godinho, S. Žumer, Sensing surface morphology of biofibers by decorating spider silk and cellulosic filaments with nematic microdroplets. Proc. Natl. Acad. Sci. **113**, 1174 (2016)
11. M. Nikkhou, M. Škarabot, S. Čopar, M. Ravnik, S. Žumer, I. Muševič, Light-controlled topological charge in a nematic liquid crystal. Nat. Phys. **11**, 183 (2015)
12. D. Svenšek, S. Žumer, Hydrodynamics of pair-annihilating disclination lines in nematic liquid crystals. Phys. Rev. E **66**, 21712 (2002)
13. G. Tóth, C. Denniston, J.M. Yeomans, Hydrodynamics of topological defects in nematic liquid crystals. Phys. Rev. Lett. **88**, 105504 (2002)
14. A. Sengupta, U. Tkalec, M. Ravnik, J. Yeomans, C. Bahr, S. Herminghaus, Liquid crystal microfluidics for tunable flow shaping. Phys. Rev. Lett. **110**, 48303 (2013)
15. V.M.O. Batista, M.L. Blow, M.M.T. da Gama, The effect of anchoring on the nematic flow in channels. Soft Matter **11**, 4674 (2015)
16. A. Sengupta, C. Bahr, S. Herminghaus, Topological microfluidics for flexible micro-cargo concepts. Soft Matter **9**, 7251 (2013)
17. Y.J. Na, T.Y. Yoon, S. Park, B. Lee, S.D. Lee, Electrically programmable nematofluidics with a high level of selectivity in a hierarchically branched architecture. ChemPhysChem **11**, 101 (2010)
18. J.G. Cuennet, A.E. Vasdekis, D. Psaltis, Optofluidic-tunable color filters and spectroscopy based on liquid-crystal microflows. Lab Chip **13**, 2721 (2013)
19. M.C. Marchetti, J.F. Joanny, S. Ramaswamy, T.B. Liverpool, J. Prost, M. Rao, R.A. Simha, Hydrodynamics of soft active matter. Rev. Mod. Phys. **85**, 1143 (2013)
20. T. Araki, M. Buscaglia, T. Bellini, H. Tanaka, Memory and topological frustration in nematic liquid crystals confined in porous materials. Nat. Mater. **10**, 303 (2011)
21. I. Muševič, *Liquid Crystal Colloids* (Springer, New York, 2017)
22. I.I. Smalyukh, Y. Lansac, N.A. Clark, R.P. Trivedi, Three-dimensional structure and multistable optical switching of triple-twisted particle-like excitations in anisotropic fluids. Nat. Mater. **9**, 139 (2010)
23. A. Sengupta, S. Herminghaus, C. Bahr, Liquid crystal microfluidics: surface, elastic and viscous interactions at microscales. Liq. Cryst. Rev. **2**, 73 (2014)
24. M. Ravnik, J. Yeomans, Confined active nematic flow in cylindrical capillaries. Phys. Rev. Lett. **110**, 26001 (2013)
25. T.C. Lubensky, D. Pettey, N. Currier, H. Stark, Topological defects and interactions in nematic emulsions. Phys. Rev. E **57**, 610 (1998)
26. M. Ravnik, M. Škarabot, S. Žumer, U. Tkalec, I. Poberaj, D. Babič, N. Osterman, I. Muševič, Entangled nematic colloidal dimers and wires. Phys. Rev. Lett. **99**, 247801 (2007)
27. U. Tkalec, M. Ravnik, S. Čopar, S. Žumer, I. Muševič, Reconfigurable knots and links in chiral nematic colloids. Science **333**, 62 (2011)
28. B. Senyuk, Q. Liu, S. He, R.D. Kamien, R.B. Kusner, T.C. Lubensky, I.I. Smalyukh, Topological colloids. Nature **493**, 200 (2013)
29. B.G.G. Chen, P.J. Ackerman, G.P. Alexander, R.D. Kamien, I.I. Smalyukh, Generating the Hopf fibration experimentally in nematic liquid crystals. Phys. Rev. Lett. **110**, 237801 (2013)
30. J. Dontabhaktuni, M. Ravnik, S. Žumer, Quasicrystalline tilings with nematic colloidal platelets. Proc. Natl. Acad. Sci. **111**, 2464 (2014)
31. E. Pairam, J. Vallamkondu, V. Koning, B.C. van Zuiden, P.W. Ellis, M.A. Bates, V. Vitelli, A. Fernandez-Nieves, Stable nematic droplets with handles. Proc. Natl. Acad. Sci. **110**, 9295 (2013)

32. S. Čopar, S. Žumer, Nematic Braids: topological invariants and rewiring of disclinations. Phys. Rev. Lett. **106**, 177801 (2011)
33. G.P. Alexander, B.G.G. Chen, E.A. Matsumoto, R.D. Kamien, Colloquium: disclination loops, point defects, and all that in nematic liquid crystals. Rev. Mod. Phys. **84**, 497 (2012)
34. P.J. Ackerman, I.I. Smalyukh, Static three-dimensional topological solitons in fluid chiral ferromagnets and colloids. Nat. Mater. **16**, 426 (2017)
35. P. Oswald, P. Pieranski, *Nematic and Cholesteric Liquid Crystals* (Taylor & Francis, Boca Raton, 2005)
36. H. Kikuchi, M. Yokota, Y. Hisakado, H. Yang, T. Kajiyama, Polymer-stabilized liquid crystal blue phases. Nat. Mater. **1**, 64 (2002)
37. H.J. Coles, M.N. Pivnenko, Liquid crystal 'blue phases' with a wide temperature range. Nature **436**, 997 (2005)
38. J.A. Kelly, M. Giese, K.E. Shopsowitz, W.Y. Hamad, M.J. MacLachlan, The development of chiral nematic mesoporous materials. Acc. Chem. Res. **47**, 1088 (2014)
39. G.J. Vroege, H.N.W. Lekkerkerker, Phase transitions in lyotropic colloidal and polymer liquid crystals. Rep. Prog. Phys. **55**, 1241 (1992)
40. I. Dierking, S. Al-Zangana, Lyotropic liquid crystal phases from anisotropic nanomaterials. Nanomaterials **7**, 305 (2017)
41. D.Y. Kim, S.I. Lim, D. Jung, J.K. Hwang, N. Kim, K.U. Jeong, Self-assembly and polymer-stabilization of lyotropic liquid crystals in aqueous and non-aqueous solutions. Liq. Cryst. Rev. **5**, 34 (2017)
42. L. van 't Hag, S.L. Gras, C.E. Conn, C.J. Drummond, Lyotropic liquid crystal engineering moving beyond binary compositional space – ordered nanostructured amphiphile self-assembly materials by design. Chem. Soc. Rev. **46**, 2705 (2017)
43. F.C. Frank, I. Liquid crystals. On the theory of liquid crystals. Discuss. Faraday Soc. **25**, 19 (1958)
44. S. Chandrasekhar, *Liquid Crystals* (Cambridge, Cambridge, 1992)
45. K. Schiele, S. Trimper, On the elastic constants of a nematic liquid crystal. Phys. Status Solidi **118**, 267 (1983)
46. G.P. Chen, H. Takezoe, A. Fukuda, Determination of K_i (i = 1-3) and μ_j (j = 2-6) in 5CB by observing the angular dependence of Rayleigh line spectral widths. Liq. Cryst. **5**, 341 (1989)
47. N.V. Madhusudana, R. Pratibha, Elasticity and orientational order in some cyanobiphenyls: Part IV. Reanalysis of the data. Mol. Cryst. Liq. Cryst. **89**, 249 (1982)
48. M. Ravnik, S. Žumer, Landau-de Gennes modelling of nematic liquid crystal colloids. Liq. Cryst. **36**, 1201 (2009)
49. B. Jerome, Surface effects and anchoring in liquid crystals. Rep. Prog. Phys. **54**, 391 (1991)
50. M. Nobili, G. Durand, Disorientation-induced disordering at a nematic-liquid-crystal-solid interface. Phys. Rev. A **46**, R6174 (1992)
51. L.M. Blinov, A.Y. Kabayenkov, A.A. Sonin, Invited lecture. Experimental studies of the anchoring energy of nematic liquid crystals. Liq. Cryst. **5**, 645 (1989)
52. J.B. Fournier, P. Galatola, Modeling planar degenerate wetting and anchoring in nematic liquid crystals. Eur. Lett. **72**, 403 (2005)
53. P.J. Collings, M. Hird, *Introduction to Liquid Crystals: Chemistry and Physics* (Taylor and Francis, London, 1997)
54. V. Fréedericksz, A. Repiewa, Theoretisches und Experimentelles zur Frage nach der Natur der anisotropen Flüssigkeiten. Z. Phys. **42**(7), 532 (1927)
55. B.J. Frisken, P. Palffy-Muhoray, Freedericksz transitions in nematic liquid crystals: The effects of an in-plane electric field. Phys. Rev. A **40**(10), 6099 (1989)
56. M. Kleman, *Points, Lines and Walls* (Wiley, New York, 1983)
57. M. Kleman, O.D. Lavrentovich, Topological point defects in nematic liquid crystals. Philos. Mag. **86**, 4117 (2006)
58. N. Mermin, The topological theory of defects in ordered media. Rev. Mod. Phys. **51**, 591 (1979)

59. H.R. Trebin, The topology of non-uniform media in condensed matter physics. Adv. Phys. **31**, 195 (1982)
60. A. Rapini, Umbilics : static properties and shear-induced displacements. J. Phys. Fr. **34**, 629 (1973)
61. I. Dierking, M. Ravnik, E. Lark, J. Healey, G.P. Alexander, J.M. Yeomans, Anisotropy in the annihilation dynamics of umbilic defects in nematic liquid crystals. Phys. Rev. E **85**, 21703 (2012)
62. P. Pieranski, B. Yang, L.J. Burtz, A. Camu, F. Simonetti, Generation of umbilics by magnets and flows. Liq. Cryst. **40**, 1593 (2013)
63. M.G. Clerc, E. Vidal-Henriquez, J.D. Davila, M. Kowalczyk, Symmetry breaking of nematic umbilical defects through an amplitude equation. Phys. Rev. E **90**, 12507 (2014)
64. M. Kléman, L. Michel, Spontaneous breaking of Euclidean invariance and classification of topologically stable defects and configurations of crystals and liquid crystals. Phys. Rev. Lett. **40**, 1387 (1978)
65. G.E. Volovik, Relationship between molecule shape and hydrodynamics in a nematic substance. JETP Lett. **31**, 273 (1980)
66. S. Hess, Irreversible thermodynamics of nonequilibrium alignment phenomena in molecular liquids and in liquid crystals. Z. Naturforsch. Pt. A **30**, 728 (1975)
67. P. Olmsted, P. Goldbart, Theory of the nonequilibrium phase transition for nematic liquid crystals under shear flow. Phys. Rev. A **41**, 4578 (1990)
68. H. Stark, T.C. Lubensky, Poisson-bracket approach to the dynamics of nematic liquid crystals. Phys. Rev. E **67**, 231 (2003)
69. A.M. Sonnet, P.L. Maffettone, E.G. Virga, Continuum theory for nematic liquid crystals with tensorial order. J. Non-Newton. Fluid. **119**, 51 (2004)
70. C. Denniston, E. Orlandini, J. Yeomans, Lattice Boltzmann simulations of liquid crystal hydrodynamics. Phys. Rev. E **63**, 56702 (2001)
71. R.A. Simha, S. Ramaswamy, Hydrodynamic fluctuations and instabilities in ordered suspensions of self-propelled particles. Phys. Rev. Lett. **89**, 58101 (2002)
72. R. Voituriez, J.F. Joanny, J. Prost, Spontaneous flow transition in active polar gels. Eur. Lett. **70**, 404 (2005)
73. T. Sanchez, D.T.N. Chen, S.J. DeCamp, M. Heymann, Z. Dogic, Spontaneous motion in hierarchically assembled active matter. Nature **491**, 431 (2012)
74. P. Guillamat, Ž. Kos, J. Hardoüin, J. Ignés-Mullol, M. Ravnik, F. Sagués, Active nematic emulsions. Sci. Adv. **4**, eaao1470 (2018)
75. T. Vicsek, A. Czirók, E. Ben-Jacob, I. Cohen, O. Shochet, Novel type of phase transition in a system of self-driven particles. Phys. Rev. Lett. **75**, 1226 (1995)
76. T. Gao, R. Blackwell, M.A. Glaser, M.D. Betterton, M.J. Shelley, Multiscale polar theory of microtubule and motor-protein assemblies. Phys. Rev. Lett. **114**, 242S (2015)
77. H.H. Wensink, J. Dunkel, S. Heidenreich, K. Drescher, R.E. Goldstein, H. Lowen, J.M. Yeomans, Meso-scale turbulence in living fluids, Proc. Natl. Acad. Sci. **109**, 14308 (2012)
78. D.W. Berreman, Liquid-crystal twist cell dynamics with backflow. J. Appl. Phys. **46**, 3746 (1975)
79. C.Z. van Doorn, Dynamic behavior of twisted nematic liquid-crystal layers in switched fields. J. Appl. Phys. **46**, 3738 (1975)
80. N. Éber, P. Salamon, Á. Buka, Electrically induced patterns in nematics and how to avoid them. Liq. Cryst. Rev. **4**, 101 (2016)
81. K.A. Takeuchi, M. Sano, Universal fluctuations of growing interfaces: evidence in turbulent liquid crystals. Phys. Rev. Lett. **104**, 230601 (2010)
82. O. Wiese, D. Marenduzzo, O. Henrich, Microfluidic flow of cholesteric liquid crystals. Soft Matter **12**, 9223 (2016)
83. Z. Dogic, S. Fraden, Ordered phases of filamentous viruses. Curr. Opin. Colloid Interface Sci. **11**, 47 (2006)
84. E. Guyon, P. Pieranski, Poiseuille flow instabilities in nematics. J. Phys. Colloq. **36**, C1 (1975)

85. E. Dubois-Violette, P. Manneville, in *Pattern Formation in Liquid Crystals*, ed. by A. Buka, L. Kramer (Springer, New York, 1996), pp. 91–163
86. S.P. Thampi, R. Golestanian, J.M. Yeomans, Driven active and passive nematics. Mol. Phys. **113**, 2656 (2015)
87. L. Giomi, Ž. Kos, M. Ravnik, A. Sengupta, Cross-talk between topological defects in different fields revealed by nematic microfluidics. Proc. Natl. Acad. Sci. **114**, E5771 (2017)
88. D. Kang, J. Maclennan, N. Clark, A. Zakhidov, R. Baughman, Electro-optic behavior of liquid-crystal-filled silica opal photonic crystals: effect of liquid-crystal alignment. Phys. Rev. Lett. **86**, 4052 (2001)
89. F. Serra, K.C. Vishnubhatla, M. Buscaglia, R. Cerbino, R. Osellame, G. Cerullo, T. Bellini, Topological defects of nematic liquid crystals confined in porous networks. Soft Matter **7**, 10945 (2011)
90. T. Araki, Dynamic coupling between a multistable defect pattern and flow in nematic liquid crystals confined in a porous medium. Phys. Rev. Lett. **109**, 257801 (2012)
91. J. Aplinc, S. Morris, M. Ravnik, Porous nematic microfluidics for generation of umbilic defects and umbilic defect lattices. Phys. Rev. Fluids **1**, 23303 (2016)
92. Ž. Kos, M. Ravnik, S. Žumer, Nematodynamics and structures in junctions of cylindrical micropores. Liq. Cryst. **44**, 2161 (2017)
93. H. Stark, Physics of colloidal dispersions in nematic liquid crystals. Phys. Rep. **351**, 387 (2001)
94. J.S. Lintuvuori, K. Stratford, M.E. Cates, D. Marenduzzo, Colloids in cholesterics: size-dependent defects and non-Stokesian microrheology. Phys. Rev. Lett. **105**, 178302 (2010)
95. P. Poulin, H. Stark, T.C. Lubensky, D.A. Weitz, Novel colloidal interactions in anisotropic fluids. Science **275**, 1770 (1997)
96. E. Terentjev, Disclination loops, standing alone and around solid particles, in nematic liquid crystals. Phys. Rev. E **51**, 1330 (1995)
97. S. Ramaswamy, R. Nityananda, V.A. Raghunathan, J. Prost, Power-law forces between particles in a nematic. Mol. Cryst. Liq. Cryst. **288**(1), 175 (1996)
98. J.i. Fukuda, Liquid crystal colloids: a novel composite material based on liquid crystals. J. Phys. Soc. Jpn **78**(4), 41003 (2009)
99. M. Yada, J. Yamamoto, H. Yokoyama, Direct observation of anisotropic interparticle forces in nematic colloids with optical tweezers. Phys. Rev. Lett. **92**, 185501 (2004)
100. M. Škarabot, M. Ravnik, S. Žumer, U. Tkalec, I. Poberaj, D. Babič, N. Osterman, I. Muševič, Two-dimensional dipolar nematic colloidal crystals. Phys. Rev. E **76**, 51406 (2007)
101. P. Poulin, D.A. Weitz, Inverted and multiple nematic emulsions. Phys. Rev. E **57**, 626 (1998)
102. P. Poulin, V. Cabuil, D. Weitz, Direct Measurement of colloidal forces in an anisotropic solvent. Phys. Rev. Lett. **79**, 4862 (1997)
103. U. Tkalec, M. Ravnik, S. Žumer, I. Muševič, Vortexlike topological defects in nematic colloids: chiral colloidal dimers and 2D crystals. Phys. Rev. Lett. **103**, 127801 (2009)
104. U. Tkalec, M. Ravnik, S. Čopar, S. Žumer, I. Muševič, Reconfigurable knots and links in chiral nematic colloids. Science **333**(6038), 62 (2011)
105. N.A. Clark, A.J. Hurd, B.J. Ackerson, Single colloidal crystals. Nature **281**(5726), 57 (1979)
106. T. Hegmann, H. Qi, V.M. Marx, Nanoparticles in liquid crystals: synthesis, self-assembly, defect formation and potential applications. J. Inorg. Organomet. Polym. Mater. **17**(3), 483 (2007)
107. L.S. Hirst, J. Kirchhoff, R. Inman, S. Ghosh, L.C. Chien, in *Proc SPIE*, vol. 7618 (2010), pp. 76,180F–1
108. I. Muševič, M. Škarabot, Self-assembly of nematic colloids. Soft Matter **4**, 195 (2008)
109. U. Ognysta, A. Nych, V. Nazarenko, M. Škarabot, I. Muševič, Design of 2D binary colloidal crystals in a nematic liquid crystal. Langmuir **25**, 12092 (2009)
110. M. Škarabot, M. Ravnik, S. Žumer, U. Tkalec, I. Poberaj, D. Babič, I. Muševič, Hierarchical self-assembly of nematic colloidal superstructures. Phys. Rev. E **77**, 61706 (2008)
111. C. Peng, T. Turiv, Y. Guo, S.V. Shiyanovskii, Q.H. Wei, O.D. Lavrentovich, Control of

colloidal placement by modulated molecular orientation in nematic cells. Sci. Adv. **2**, e1600932 (2016)
112. N. Silvestre, P. Patricio, M.T. da Gama, Key-lock mechanism in nematic colloidal dispersions. Phys. Rev. E **69**(6), 061402 (2004)
113. Z. Eskandari, N. Silvestre, M.T. da Gama, M. Ejtehadi, Particle selection through topographic templates in nematic colloids. Soft Matter **10**(48), 9681 (2014)
114. M. Cavallaro, M.A. Gharbi, D.A. Beller, S. Čopar, Z. Shi, T. Baumgart, S. Yang, R.D. Kamien, K.J. Stebe, Exploiting imperfections in the bulk to direct assembly of surface colloids. Proc. Natl. Acad. Sci. **110**, 18804 (2013)
115. Y. Luo, F. Serra, D.A. Beller, M.A. Gharbi, N. Li, S. Yang, R.D. Kamien, K.J. Stebe, Around the corner: colloidal assembly and wiring in groovy nematic cells. Phys. Rev. E **93**, 32705 (2016)
116. C.P. Lapointe, T.G. Mason, I.I. Smalyukh, Shape-controlled colloidal interactions in nematic liquid crystals. Science **326**, 1083 (2009)
117. S. Park, Q. Liu, I.I. Smalyukh, colloidal surfaces with boundaries, apex Boojums, and nested elastic self-assembly of nematic colloids. Phys. Rev. Lett. **117**, 277801 (2016)
118. B. Senyuk, Q. Liu, P.D. Nystrom, I.I. Smalyukh, Repulsion–attraction switching of nematic colloids formed by liquid crystal dispersions of polygonal prisms. Soft Matter **13**, 7398 (2017)
119. X. Li, J.C. Armas-Pérez, J.P. Hernández-Ortiz, C.G. Arges, X. Liu, J.A. Martínez-González, L.E. Ocola, C. Bishop, H. Xie, J.J. de Pablo, P.F. Nealey, Directed self-assembly of colloidal particles onto nematic liquid crystalline defects engineered by chemically patterned surfaces. ACS Nano **11**, 6492 (2017)
120. T. Yamamoto, Y. Tabe, H. Yokoyama, Photochemical transformation of topological defects formed around colloidal droplets dispersed in azobenzene-containing liquid crystals. Colloid Surf. A **334**, 155 (2009)
121. Q. Liu, P.J. Ackerman, T.C. Lubensky, I.I. Smalyukh, Biaxial ferromagnetic liquid crystal colloids. Proc. Natl. Acad. Sci. **113**, 10479 (2016)
122. M. Ravnik, S. Čopar, S. Žumer, Particles with changeable topology in nematic colloids. J. Phys. Condens. Matter **27**, 354111 (2015)
123. S.M. Hashemi, U. Jagodič, M.R. Mozaffari, M.R. Ejtehadi, I. Muševič, M. Ravnik, Fractal nematic colloids. Nat. Commun. **8**, 14026 (2017)
124. P.C. Mushenheim, R.R. Trivedi, H.H. Tuson, D.B. Weibel, N.L. Abbott, Dynamic self-assembly of motile bacteria in liquid crystals. Soft Matter **10**(1), 88 (2014)
125. O.D. Lavrentovich, Active colloids in liquid crystals. Curr. Opin. Colloid Interface Sci. **21**, 97 (2016)

4

Janus Particles: Synthetic Methods and their Applications

Andrei Honciuc

Acronyms

3-TSPM	3-(trimethoxysilyl)propyl methacrylate
APS	(3-Aminopropyl)triethoxysilane
CA	contact angle
DIW	de-ionised water
HLB	hydrophilic–lyophilic balance
HP	homogeneous particle
IBA	isobornyl acrylate
IFT	interfacial tension
JP	Janus particle
MMA	methyl methacrylate
NP	nanoparticle
OTS	n-octadecyltrichlorosilane
P(3-TSPM)	poly(3-(trimethoxysilyl)propyl methacrylate)
PA	polyacrylate
PAA	poly(acrylic acid)
PI	polyisoprene
PMMA	poly(methyl methacrylate)
pNA	poly[N-isopropylacrylamide-co-(acrylic acid)]
PPA	poly(propargyl-acrylate)
PPFBEM	poly(2-(perfluorobutyl)ethyl methacrylate)
PPy	polypyrrole
PS	polystyrene

A. Honciuc (✉)
Institute of Chemistry and Biotechnology, Zurich University of Applied Sciences, Waedenswil, Switzerland
e-mail: andrei.honciuc@zhaw.ch

PS-PDIPAEMA	polystyrene-poly[2-(diisopropylamino)ethyl methacrylate
PtBA	poly(tert-butyl acrylate)

4.1 Introduction

Janus particles (JPs) can be generally defined as asymmetric particles with at least two surface regions or bulk composition differing in their physicochemical properties. JPs can adopt different shapes, for example, perfectly spherical with two hemispheres having different surface properties as depicted in Fig. 4.1, or they can adopt different shapes, such as snowman [1], or dumbbell, hybridised-like [2] orbitals, mushroom [3] with clear geometrical and topological asymmetries. Under the same "Janus" category other particles, such as raspberry, [4] rods or discs [5] have also been included and generally all asymmetric particles, as long as there is a difference in composition or surface properties on the same particle but on distinctive regions. Probably the most typical shape of a Janus particle is that of a dumbbell or snowman, Fig. 4.1. Unlike homogeneous particles (HPs) the JPs have some interesting properties and exhibit extra-functionality conferred by their asymmetry. One example of such functionality is amphiphilicity. Due to the inherent polarity contrast between two surface regions, they resemble surfactants with one polar side and one-less polar; therefore, the JPs are promising as "solid

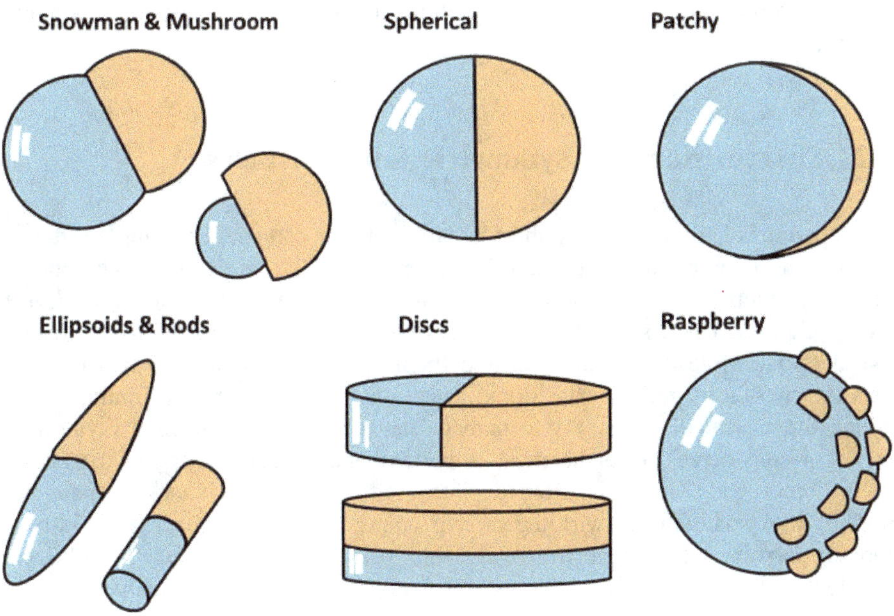

Fig. 4.1 Types of Janus particles

state amphiphiles" or the next generation of amphiphiles. But unlike surfactants, generally small molecules, or low molecular weight polymers, JPs exhibit some significant differences. First, because they are solid-state particles and due to their size, they have large interfacial attachment energies, on the order of thousands of kT (parameter that scales with the R^2, where R is the radius of the particle), meaning that once adsorbed at the interface they remain trapped and secondly their diffusion through the liquid is much slower. This can be an advantage because JPs can be used as emulsifiers of oils and water and create ultrastable Pickering emulsions. Pickering emulsions can also be generated with HPs but it has been shown that due to their amphiphilicity the JPs are several times more interfacially active and thus superior in such applications. Furthermore, the JPs are also active at the air–water interface and this makes them attractive as stabilisers of air bubbles and foams. Their size may also bring further advantages, for example, they can be used as carriers of actives (small molecules serving as pharmaceutically active ingredients), or smart catalysts moving in concentration gradients and even nanomotors for transportation of "heavy" cargo, which molecular surfactants cannot do. JPs can also be regarded as building blocks of matter that can self-assemble to give rise to suprastructures. JPs can be multifunctional because they can carry different properties on each lobe; this is especially attractive to creating new multifunctional materials where the surface and bulk-like properties can be combined to obtain unexpected functionalities. For example, it has been demonstrated that conductivity and surface polarity of snowman-type JPs can be tuned by changing the lobe ratio between a semiconductive lobe and an electrically insulating lobe [6]. This opens up the path to new multifunctional materials made from Janus building block that carry on the different lobes different functionalities, optic, magnetic, surface functional groups, etc.

4.2 Short History of Asymmetric Janus Particles

Janus particles have a relatively short history and only came recently into existence through the imagination of a handful of scientists in the late 1980s. The concept of an asymmetric and amphiphilic particle was put forward by Casagrande, Veysié and de Gennes. The first synthesis of micron-sized JPs was attributed to the former two authors, while de Gennes baptised them after the Roman god Janus. Janus was the two-faced god of transitions, gates, passages and new beginnings, after which the month of January was named. Janus is a god in Roman mythology and was considered to be the most important one because it provided passage to the other gods. Janus is often depicted with two faces of an old man, but initially this was depicted with the face of young man looking back and old man looking forward, very suggestive of the time passage. In his famous Nobel Prize talk [7] de Gennes also called the asymmetric particle Janus grains. Interestingly, the amphiphilic nature of the Janus glass beads produced by Casagrande et al. [8] could be probed directly for the first time by observing from photographs of breath-patterns "figures de soufflé" clearly showing droplets of water pearling up on the hydrophobic side but making a contiguous film on the hydrophilic side. A few years earlier, in 1985, Grünning et al. [9] filed a patent claiming a procedure for the preparation of amphiphilic particles from 100 μm hollow beads, with the exterior surface hydrophobised and hydrophilic side remaining in the interior, which were then crushed to produce irregularly shaped glass shards that were amphiphilic; this procedure was later detailed by one of the inventors, Rosmmy in 1998 [10]. Remarkably, right from the beginning the inventors proposed the use of such particles as surface active products, for emulsification, production of foams and deployment in tertiary oil recovery. This was the first time when amphiphilic particles were produced in large amounts. In 1991 Chen et al. [11] synthesised biphasic snowman type polymeric particles by seeded emulsion polymerisation and phase separation, but were probably unaware of the fact that they synthesised the first polymeric JPs. Within the 1990s to 2000s the scientific community did not seem to catch interest in these particles. It was then only much later, in the early of the first decade of the twenty-first century when the new synthetic routes were published. This was in part triggered by the theoretical work of Ondarçuhu [12] and later by Binks and Fletcher [13] who showed by calculation that the interfacial activity of spherical JPs at their maximum amphiphilicity (highest polarity contrast), in terms of interfacial desorption energy should be up to three times larger than that of HPs and much larger for snowman or dumbbell JPs. However, the desorption energy only is not a proper gauge for measuring the interfacial activity, but rather the ability to lower the interfacial tension, as it will be discussed later. Since then the research on Janus particles grew exponentially, estimated from the increasing number of publications each year. Initially the JPs could only be produced in very low amounts and the synthetic challenges were preventing their use in applications. Currently JPs can be produced in large amounts and therefore, their use in new applications is being explored, such as emulsifiers, catalysts, foam stabilisers, polymer blend compatibilisers, amphiphiles, building-blocks, etc. It turns out the Janus is not only reserved for synthetic particles but also naturally occurring proteins, such as the hydrophobins produced by fungi, HFBII from *Trichoderma Reesei* that is nearly globular with a 3 nm diameter and 7.2 kDa [14]. Interestingly, it has been shown that HFBII is an excellent foam stabiliser and is responsible for beer gushing. Beer gushing is the phenomenon of beer foam gushing out of the bottle when opened or mechanically shocked by hitting its bottom to the table, completely emptying the bottle [15]. The HFBII ends up in malt by fungi infection.

4.3 General Synthetic Routes

JNPs can be prepared by chemical or physical methods. Next we give a short overview of the methods used in the synthesis of JPs.

4.3.1 Masking and Asymmetric Modification

This preparation method relies on a simple concept, in a first step one side of a spherical particle is masked by a protective layer and in a second step chemical or physical modification is performed on the unmasked part followed by the removal of the masking layer. In this way the original HP particle is now a JP particle because it has different surface properties, even though its bulk composition remains the same. While the concept of the method is simple its implementation and scalability can be an issue. The masking and modification method was first used by Casagrande et al. [8] who used varnish to mask half of the micron-sized spherical glass beads and performed silanisation/hydrophobisation on the other half of the spherical bead to produce JPs. Instead of applying a varnish, a spherical particle can be deposited on a flat surface or self-assembled in a monolayer followed by the deposition, via evaporation, of a metal on the exposed part, Fig. 4.2. Due to the masking, the metal will only deposit on one side of the particle facing away from the flat substrate. In this way hybrid polymer/metallic, silica/metallic JPs can be made. This is a highly effective method to produce JPs with tunable optical or electric properties as the thickness of the metal layer deposited can be precisely controlled; Kawaguchi et al. [16] applied this method to produce JPs with a functional gold surface for control of surface plasmon resonance and thus the colour of the particles with potential applications in paper displays. Other patchy JPs with tunable optical properties were successfully prepared by Composto et al. [17]. Bimetallic Co/Ni Ag/Au, Ni/Au Janus particles could be made using the same method by Carroll et al. [18] from a single layer of self-assembled silica beads on a substrate in two steps:

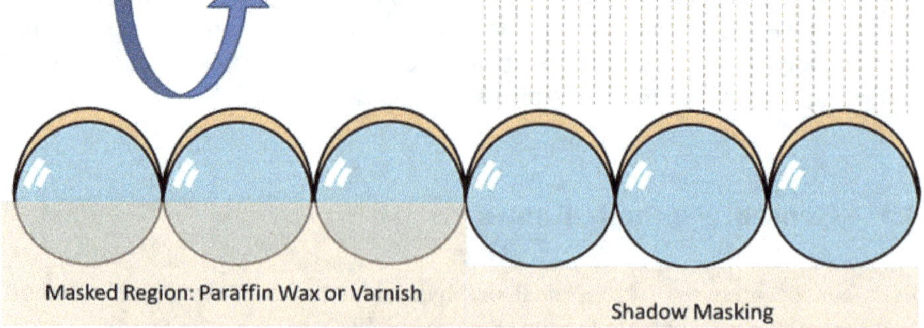

Fig. 4.2 Masking and surface modification techniques for manufacturing of Janus particles

first the particles were coated with one metal using e-beam evaporation, then the beads were inverted and coated with another metal. The preparation of JPs from 2D layers is highly effective but not scalable to produce large amounts. In this context Granick's group [19, 20] has succeeded in producing gram-scale amount of JPs by first taking fused silica particles HPs (800 nm and 1.5 μm in diameter) and use them to emulsify molten wax at high temperature. Then the obtained o/w emulsion was cooled to room temperature to obtain solid wax colloidosomes that have HPs trapped/embedded on their surface. Because the HPs were half protected by the wax they could be chemically reacted with APS on the water exposed part, then after dissolving the wax colloidosome they could be hydrophobised on the other side with OTS. In this way gram-scale amounts of JPs could be obtained. Suzuki et al. [21] used the same strategy to prepare large amounts of JPs from thermo-responsive pNA microgels by first trapping them at the heptane/water via emulsification to create o/w Pickering emulsions. Once at the interface of the oil droplets the microgel particles were further reacted with water soluble only ethylenediamine (ED) and 1-ethyl-3-(3-dimethylaminopropyl)-carbodiimide hydrochloride (EDC) and amine groups were introduced this way at the surface of the water exposed gel particles. Smaller Au nanoparticles could be attached only on the $-NH_2$ rich side of the stimuli-responsive Janus microgel particles. Fujimoto et al. [22] prepared Janus microspheres at the solid–liquid interface by allowing poly(methacrylic acid-co-nitrophenyl acrylate) microspheres to interact with a substrate on which human immunoglobulin (IgG) was previously adsorbed. After the microspheres settled on the flat substrate, NaOH was added to activate the ester bonds by cleaving the p-nitrophenol. The part of the microsphere touching the IgG-substrate reacted with the amine and thiol bonds of the IgG molecule, while the other side remained unmodified.

4.3.2 Seeded Emulsion Polymerisation and Phase Separation

The preparation of JPs by seeded emulsion polymerisation and phase separation of polymers was first performed by Chen et al. [11] in 1991, although probably unaware of the fact that these were being baptised Janus particles by P. de Gennes in the same year. The procedure is simple in theory, and starts with monodisperse seed latex polystyrene (PS) particles. Their methods were later revived and extended to a multitude of different biphasic polymeric Janus particles. The method consists of first preparing polymeric latex particles in the nanometre range, such as PS. Then these are re-dispersed in water and a second monomer (partially soluble in water) is added, such as MMA or 3-TSPM. Emulsification of the monomers is necessary for the swelling of the latex particles to succeed. Then the mixture is given some time for the swelling of the seed particle by the monomer to take place, eventually the polymerisation is started. As the monomer polymerises the newly created polymer, due to incompatibility with PS, bulges out from the seed particles leading to the creation of a second lobe and the formation of snowman

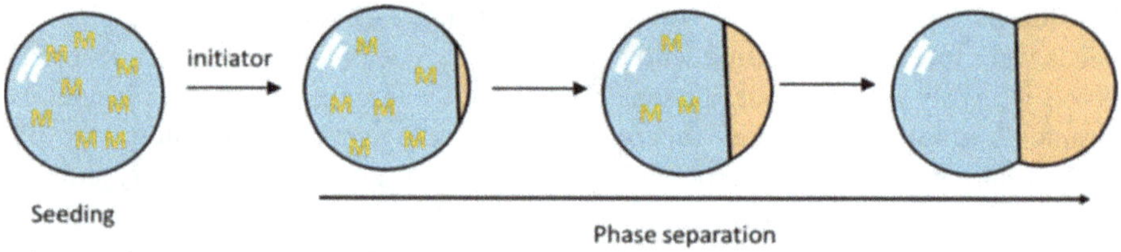

Fig. 4.3 Cartoon depicting the preparation of JPs via seeded emulsion polymerisation and phase separation method. The second lobe grows as the monomer (M) from the reservoir is being consumed

or dumbbell type biphasic Janus particles as depicted in Fig. 4.3. The degree of phase separation as well as the final geometry of the obtained JPs depends on the Flory–Huggins interaction parameter, wettability between the two polymers, cross-linking degree, phase separation kinetics. Wettability between the polymers can also be influenced by initiator type, addition of surfactant, etc. [11]. Several different types of polymeric JPs could be made this way and with precisely tuned reaction conditions snowman-type particles could be produced. JPs with hard and soft lobes PtBA/PS JPs were prepared by Bon et al. [23], Daeyeon Lee et al. [24] prepared PS/PPA JPs that can be subsequently modified via thiol-yne click reactions, Sun et al. [25] obtained PS/PMMA with a hollow PS lobe, Hoffmann et al. [26] prepared PMMA/PS JPs from PMMA seeds, Weitz et al. [27] synthesised PS/PMMA and PS/PtBA seeds. Note the convention we have used, the initial seed particles are first written followed by the second lobe.

4.3.3 Microfluidic and Capillary Electro-Jetting Methods

Multi-compartmented or multiphasic particles can be fabricated by microfluidic as well as electro-jetting processes [28]. Droplet microfluidics refer to the preparation and manipulation of discrete micron-sized droplets, double emulsions droplets, microbubbles, etc., and the fabrication of polymeric JP by this method has been extensively reviewed [29]. Black and white bicolored JPs with electrical anisotropy were synthesised for the first time by Torii et al. [30] using a microfluidic co-flow system. Pigments of carbon black and titanium oxide were dispersed in IBA then were separately introduced into the Y-junction at the same volumetric flow rate to form a two-colour stream followed by break-up into droplets due to surface tension further down into the stream channel. The obtained Janus droplets were polymerised outside of the microfluidic system by heating, but in principle polymerisation can also be done in the microfluidic channel by UV exposure. Multicoloured JPs with electrical anisotropy can be used in panel displays where the colour switch can be done by changing the orientation of the JPs and this can be actuated between two electrodes by applying voltage.

The electrodynamic co-jetting methods consist of flowing two or more different polymer solutions through a bi-compartmented metal capillary, while maintaining a laminar flow, at the apex of the capillary tip the polymer solution comes together to form a droplet to which a high electrical voltage is applied. The application of the electric field causes the solutions to form a Taylor cone, which creates a spray of individual droplets accelerating toward the counter electrode. During this, the solvents from the droplet evaporate, resulting in polymeric nanoparticles that are collected on the surface of the counter electrode. In this method due to the rapid evaporation of the solvents, and because the process is very fast, the polymers touching do not have sufficient time to mix, the resulting nanoparticles remain bi-compartmented. These methods have the great advantage of high-throughput to prepare large amounts of micrometre sized multiphasic Janus particles. Preparing of monodispersed nano-sized particle is more challenging; however, Lahann et al. [31] succeeded with the synthesis of PMMA/PtBMA biphasic Janus nanoparticles having diameters $d = 172 \pm 28$ nm and tri-phasic [32] from poly(ethylene oxide), poly(acrylic acid) and poly(acrylamide-*co*-acrylic acid). Furthermore the individual polymer phases can be independently loaded with biomolecules or selectively modified with model ligands [33]. The resulting JPs are generally non-crosslinked, but it is possible to crosslink these by using a photoinitiator and by UV exposure after immediate generation of the oil droplets can lead to cross-linking of the polymers [34].

4.3.4 Polymer Co-precipitation and Phase Separation

One of the most simple and feasible pathways to synthesise JPs is the phase separation of polymeric solutions, such as A/B homopolymer/homopolymer and AB/C copolymer/homopolymer blends dissolved under confinement followed by the evaporation of the solvent. This method usually involves an oil-in-water (o/w) emulsion system working as the confinement system, in which oil droplets comprise two incompatible polymers with a large Flory–Huggins interaction parameter, such as PS and PMMA, dissolved in a common solvent [35]. After evaporation, the solvent leaves behind solid well-defined particles, with the two separated polymer phases inside. Deng et al. [36] were able to prepare JPs with hierarchical structures from AB/C polymer blends. The preparation of polymeric Janus nanoparticles by this method involves first dissolving multiple, chemically distinct polymers in a mutually favourable solvent and gradually altering the solubility character of the solution until the polymer molecules co-precipitate as particles and different polymers phase separate. Priestley et al. [37] succeeded in scaling up preparation of JPs and multi-compartmented particles by co-precipitation and phase separation method from dissimilar polymers to up to 1400 kg/day by designing a so-called confined impinging jet mixer, a technique they called termed flash nanoprecipitation (FNP). In the FNP method, the PS and PI (with a Flory–Huggins interaction parameter $\chi_{PS\text{-}PI} = 0.07$) were dissolved at a certain ratio in a common solvent

THF and injected in one of the arms of a fluidic device, in the same time the "anti-solvent", DIW, was flown through a second arm of a fluidic device. The two arms converged into a single one called the mixing region where the PI-PS co-precipitated and phase separated as the solvent rapidly exchanges with an "anti-solvent". By changing the polymer feed concentration from 0.1 to 1.0 mg/mL they could systematically increase the size of the Janus nanocolloids from $\simeq$125 to 540 nm in diameter. Anisotropy of the produced particles could also be changed by altering the PS–PI polymer ratio from 1:4 to 4:1 to produce multifaceted colloids. The later technology proves to be highly versatile, with perhaps the only disadvantage is that starting already from polymer chains the co-precipitation and phase separation methods do not offer the possibility for polymer cross-linking restricting the use of the obtained JPs, for example, as stabilisers in Pickering emulsions in which case the JPs would likely disintegrate/dissolve upon interaction with the oil.

4.4 Tuning the Surface Polarity in JPs

Unlike the HPs whose surface polarity can be changed only by chemical means, tuning the surface polarity of the JPs can be done in a gradual and predictive way by adjusting the geometric ratio or the surface area between the lobes having different polarities. In this way, homologous series of JPs can be created, in analogy to homologous series of molecular surfactants, for example, by increasing the amount of the monomer to seed latex particles in seeded emulsion polymerisation, Fig. 4.4.

Wu and Honciuc [1] have synthesised a homologous series of PS/P(3-TSPM) JPs by changing the volume of 3-TSPM monomer to the PS seed NPs, see Fig. 4.4, and demonstrated that the PS lobe is the less polar than the P(3-TSPM) one; by increasing the size of the more polar P(3-TSPM) they could achieve polarity inversion in the homologous series purely by geometric means when the size of the P(3-TSPM) became larger than that of the PS lobe. The polarity inversion in the homologous series could be demonstrated from heptane–water emulsification experiments, whereas the JPs with smallest P(3-TSPM) lobe have high affinity to heptane, the largest P(3-TSPM) lobe JPs have higher affinity to water and as a consequence an inversion of the emulsion phase from w/o to o/w occurs at the middle of the homologous series. But the aspect ratio and implicitly the polarity of the JPs can also be tuned dynamically by stimuli, such as in stimuli-responsive particles that change both their geometry and polarity. One such example are the shape-changing and pH-responsive particles PtBA/PS produced in Lee's group [38] which upon cleaving the -tBA group by hydrolysis in the PtBA lobe, pH-responsive PA/PS JPs are obtained. At high pH the carboxyl groups are mostly ionised, resulting in large intake of water and swelling of the PA lobe; as a consequence, the particles become polar, have a good affinity to water and form o/w emulsions. At low pH, below the pKa value, the -COOH groups are protonated, non-ionic and

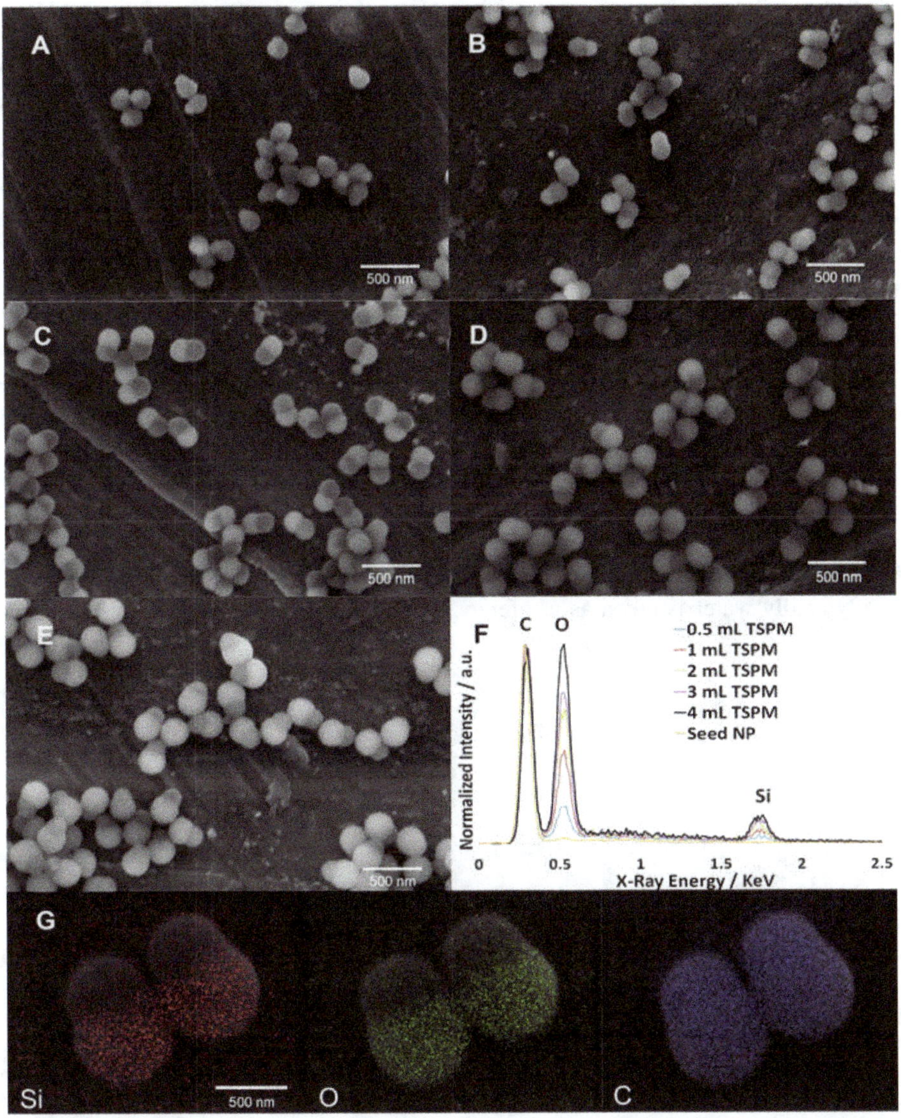

Fig. 4.4 SEM images of PS/P(3-TSPM) JPs with progressively enlarged P(3-TSPM) lobe (light-grey/white) from the same seed PS NPs (dark-grey). **(a)**–**(e)** JPs with progressively larger lobes obtained for a volume of 3-TSPM monomer **(a)** 0.5 mL, **(b)** 1 mL, **(c)** 2 mL, **(d)** 3 mL and **(e)** 4 mL added to 1 g of PS seed NPs; **(f)** EDX spectra, normalised with respect to the reference carbon peak of the PS seed NPs. **(g)** EDX mapping of "2 mL TSPM" JNPs obtained from larger seed PS NPs, 320 ± 5 nm diameter, showing asymmetric distribution of oxygen, silicon elements, namely a higher concentration in the P(3-TSPM) lobe in contrast to a symmetric distribution of carbon in both Janus lobes. Reprinted with permission from Ref. [1].

there is no water intake of the lobe; the geometry of the lobe remains rigid and the particles have higher affinity to the oil phase forming w/o emulsions.

Surface Polarity Contrast Between Lobes: Quantification of Amphiphilicity
The JPs are amphiphilic because of the inherent surface polarity contrast between the lobes. The concept of amphiphilicity is however understood in a qualitative way and it denotes the ability of the amphiphile to adsorb and partition at the oil/water or air/water interfaces.

The earlier theoretical work of Ondarçuhu [12] already set the framework for estimating the amphiphilicity balance of spherical JP adsorbed at an interface (oil–water or air–water) by measuring its contact angle with reference to one of the phases, usually the water phase. His geometrical model was parameterised as to include the angle α that denotes the position of the boundary between the apolar and polar regions of the Janus lobes, see Fig. 4.5. The contact angle of the JP with the interface is given by β. For a perfectly spherical JP α is almost a measure of its amphiphilicity, because zero amphiphilicity (corresponding to homogeneous particles) corresponds to either $\alpha = 0$ or 180°. Strongest amphiphilicity is expected when $\alpha = 90°$. Two additional parameters were introduced, namely θ_A and θ_P which are the contact angle of one of the phases (depends on the chosen reference but typically water is taken as a reference) with each of the lobes, the apolar and

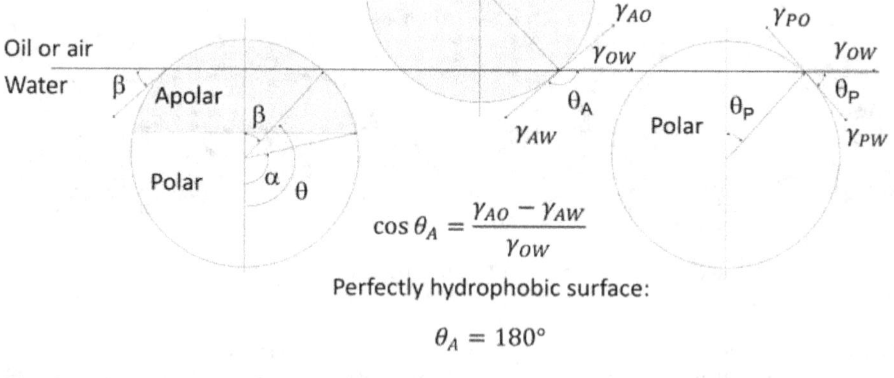

Fig. 4.5 Model describing a spherical JP, apolar HP and polar HP at the oil–water interface. The parameters can be defined as follows: α keeps track of the position of the boundary between the apolar and polar regions of the JP, whereas β keeps track of the position of the oil–water interface relative to the particle centre and it represents the contact angle with the water phase. The angle θ_A is the water contact angle of the apolar HP corresponding to the apolar region of the JP, the θ_P is the water contact angle of the HP corresponding to the polar region of the JP

the polar, respectively. These angles can be better understood if HPs corresponding to each of the JP lobes are depicted at the interface as in Fig. 4.5. For JPs no amphiphilicity is expected when $\theta_A - \theta_P = 0°$ and the strongest amphiphilicity for $\theta_A - \theta_P = 180°$, meaning that one lobe has a perfectly hydrophobic surface and the other a perfectly hydrophilic surface, respectively.

Later Jiang and Granick [39] introduced the concept of Janus balance or "J-value" to effectively quantify the amphiphilicity as the dimensionless ratio of work to transfer an amphiphilic JP from the oil–water interface into the oil phase, normalised by the work needed to move it into the water phase:

$$J = \frac{\sin^2\alpha + 2\cos\theta_P(\cos\alpha - 1)}{\sin^2\alpha + 2\cos\theta_A(\cos\alpha + 1)}, \tag{4.1}$$

where the angle α, θ_A and θ_P have the same meaning as those depicted in Fig. 4.5. The above equation shows that Janus balance depends on the relative areas of hydrophilic and hydrophobic lobe, quantified by α and on the hydrophobicity of the two sides, quantified by θ_A and θ_P. When θ_A and θ_P are fixed, J increases as α increases (because $\cos\theta_P < 0$) meaning a larger hydrophilic area. When α is fixed, J increases when θ_A and θ_P increase, which corresponds to the hydrophilic part becoming more hydrophilic or the hydrophobic part becoming less hydrophobic. The larger the magnitude of J, the more hydrophilic is the JP, which follows the same trend as that of the HLB[40] for surfactant molecules: larger HLB meaning a higher affinity for water. The J value can be therefore calculated from the interfacial contact angle and the geometry of Janus particles. However, the above model has two caveats: first the model was deduced by assuming a perfect orientation of the particle at the oil/water interface, i.e. the Janus axis perpendicular to the interface, and second it assumed a perfectly spherical particle, but α loses its meaning for a snowman, dumbbell or any other shape of the JP and the problem has to be re-parameterised. Therefore, the above model is not generally applicable. Based on these calculations Binks and Fletcher [13] have shown that the interfacial activity of a JP can be up to three times larger than that of an HP. This is lately taken, mistakenly, as an upper limit of what JPs can achieve but in fact the interfacial activity of these dual particles can be significantly larger than a factor of three for other geometries, as shown in simulations by Gao et al. [41]. However, it would be more useful if amphiphilicity can be discussed quantitatively and can be measured.

In a different approach to quantify the amphiphilicity of any shape of JPs Honciuc et al. [1, 6, 42] have proposed the direct calculation of the HLB balance of a JP using Griffin's approach [43], that is, the same model used to calculate the HLB balance for a homologous series of surfactants [40] and adapted it for JPs:

$$\mathrm{HLB} = 20\frac{A_{\mathrm{Polar}}F_1}{A_{\mathrm{Polar}}F_1 + A_{\mathrm{Apolar}}F_2}, \tag{4.2}$$

where A_{Polar} is the area of the polar lobe, A_{Apolar} is the area of the non-polar lobe and in addition we have introduced the weighting factors F_i $(i = 1, 2)$ accounting for the "degree" of polarity of the lobes. The original approach of Griffin for surfactants did not account for the polarity of the surfactant moieties, but only considers their relative molecular weights $20M_{w(polar)}/M_{w(molecule)}$. The above equation takes the value of 20 for $F_2 = 0$ and 0 for $F_1 = 0$, which are two limiting situations: strongly polar and apolar particles, respectively, with no amphiphilicity. On the other hand a value of F_1= 1 (hypothetical 100% polar surface) and F_2= 1 (hypothetical 100% non-polar surface) assumes an "ideal" polarity contrast between the two surface regions, see Fig. 4.6a, and thus, the HLB is decided by the geometry of the lobes, i.e. their aspect ratio. The polarity weighting factors F can be calculated from the ratio between the polar and apolar or dispersive surface energy components for each of the Janus lobes, as depicted in Fig. 4.6:

$$F_1 = \frac{\gamma_1^p}{\gamma_1^p + \gamma_1^d} \quad \text{and} \quad F_2 = \frac{\gamma_2^d}{\gamma_2^p + \gamma_2^d}, \tag{4.3}$$

where the small Greek gammas are the surface energies and the superscripts "p" and "d" indicate the polar and dispersive or apolar surface energy components of the corresponding Janus lobes, subscripts 1-polar lobe and 2-apolar lobe. In practice, to determine F one must know the surface energy and its polar and dispersive components which is not trivial. Recently, Mihali and Honciuc [6] have measured the surface energy and the polar/disperse components of each Janus lobes in a homologous series of semiconducting PPy/P(3-TSPM) JPs with increasing size of the polar lobe and with these values they have calculated the corresponding weighting factors F and subsequently the HLB values, these are given in Table 4.1.

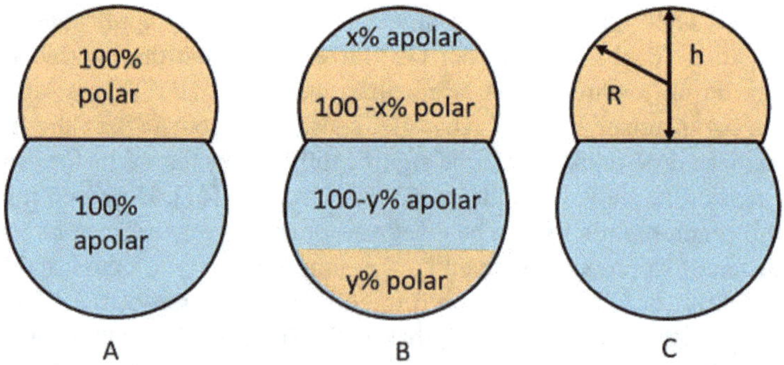

Fig. 4.6 (**a**) Hypothetical amphiphilic dumbbell Janus particle displaying an ideal polarity contrast between a purely dispersive surface and "purely polar" surface of the two Janus lobes, where the polarity factors are $F_1 = F_2 = 1$; (**b**) the more realistic representation of a snowman Janus particle with the surface polarity of the lobes departing from ideality and whose polarity factors, F_1 and F_1 can be calculated with Eq. (4.3); (**c**) the parameters of a JNP used to calculate areas of the lobes, radius R and height of the lobe h

Table 4.1 Aspect ratio and HLB values in the homologous series of JNPs

JPs in increasing order of the (3-TSPM) polar lobe	[a]Area × 1000/nm^2, PPy-lobe	[a]Area × 1000/nm^2, P(3-TSPM)-lobe	Aspect ratio P(3-TSPM) /PPy	[b]HLB number	[c]F_1	[c] F_2	[d]HLB number (weighted)
PPy/P(1 mL 3-TSPM)	177.6	83.2	0.5	6	0.93	0.92	6
PPy/P(2 mL 3-TSPM)	211.8	172.9	0.8	9	0.93	0.92	9
PPy/P(3 mL 3-TSPM)	164.1	318.9	1.9	13	0.93	0.92	13
PPy/P(4 mL 3-TSPM)	152.7	348.7	2.3	14	0.93	0.92	14

[a]The areas of the lobes were calculated from the equation $A = 2\pi Rh$, where h is the height of the Janus lobe and R is its radius, Fig. 4.6c
[b]Values calculated with Eq. (4.2) with $F_1 = F_2 = 1$
[c]F_1 and F_2 were calculated with Eq. (4.2)
[d]Values calculated with Eq. (4.2)
From Ref. [6].

Interestingly the HLB values calculated for the JPs excluding the weighting factors, in the 5th column of Table 4.1, are similar with those calculated after taking into account the degree of polarity of the lobes, in the 8th column of Table 4.1. That is because the measured F-values are close to unity and thus the JPs have an almost ideal polarity contrast. Calculating HLB value this way for a homologous series of JPs has the advantage of being able to predict the behaviour of JPs with respect to their ability to act as w/o or o/w emulsifiers, discussed later. HLB range, taking values from 1 to 20, is more readily understood by scientists working with surfactants. For example, amphiphiles with values below 10 on the Griffin's scale are good w/o emulsifiers (good affinity to the oil phase), while those with HLBs above 10 are good o/w emulsifiers, which was clearly shown by the emulsification experiments [1]. We believe this method to quantify the Janus balance is universally applicable because it makes no assumptions about the particle geometry, orientation or its position at interface.

4.5 Interfacial Activity and Adsorption at Interfaces

It is well known that particles can spontaneously adsorb at liquid–liquid and air–liquid interfaces and can thus lower the system's Gibbs free energy, which translates into the reduction of the interfacial IFT. By monitoring the decrease of the interfacial tension with time, i.e. the dynamic surface tension, usually with pendant drop tensiometry, one can obtain information about JPs' interfacial adsorption kinetics. The bulk diffusivity of any particles obeys the Stokes diffusion law. For particles with appropriate wettability the bulk-to-surface diffusion may lead to the adsorption and attachment at the interface if: (a) there is no strong electrostatic repulsion interaction between the interface and the particle (image charge repulsion) and (b) the energy costs related to the surface dehydration and re-solvation of the surface by the next solvent are not too high.

The adsorption kinetics of particle adsorption may operate in different regimes, diffusion limited, activation energy limited or a combination of both. Therefore, it is expected that the decrease in IFT vs. time is slower for larger particles as compared to the smaller ones. Furthermore, according to the time evolution of the interfacial tension, the adsorption is characterised by three adsorption stages, depicted in Fig. 4.7: (I) the free diffusion of some particles to the interface (II) continuous adsorption of Janus particles to form domains at the interface, and (III) particle packing and rearrangement in compact domains/islands [41, 44].

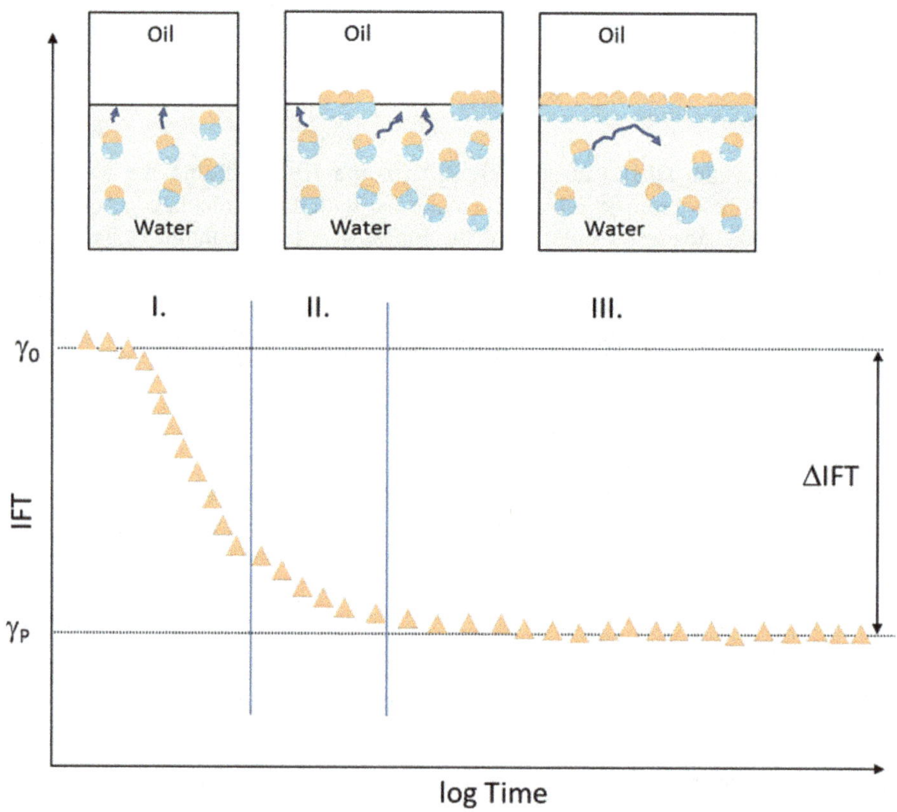

Fig. 4.7 Cartoon depicting the time evolution of the IFT and the three main stages of JPs adsorption: (I) JPs adsorb at a pristine oil–water interface, the diffusion from bulk-to-surface may be the limiting step, but also a large activation energy barrier to adsorption, (II) continuous adsorption of JPs at the interface and formation of domains and 2D islands and (III) full occupation of the interface by JPs, particle–particle repulsive interactions contribute to increasing the energy barrier to adsorption and interfacial re-organisation of the monolayer

The magnitude of ΔIFT, which is the distance between the starting value at $t = 0$ and t_{plateau} at which IFT does not decrease anymore, Fig. 4.7, indicates how effective the JPs are at lowering the interfacial tension. The magnitude of ΔIFT encompasses several phenomena, such as the ability of the particles to pack in a compact monolayer, surface and interfacial energy of the particles with both phases and the particle–particle lateral interactions. Hypothetically, the JPs can lower the IFT to almost zero if their interfacial energy with both phases is zero, perfect affinity for each lobe for the oil and water phases, and have the ability to pack compactly leaving no free oil–water interface. It has been predicted [13] and experimentally demonstrated that JPs have a significantly higher ability to lower the IFT than the HPs [45, 46]. The reason of that is that JPs constituted of two parts can achieve better solvent-JP and water-JP compatibility (that translate in very low interfacial energies), which cannot be achieved in HPs. In the case of JPs not only the size but also the shape affects the ΔIFT. As the shape changes from sphere to disc and rod, Gao et al. [41] observed different adsorption kinetics, different packing behaviours and ultimately different ΔIFT values. For the three types of Janus particles with the same surface area, the ability to decrease the interfacial tension increases from Janus sphere to Janus disc to Janus rod. The particle-interface interaction has also been shown to play a role, for example, the particle with a large zeta potential seems to adsorb better at the interface due to an interfacial charge re-distribution due to the strong electric field of the particle that locally inverts the charge density of the air–water interface [47]. The way the JNPs can assemble at the oil–water interfaces can be affected by their aspect ratio, i.e. the geometrical packing parameter and by the polarity contrast, analogue to molecular surfactants.

4.5.1 Contact Angle and Interfacial Adsorption Energies of HPs vs. JPs

The adsorption of the JPs at interfaces can be treated purely thermodynamically, whereas only the initial and final states are taken into account regardless of the path the system takes. For example, the Gibbs free energy of JPs' adsorption at oil–water interface can be calculated by taking into account the free energy of the particle in one of the bulk phases as the initial state and the free energy of the particle at the interface between two phases. The difference between the two will be the Gibbs free energy of JPs' adsorption or minus desorption energy. For an HP this can be easily calculated, for example, at the air–water interface if we know its contact angle β with water, Fig. 4.8.

The free energy can be calculated for any particle simply by multiplying its interfacial energy also known as the energy density (mJ/m^2) by its exposed area to each liquid forming the interface and for the case in Fig. 4.8 the free energy at the interface is:

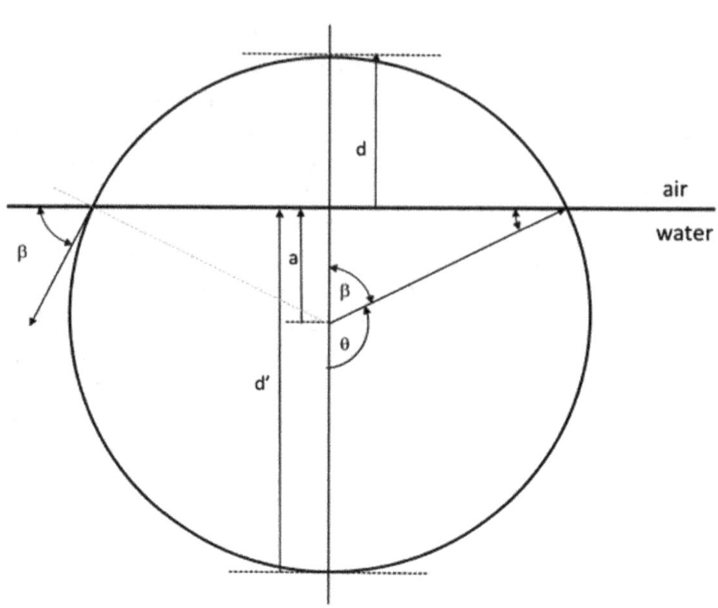

Fig. 4.8 Spherical particle adsorbed at the air–water interface. The contact angle with water β can be determined in two equivalent ways: (left) between the position of the air–water interface with the surface of the particle or (right) from the angle formed between the radius pointing at the three-phase line. The parameters a and d' are the immersion depth of the particle in water measured from the centre and apex of the particle, respectively, and d is the immersion of the particle in the second phase, air or oil

$$E_{\text{interface}} = \gamma_{(\text{HP,air})} A_{(\text{HP,air})} + \gamma_{(\text{HP,water})} A_{(\text{HP,water})} + \gamma_{(\text{air,water})} A_{(\text{circular base radius r})}. \tag{4.4}$$

The explicit expressions of the areas are:

$$A_{(\text{HP,air})} = 2\pi R(R-a) = 2\pi R[R - R\sin(\pi/2-\beta)] = 2\pi R^2(1-\cos\beta) \tag{4.5}$$

$$A_{(\text{HP,water})} = 2\pi R(R+a) = 2\pi R[R + R\sin(\pi/2-\beta)] = 2\pi R^2(1+\cos\beta) \tag{4.6}$$

$$A_{(\text{circular base radius r})} = \pi r^2 = \pi[R\cos(\pi/2-\beta)]^2 = \pi R^2\sin^2\beta, \tag{4.7}$$

where the last expression accounts for the area of the excluded air–water interfacial area.

Finally, the free energy of the HP at the interface is:

$$E_{\text{interface}} = \gamma_{(\text{HP,air})} 2\pi R^2(1-\cos\beta) + \gamma_{(\text{HP,water})} 2\pi R^2(1+\cos\beta) + \gamma_{(\text{air,water})} \pi R^2\sin^2\beta. \tag{4.8}$$

The Gibbs free energy of a single HP completely immersed in water will be:

$$E_{\text{water}} = \gamma_{(\text{HP},\text{water})} 4\pi R^2. \tag{4.9}$$

Therefore, the total interfacial adsorption energy only as a function of the contact angle β will be:

$$\begin{aligned} E_{\text{water}} - E_{\text{interface}} &= \gamma_{(\text{HP},\text{air})} 2\pi R^2 (1 + \cos\beta) - \gamma_{(\text{HP},\text{water})} 2\pi R^2 (1 + \cos\beta) \\ &\quad - \gamma_{(\text{air},\text{water})} \pi R^2 \sin^2\beta. \end{aligned} \tag{4.10}$$

Pieranski [48] was among the pioneers to attempt calculating the particle energies at interfaces and the equations proposed were as a function of the HPs immersion depth "a", Fig. 4.8, which is the distance from the centre of the particle to the interface. This can obviously also be determined directly without the need to measure the contact angle directly from SEM of cryogenised particles trapped at interfaces [49]. The above expression as a function of "a" would be:

$$\begin{aligned} E_{\text{interface}} &= \gamma_{(\text{Particle},\text{air})} 2\pi R (R - a) + \gamma_{(\text{Particle},\text{water})} 2\pi R (R + a) \\ &\quad + \gamma_{(\text{air},\text{water})} \pi (R^2 - a^2), \end{aligned} \tag{4.11}$$

which after rearrangement becomes:

$$\begin{aligned} E_{\text{interface}} &= \gamma_{(\text{Particle},\text{air})} 2\pi R^2 (1 - a/R) + \gamma_{(\text{Particle},\text{water})} 2\pi R^2 (1 + a/R) \\ &\quad + \gamma_{(\text{air},\text{water})} \pi R^2 [1 - (a/R)^2]. \end{aligned} \tag{4.12}$$

The depth a is related to the interfacial tension via the Young–Dupré contact angle:

$$\cos\beta = \frac{a}{R} = \frac{\gamma_{(\text{HP},\text{air})} \pm \gamma_{(\text{HP},\text{water})}}{\gamma_{(\text{HP},\text{water})}}. \tag{4.13}$$

The sign in parenthesis is the negative for $\beta < 90°$ and positive for $\beta > 90°$.

Very often in different works one finds the following expression as the energy with which a small particle of radius R is held at the air–water or oil–water interfaces[50]:

$$E_{\text{interface}} = \gamma_{(\text{air},\text{water})} \pi R^2 [1 \pm \cos\beta]^2, \tag{4.14}$$

where the sign in parenthesis is the negative for $\beta < 90°$ and positive for $\beta > 90°$. This form of the last equation results from inserting the expression of $\cos\beta$ in Eq. (4.13), defined by the Young–Dupré expression, into the $E_{\text{interface}}$ expression Eq. (4.11) obtaining[51]:

$$
\begin{aligned}
E_{\text{interface}} =& \gamma_{(\text{HP,air})} 2\pi R^2(1-\cos\beta) + \gamma_{(\text{HP,water})} 2\pi R^2(1+\cos\beta) \\
&+ \gamma_{(\text{air,water})} \pi R^2 \sin^2\beta \\
=& 2\pi R^2 \left[\gamma_{(\text{HP,air})} + \gamma_{(\text{HP,water})} + \cos\beta(\gamma_{(\text{HP,water})} - \gamma_{(\text{HP,air})})\right] \\
&+ \gamma_{(\text{air,water})} \pi R^2 \sin^2\beta \\
=& 2\pi R^2 \left[\gamma_{(\text{HP,air})} + 2\gamma_{(\text{HP,water})} - \gamma_{(\text{HP,water})} - \gamma_{(\text{air,water})} \cos^2\beta\right] \\
&+ \gamma_{(\text{air,water})} \pi R^2 \sin^2\beta \\
=& 4\pi R^2 \gamma_{(\text{HP,water})} + 2\pi R^2 \left[\gamma_{(\text{air,water})} \cos\beta - \gamma_{(\text{air,water})} \cos^2\beta\right] \\
&+ \gamma_{(\text{air,water})} \pi R^2 \sin^2\beta.
\end{aligned} \tag{4.15}
$$

Keeping in mind that $\sin^2\beta = 1 - \cos^2\beta$,

$$
\begin{aligned}
E_{\text{interface}} =& 4\pi R^2 \gamma_{(\text{Particle,water})} - \pi R^2 \gamma_{(\text{air,water})} \\
&\times \left[1 - 2\cos\beta + 2\cos^2\beta - \cos^2\beta\right] \\
=& 4\pi R^2 \gamma_{(\text{Particle,water})} - \pi R^2 \gamma_{(\text{air,water})} (1-\cos\beta)^2,
\end{aligned} \tag{4.16}
$$

where the first term in the last equation is the interfacial energy of a sphere completely immersed in water; therefore, the last term is the change in interfacial energy of the particle attached to the air–water interface, i.e. the final state. For example, if $\gamma_{(\text{air,water})} = 75\,\text{mJ/m}^2$ and $R = 1\,\mu\text{m}$ and $\beta = 90°$, then the energy holding the particle at the interface at room temperature would be 2.4×10^{-13} J $\approx 5.7 \times 10^7$ kT, where 1 kT $= 4.114 \times 10^{-21}$ J at 298 K.

In order to calculate the Gibbs free energy of JPs adsorption at air–water interface, consisting of two hemispheres, one apolar (A) and one polar (P), one needs to calculate the energy difference between the particle entirely immersed in water and that adsorbed at air–water interface. This can be done following strictly the Ondarçuhu [12] nomenclature, which assumes a perfectly spherical JP and its adsorption at the interface occurs via partial dehydration of the surface of one lobe, as depicted in Fig. 4.9.

A. Case of partial de-wetting of the A-lobe, $\theta > \alpha$ and the general case $\alpha \neq \pi/2$.

In order to solve this, the expression of the energy for each interface must be found. Next, we find the expressions of $E_{(\text{A,air})}$, $E_{(\text{A,water})}$, $E_{(\text{P,water})}$ and $E_{(\text{air,water})}$ as a function of contact angle,

$$
E_{(\text{A,air})} = \gamma_{(\text{A,air})} A_{(\text{A,air})} = \gamma_{(\text{A,air})} 2\pi R^2 (1+\cos\theta), \tag{4.17}
$$

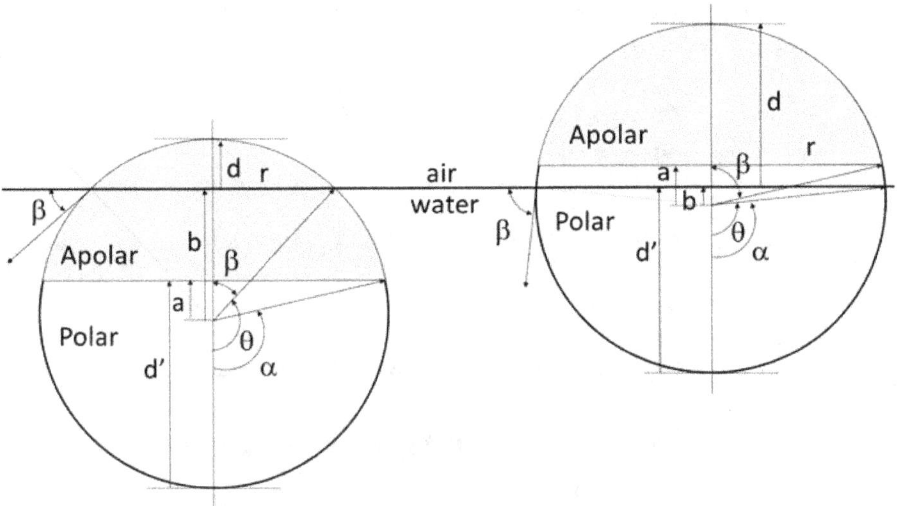

Fig. 4.9 A perfectly spherical JP adsorbed at the oil–water interface. (Left) Case of partial de-wetting of the A-lobe, for $\theta > \alpha$ and the general case $\alpha \neq \pi/2$. (Right) Case of partial de-wetting of the P-lobe, $\theta < \alpha$ and the general case $\alpha \neq \pi/2$

where the area is given by

$$\begin{aligned} A_{(\mathrm{A,air})} &= 2\pi Rd = 2\pi R(R-b) \\ &= 2\pi R^2[1-\sin(\pi/2-\beta)] = 2\pi R^2[1-\sin(\pi/2-\pi+\theta)] \\ &= 2\pi R^2[1-\sin(-\pi/2+\theta)] = 2\pi R^2[1+\sin(\pi/2-\theta)] \\ &= 2\pi R^2(1+\cos\theta), \end{aligned} \tag{4.18}$$

and $\sin(-\pi/2+\theta) = -\cos\theta$

$$E_{(\mathrm{A,water})} = \gamma_{(\mathrm{A,water})} A_{(\mathrm{A,water})} = \gamma_{(\mathrm{A,water})} 2\pi R^2(\cos\alpha - \cos\theta), \tag{4.19}$$

where

$$b = R\sin(-\pi/2+\theta) = -R\sin(\pi/2-\theta) = -R\cos\theta \tag{4.20}$$

$$a = R\sin(-\pi/2+\alpha) = -R\sin(\pi/2-\alpha) = -R\cos\alpha \tag{4.21}$$

$$A_{(\mathrm{A,water})} = 2\pi R^2(\cos\alpha - \cos\theta). \tag{4.22}$$

The surface area of the zone, excluding the top and bottom bases is: $A_{(\mathrm{A,water})} = 2\pi Rh$, where $h = b - a$ and is the height of the zone.

$$E_{(\mathrm{P,water})} = \gamma_{(\mathrm{P,water})} A_{(\mathrm{P,water})} = \gamma_{(\mathrm{P,water})} 2\pi R^2(1 - \cos\alpha), \tag{4.23}$$

where the area is:

$$A_{(\mathrm{P,water})} = 2\pi Rd' = 2\pi R(R + a) = 2\pi R^2(1 - \cos\alpha) \tag{4.24}$$

$$E_{(\mathrm{air,water})} = \gamma_{(\mathrm{air,water})} A_{(\text{circular base radius r})} = \gamma_{(\mathrm{air,water})} \pi R^2 \sin^2\theta. \tag{4.25}$$

Therefore, the total Gibbs free energy of the JP at the interface is:

$$\begin{aligned} E = 2\pi R^2 \Big[&\gamma_{(\mathrm{A,air})}(1 + \cos\theta) + \gamma_{(\mathrm{A,water})}(\cos\alpha - \cos\theta) \\ &+ \gamma_{(\mathrm{P,water})}(1 - \cos\alpha) - \gamma_{(\mathrm{air,water})}\frac{1}{2}\sin^2\theta \Big]. \end{aligned} \tag{4.26}$$

This equation has several limiting cases, for example, if the Janus lobe is a perfect hemisphere, then $\alpha = \pi/2$ and the above equation becomes:

$$\begin{aligned} E = 2\pi R^2 \Big[&\gamma_{(\mathrm{A,air})}(1 + \cos\theta) - \gamma_{(\mathrm{A,water})}\cos\theta + \gamma_{(\mathrm{P,water})} \\ &- \gamma_{(\mathrm{air,water})}\frac{1}{2}\sin^2\theta \Big]. \end{aligned} \tag{4.27}$$

Also, if instead of the parameters α and θ we use instead only the contact angle β, where $\theta = \pi - \beta$, then the equation becomes:

$$\begin{aligned} E_{\mathrm{interface}} = 2\pi R^2 \Big[&\gamma_{(\mathrm{A,air})}(1 - \cos\beta) + \gamma_{(\mathrm{A,water})}\cos\beta \\ &+ \gamma_{(\mathrm{P,water})} + \gamma_{(\mathrm{air,water})}\sin^2(\beta)/2 \Big]. \end{aligned} \tag{4.28}$$

B. Case of total de-wetting of the A-lobe and partial de-wetting of the lobe P, $\theta < \alpha$ and the general case $\alpha \neq \pi/2$.

$$E_{(\mathrm{A,air})} = \gamma_{(\mathrm{A,air})} A_{(\mathrm{A,air})} = \gamma_{(\mathrm{A,air})} 2\pi R^2(1 + \cos\alpha), \tag{4.29}$$

where

$$\begin{aligned} A_{(\mathrm{A,air})} &= 2\pi R(R - a) = 2\pi R\left[R - R\sin(\pi/2 - \beta)\right] \\ &= 2\pi R\left[R - R\sin(\pi/2 - \pi + \alpha)\right] \\ &= 2\pi R\left[R + R\sin(\pi/2 - \alpha)\right] = 2\pi R^2(1 + \cos\alpha) \end{aligned} \tag{4.30}$$

$$E_{(\mathrm{P,air})} = \gamma_{(\mathrm{P,air})} A_{(\mathrm{P,air})} = \gamma_{(\mathrm{P,air})} 2\pi R^2(\cos\theta - \cos\alpha), \tag{4.31}$$

where

$$R - b = R - R\sin(\theta - \pi/2) = R + R\sin(\pi/2 - \theta) = R + R\cos(\theta) \tag{4.32}$$

$$R - a = R + R\sin(\pi/2 - \alpha) = R + R\cos(\alpha) \tag{4.33}$$

$$A_{(\mathrm{A,water})} = 2\pi R^2(\cos\theta - \cos\alpha) \tag{4.34}$$

$$\begin{aligned} E_{(\mathrm{P,water})} &= \gamma_{(\mathrm{P,water})} 2\pi R^2(1 - \cos\theta) \\ A_{(\mathrm{P,water})} &= 2\pi R d' = 2\pi R(R + b) = 2\pi R^2(1 - \cos\theta) \end{aligned} \tag{4.35}$$

$$\begin{aligned} E_{(\mathrm{air,water})} &= \gamma_{(\mathrm{air,water})} A_{(\mathrm{circular\ base\ radius\ r})} = \gamma_{(\mathrm{air,water})} \pi R^2 \sin^2\theta \\ A_{(\mathrm{circular\ base\ radius\ r})} &= \pi r^2 = \pi R^2 \cos^2(\pi/2 - \pi + \theta) = \pi R^2 \sin^2\theta. \end{aligned} \tag{4.36}$$

Therefore, the total Gibbs free energy of the particle at the interface is:

$$\begin{aligned} E_{\mathrm{interface}} = 2\pi R^2\Big[&\gamma_{(\mathrm{A,air})}(1 + \cos\alpha) + \gamma_{(\mathrm{A,water})}(\cos\theta - \cos\alpha) \\ &+ \gamma_{(\mathrm{P,water})}(1 - \cos\theta) - \gamma_{(\mathrm{air,water})}(\sin^2\theta)/2\Big]. \end{aligned} \tag{4.37}$$

The surface energy of the Janus particles located completely in air(or oil) $E_{\mathrm{air(oil)}}$ or water E_{water} can be easily found by setting the angle β in the particle Gibbs free energy equations to 0° or 180°, respectively. Further, Binks and Fletcher [13] also suggested that the interfacial activity of the particle should be evaluated by the magnitude of its energy of interfacial detachment that should be equal to $E_{\mathrm{water}} - E_{\mathrm{interface}}$ but this can be debated. Because the above energies scale with R^2, this criterion of evaluating the interfacial activity of the particle cannot be used to compare particles of different size as the larger particle or even large particle, let's say a tennis ball would be the most interfacially active particle according to this framework of thinking. Instead we proposed that the measurable effects in the drop of the surface tension of the liquid or interfacial tension of oil–water should be instead used as criterion for estimating the interfacial activity of particles. In addition, this method is difficult to apply for JPs that are dumbbell shaped, for example, because it is difficult to keep track of the position of the long axis, and long axis tilt with the respect of surface normal, but in principle it can be done. However, determining accurately the contact angle on particles it is very difficult and therefore Ondarçuhu's model is difficult to apply in practice. Therefore, determining the Gibbs free energy of a particle at interface and especially of JPs is very difficult.

4.5.2 Inter-Particle Interaction at Interfaces vs. Lowering the Interfacial Tension

Once adsorbed at the oil–water and air–water interfaces the HPs are capable of interaction with other particles via electrostatic forces, van der Waals forces but also capillary action due to the interface deformation that lead to repulsive and attractive interactions. Because the latter is particularly true for the case of particles that strongly deform the interface, or very rough particles that pin the interface line, due to pinning the interface at a rough particle surface, for smooth colloids this is rarely the case. Capillary interactions occur especially when large particles, depending on their surface roughness, wettability by the liquid and buoyancy effect lead to interfacial deformation, by changing the local curvature of the interface. The effect of the capillary interaction can be attractive or repulsive depending on the local curvature of the interface [52]. The capillary interaction forces between particles bound at interfaces have been fully discussed by Kralchevsky and Nagayama [53]. This is also known as Cheerios effect, where the grains of cereals floating on the milk clump together exactly due to this effect, deformation of the milk–air interface.

On the other hand, the surface charge on the JPs' surface arises due to the ionisation of the charged groups in water. The interaction between particles is generally well described by the Derjaguin–Landau–Verwey–Overbeek (DLVO) theory. The expression for the overall electrostatic pair-potential interaction between two point charges at the air–water interface was first derived by Stillinger [54] using the Debye–Hückel theory. In non-aqueous solvents colloidal particles can also carry charge, but in non-aqueous solvents this is typically much less than in water due to lack of ionization of surface functional groups, therefore it is considered less important in the stabilization of colloids. Therefore, upon adsorption the part of the particle immersed in the water phase will remained charged, while the other in the non-polar solvent or air will be neutral or less charged which leads to an asymmetric formation of the double layer. This leads to the formation of a dipole with its vector oriented perpendicular at the interface. The parallel orientation of the dipolar interfacial vectors oriented in the same way leads to repulsion. Therefore the overall contribution to the interaction pair-potential $U(r)$ are the repulsive double-layer interactions at the short range and dipole–dipole repulsion at longer range [55]. The magnitude of the dipole–dipole interaction is critical for particle assembly and ordered 2-D crystal monolayer formation at the interface as noted by Pieranski [48]. Van der Waals interaction may also play a role but its effects are negligibly small compared to the electrostatic repulsion.

With respect to interfacial tension or surface tension modification by particles the adsorption of the HPs at the interface is expected to have some measurable effects. In a recent study [56] the interfacial activity of simple non-amphiphilic silica nanoparticles at air–water and hexadecane–water interface done with pendant drop tensiometry has shown that indeed simple homogeneous particles produce a notable effect in IFT drop upon interfacial adsorption. Silica HPs with most "favourable" wetting, close to 90° reach the interface, are able to produce the strongest effect on the IFT drop. The magnitude in IFT drop is expected to be dependent on the bulk particle concentration. Much earlier, Okubo [57] has shown the same effect on PS nanoparticle, with most noticeable effect in IFT drop occurred when the particle concentration was large enough that crystallites formation was already noticeable in the bulk phase. Notable is also the work of Johnson et al. [58–60] that studied the effects of TiO_2 nanoparticle adsorption and found a similar dependence in IFT drop with increase in HPs concentration. A measurable drop in the IFT of an air–water or oil–water interface can occur as a consequence of the weakening of the cohesion forces in the top-most interfacial layer, as it is the case for surfactants possessing an alkyl tail which sticks out into the non-polar phase or air and is only able to weakly interact with the neighbouring adsorbed particles, thus weakening the cohesion and the IFT. Striolo et al. [61] have shown that the IFT decreases significantly when the surface coverage is large enough that repulsive HP–HP interactions are expected. The existence of strong inter-particle capillary interaction may lead to an increase in the observed IFT as observed by Johnson and Dong [59].

Similar behaviour is expected for JPs, however with some differences. If the JPs consist of the polar part and an apolar part decisive for their interfacial behaviour is the polarity contrast between the Janus lobes and the role these play in JP–JP interfacial interaction. Ignoring the interfacial deformation effects and excluding the capillary interactions, the interfacial activity determined by the ΔIFT can be more dramatic then say that observed for the HPs with the same size, surface properties and composition as each of the Janus lobe. As pointed out earlier the interfacial activity of the particles cannot be estimated solely by the interfacial adsorption/desorption Gibbs free energy parameter because this scales with R^2, which becomes ambiguous when comparing particles that differ in size. Instead we propose that the interfacial activity should be evaluated by the ability to decrease the IFT of an interface—this is Janus effect. The Janus effect has been demonstrated by Fernńdez-Rodríguez et al. [31] on PMMA/PtBMA JPs microparticles fabricated by electrohydrodynamic co-jetting method and by us on PDIPAEMA/PS JPs nanoparticles [62]. Yet, another example of enhanced interfacial activity of JPs compared to the constituting HPs is that of Glaser et al. [45] which observed a significant decrease in the IFT vs. time of Au/Fe_2O_3 JPs at hexane–water interface. The ability of JPs to adsorb at interfaces and lower the IFT may find important applications in oil recovery applications [63].

4.5.3 Activation and Adsorption Energies of JPs Spontaneously Adsorbing at Interfaces

The adsorption of particles at interfaces is mostly entropically driven, the overall free energy of the system decreases due to increase in the water entropy; the ordered water layer on the surface of the particles becomes free upon particle interfacial adsorption. The dehydration and re-solvation of the surface involves however some

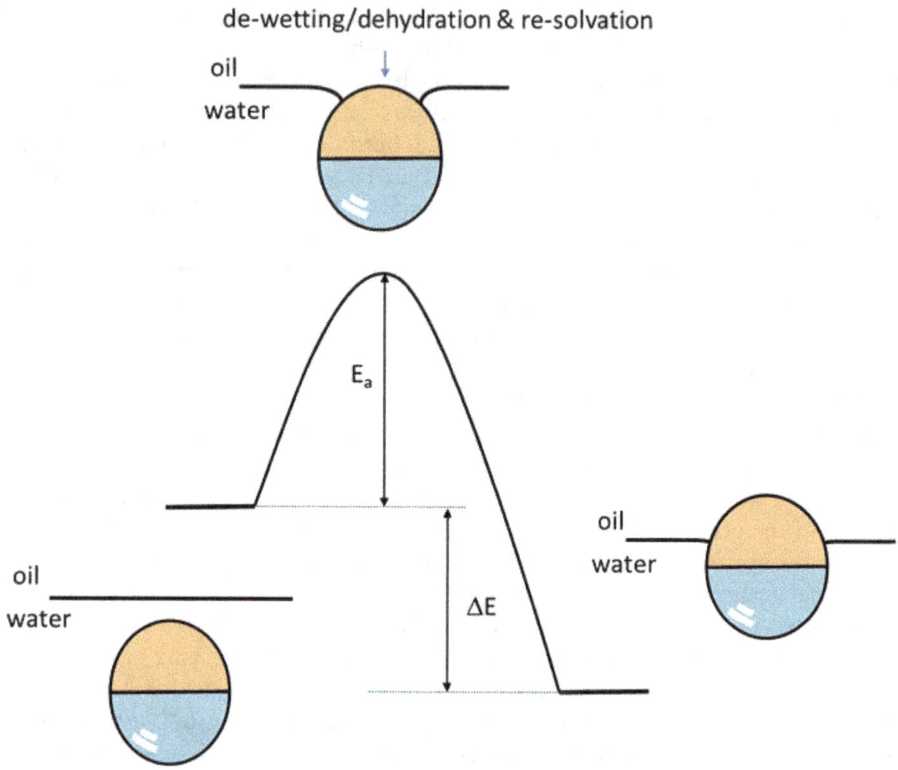

Fig. 4.10 Cartoon depicting the adsorption energy ΔE and the activation energy E_a for interfacial adsorption of JP at the oil–water. The JPs are drawn with a hypothetical orientation, long axis perpendicular to the interface

energy costs and it is one of the factors contributing to the magnitude of the activation energy barrier, Fig. 4.10. Activation energy can also arise due to particle-interface electrostatic interactions, or between the incoming particles and already adsorbed particles at the interface.

The adsorption energies of JNPs at interfaces can be calculated by first measuring the contact angle of the JPs with the solvents from the two phases, adopting a geometric model and performing the calculations as shown in Sect. 4.5.1. For complex JP geometries such as dumbbell shape or disc shape, the contact angle with each liquid phase is much more difficult to determine because it greatly increases the complexity of the geometric model in Sect. 4.5.1, adding severe uncertainties mainly due to many possible orientations of particles at the interface, the exact value of the angle of the long axis of the particle with respect to the surface. A much simpler way is to directly calculate the adsorption energy from the IFT vs. time data using pendant drop tensiometry. In pendant drop tensiometry the IFT drop with time can be monitored at oil–water interface for a long time. Same measurements can be performed at the air–water interface but have the disadvantage of the liquid evaporation leading to relatively shorter observation times. Such measurements are universally applicable to any interfacially active compound, for all types of particles and surfactants. The IFT vs. time curves data represent the starting point and opportunity to apply different kinetic models. The same kinetics models that apply to HPs [44] also apply to JPs without restrictions. The dynamic IFT measurement typically stops when ΔIFT remains constant over time, that is, a plateau equilibrium value of the interfacial tension, γ_p, has been reached, Fig. 4.7.

Bulk particle concentration can influence the magnitude of the ΔIFT until the interface becomes fully saturated. Determining the maximum ΔIFT achieved at the highest concentration of particles is an important parameter for calculating the interfacial adsorption energy of the particle. A typical evolution of the IFT vs. time function of concentration is given in Fig. 4.11 and corresponds to PS-PDIPAEMA/P(3-TSPM) JPs at heptane–water interface [62]. Notice that γ_p remains constant above a concentration of 10 mg/mL PS-PDIPAEMA/P(3-TSPM)-1 JPs meaning that the maximum ΔIFT was reached at this concentration. In addition,

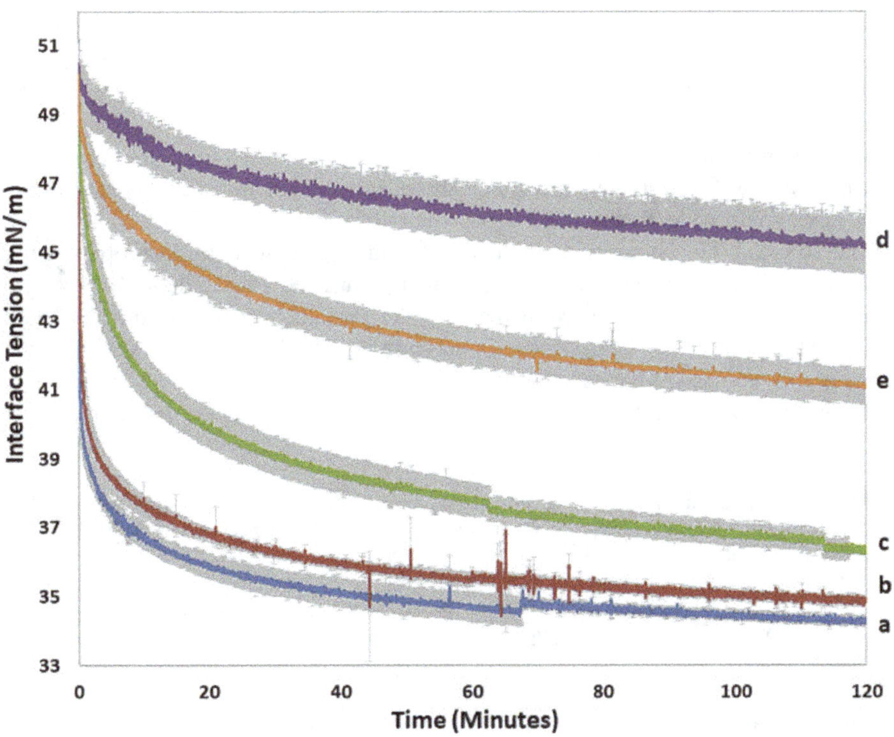

Fig. 4.11 IFT vs. time curves of the heptane–water interface in the presence of PS-PDIPAEMA/P(3-TSPM)-1 JPs at pH = 2: (**a**) 20 mg/mL, (**b**) 10 mg/mL, (**c**) 1 mg/mL, (**d**) 0.1 mg/mL and HNPs (**e**) at 10 mg/mL. Each data point is the average of three independent measurements and the error bars in grey represent the standard deviation. The data was acquired at 21 °C. Reprinted with permission from Ref. [62].

in the same figure the dynamic surface tension of JPs is compared to that of the HPs of the same composition and size as each of the Janus lobe and can be concluded that the latter are considerably less efficient than the JPs at lowering the IFT, in agreement with other similar findings of Fernández-Rodríguez et al.[31], Glaser et al. [45]. This demonstrates that the amphiphilicity of Janus particles enhances the interfacial activity of the particles.

Dinsmore et al. [64] proposed that the lowest γ_p, reached when the interface is fully saturated with particles, can be used to calculate the interfacial attachment energy ΔE:

$$\Delta E = -(\gamma_0 - \gamma_p)\pi R^2/\eta, \tag{4.38}$$

where γ_0 is the IFT for the initial concentration of particles adsorbed at the interface, Fig. 4.7, and R is the radius of the particles. Analysing the dynamic IFT measurement curves in Fig. 4.11 for the polymeric JPs adsorbing at the heptane–water interface, Wu and Honciuc [62] calculated energy of attachment ΔE using Eq. (4.38) and the obtained values are compared to the values obtained at heptane–water, toluene–water and air–water interfaces that are summarised in Table 4.2. Interestingly the ΔE values for the JPs are larger than those of HPs at the same interfaces within one order of magnitude, larger than those predicted by the calculations of Binks and Fletcher [13].

The *activation energy of adsorption* can be determined from the same IFT vs. time curves. As already mentioned the adsorption kinetics of any particle at interfaces could be diffusion controlled, energy barrier controlled or a combination of two [65–67]. The adsorption kinetics of HPs measured via pendant drop dynamic IFT measurements are typically modelled using Ward and Tordai theory [68], which considers that adsorption is controlled by the particle's concentration and bulk diffusivity followed by instantaneous adsorption at the interface. However, in the presence of an energy barrier, the adsorption at the interface is much slower than those predicted by purely diffusive models of Ward and Tordai. In order to account

Table 4.2 Activation energies of attachment of the PS-PDIPAEMA/P(3-TSPM) JPs and PS-PDIPAEMA HPs at toluene–water, heptane–water and air–water interfaces and their diffusivity, effective vs. actual

Interface	D_0 ($m^2\,s^{-1}$)	D_{eff} ($m^2\,s^{-1}$)	Ea (k_BT)	ΔE (k_BT)
JPs/toluene–water	1.65×10^{-12}	5.41×10^{-15}	5.8	-2.2×10^5
JPs/heptane–water	1.65×10^{-12}	6.24×10^{-15}	5.6	-2.9×10^5
JPs/air–water	1.65×10^{-12}	1.47×10^{-15}	7.1	-7.2×10^4
PDIPAEMA HPs/tol–water	4.94×10^{-12}	3.39×10^{-15}	7.3	-3.0×10^4
PDIPAEMA HPs/hep–water	4.94×10^{-12}	1.23×10^{-14}	6.0	-6.9×10^4
PPDIPAEMA-2 HPs/air–water	4.94×10^{-12}	1.83×10^{-15}	8.0	-2.0×10^4

Reprinted and adapted with permission from Ref. [62].

for this discrepancy Liggieri et al. [65] and Ravera et al. [66] proposed the effective diffusion model that includes an activation energy barrier. In other words not all the particles that arrive at the interface via diffusion also adsorb at the interface. Some particles that have a low kinetic energy are not able to overcome the potential barrier for surface adsorption and will diffuse back into the bulk, Fig. 4.7. The effective diffusion model enables the calculation of the activation energy barrier from the observed effective diffusion coefficient from the IFT vs. time data. The D_{eff} can be determined from the IFT vs. time data using the following equation [67]:

$$\gamma = \gamma_0 - 2N_A C_0 \Delta E \sqrt{\frac{D_{\text{eff}} t}{\pi}}, \tag{4.39}$$

where C_0 is the concentration of particles in bulk, γ_0 is the surface tension of the clean interface and ΔE is the attachment energy calculated with Eq. (4.38). By fitting the earlier portion of the IFT vs. $\sqrt{t}$ time curves one can calculate the D_{eff}. Fitting only the earlier portion of the curves is justified by the fact that the incoming particles meet a pristine interface in the first stage of adsorption, at a later time the electrostatic repulsion between the adsorbed and incoming particles dominates, Fig. 4.7 [67]. The obtained effective diffusion coefficient D_{eff} can be compared with the ones calculated from the Stokes–Einstein equation:

$$D_0 = \frac{k_B T}{6\pi \mu R}, \tag{4.40}$$

where μ is the viscosity of water and R is the hydrodynamic radius of the particle. The D_{eff} is typically much lower than the Stokes–Einstein diffusivity if an energy barrier is indeed present. Basavaraj et al. [67] obtained differences between D_{eff} *vs.* D_0 as large as three orders of magnitude for 10 nm silica particles at dodecane–water interface. The activation energy for attachment can be further calculated from the equation:

$$D_{\text{eff}} = D_0 \exp\left(\frac{-E_a}{k_B T}\right), \tag{4.41}$$

where E_a is the activation energy of attachment at interfaces. The calculated values of ΔE, γ_p, D_{eff} and D_0 and E_a for PS-PDIPAEMA/P(3-TSPM) JPs and the PS-PDIPAEMA HPs at three interfaces are compared in Table 4.2, whereas P(3-TSPM) HPs are not interfacially active. A quick inspection of the data shows that the D_{eff} effective diffusion coefficient is three orders of magnitude lower in all cases than for the Stokes diffusion coefficient calculated with Eq. (4.40), which can only be explained by the existence of an activation energy barrier. The E_a values are the largest for the air–water interface and the lowest for the adsorption at the heptane–water interface which can be explained in part by the good ability of heptane to "wet" and replace the water hydration layer from the JP surface, while at the air–water interface the high cost of JPs' surface dehydration remains uncompensated. Further, the value of ΔE for JPs than HPs, up to ten times at the heptane–water interface and up to three times at the air–water interfaces is larger than the three times upper limit predicted by the calculations of Binks and Fletcher [13]. The reason for this discrepancy may lie in the shape of the particle, which is snowman type while Binks and Fletcher have treated a perfectly spherical shaped JP.

4.6 Pickering Emulsions: Arrested JPs at Interfaces

Emulsions are mixtures of two immiscible liquids, typically oil and water, that find many applications ranging from food, cosmetics, pharmaceuticals and enhanced oil recovery. The emulsions destabilise via coalescence and Ostwald ripening resulting in phase separation, therefore enhancing their stability and extending the shelf-life of emulsion-based products represents an important topic for research and development. Pickering emulsions are emulsions stabilised by particles and have been named after the British chemist and horticulturist S.U. Pickering who first discovered them in 1907. The particles adsorb at the interface between oil–water and prevent the formed droplets from coalescing. When producing Pickering emulsion typically external energy input is required in form of mechanical stirring or ultrasonication. One advantage of the Pickering emulsions as compared to standard surfactant emulsions is their stability. Pickering emulsions can be used in a variety of applications from, drug delivery, scaffolding materials for tissue and bone growth, environmentally responsive materials, catalysis [69]. Yang et al. gave a general description of Pickering emulsions and their applications [69].

A variety of homogeneous and asymmetric nanoparticles can be used for stabilising and forming Pickering emulsions. Particles are first dispersed in one of the phases, but typically in water, and then oil is added and then high shear forces are applied, either by ultrasonication, shaking or high-power stirring. For particles that exhibit good interfacial activity the external energy input can also be lower and gentle shaking by hand may suffice for producing emulsions. It is important to note that when external energy is applied via ultrasonication or shearing forces, particles acquire high kinetic energy, easily overcome the activation energy barrier to interfacial attachment as depicted in the cartoon in Fig. 4.10. Due to the energy input one of the phases becomes dispersed forming droplets whose interface is then "bombarded" by particles and becomes quickly saturated. The particles are irreversibly trapped at the interfaces, because the desorption energy is very high, some refers to this as arrested particles at the interfaces or arrested adsorption.

Depending on the affinity to one phase or the other, oil-in-water (o/w) or water-in-oil (w/o) emulsions can be obtained. The phase of the emulsion is determined by the particles' affinity to one phase or the other according to Finkle et al. [70] and similar to the Bancroft rules [71]. For example, hydrophobic carbon black particles are more likely to form w/o emulsions than the silica particles, due to their higher affinity to the apolar phase than to water [72]. Affinity of the particle to one of the phases translates into a preferred immersion depth into one phase or the other, changing in this way the curvature of the interface toward one phase or the other, as depicted in Fig. 4.12. Affinity of a particle to the interface has to do with its

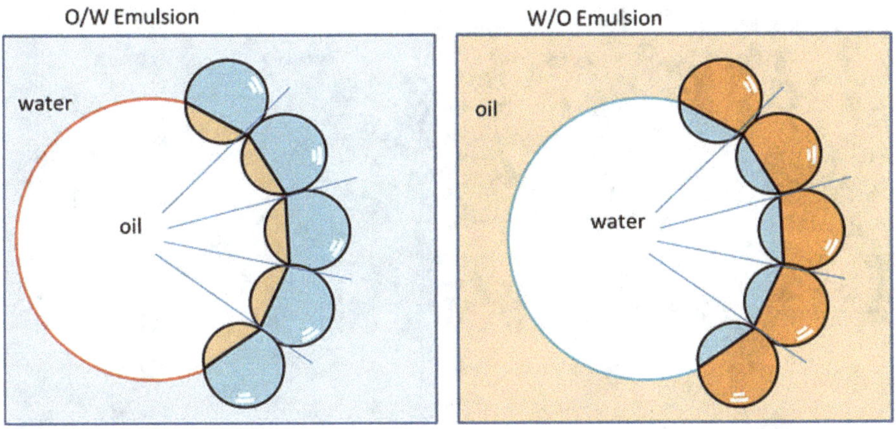

Fig. 4.12 Cartoon depicting the emulsion phase as function of the immersion depth (affinity) of a particle into the oil phase or the water phase (left) formation of o/w emulsions when the affinity of the particles is greater for water; (right) formation of w/o emulsion when the affinity of the particles is greater for oil

wettability, contact angle and eventually the immersion depth at the interface. If the immersion depth in one of the phases is stronger than the curvature of the interface will be such that the dispersed phase becomes the phase in which the particles are least immersed.

Why choosing JPs over HPs for Pickering emulsions? It has been demonstrated that the JPs are more interfacially active than the HPs due to their amphiphilicity. From the thermodynamic point of view the JPs stabilised emulsion are energetically more favourable than the HPs due to the positive line tension acting at the three-phase line oil–water–particle [73]. In addition, the surface polarity of HPs can be hard to control by surface chemical modification, often involving surface capping agents that are themselves surface active and do interfere in emulsification ability. It is for this reason why the surfactant-free JPs are more attractive than HPs in emulsification and other interfacial applications, because of the ability to precisely and gradually tune their overall polarity. Their surface energy can be varied by changing the aspect ratio between the lobes of different polarities to the desired conditions without using modifying agents like surfactants.

For example, a homologous series of five nano-sized PS/P(3-TSPM) JPs with different relative lobe sizes were tested for their emulsification ability of different volumetric ratios of heptane:water mixtures (heptane is a purely apolar liquid)[74]. Photographs of the emulsions obtained with these JPs series and the corresponding fluorescence microscopy images are presented in Fig. 4.13, whereby the top row depicts the SEM images of each particle in the homologous series; the first particle is PS HPs from which the second Janus lobe (brighter lobe) was generated, in addition the oil phase is fluorescent and the dark phase is water. The yellow line delimitates the boundary at which the emulsion phase inversion from w/o (top of the line) to o/w (bottom of the line) emulsion takes place. The horizontal yellow line depicts a transitional emulsion phase inversion that depends on the polarity of

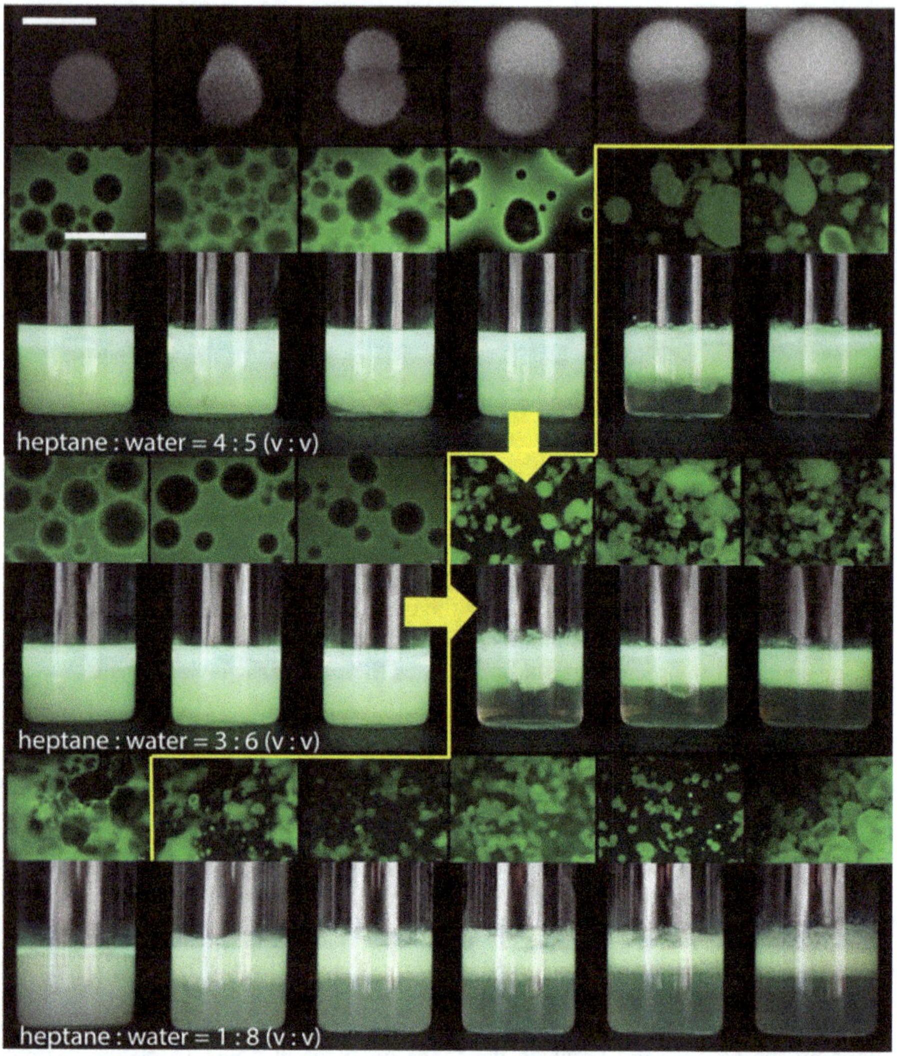

Fig. 4.13 Formulation—composition maps with photographs of emulsions in glass vials and their corresponding fluorescence microscopy images (scale bar is 400 nm) obtained with PS/P(3-TSPM) JPs. The top row depicts seed HPs and five PS/P(3-TSPM) JPs with increasing P(3-TSPM) lobe sizes (scale bar is 200 nm), while the subsequent three rows represent a different volumetric ratio of heptane to water and the six columns represent the emulsification results from each particle. The yellow line indicates the w/o and o/w emulsion phase boundary; the vertical arrow indicates the catastrophic and the horizontal the "static" transitional phase inversion. The fluorescent phase is the oil phase and the dark phase the water. Reprinted with permission from Ref. [1].

the particle, its affinity to one of the phases and eventually its immersion depth according to the cartoon in Fig. 4.12. This is the principle behind creating stimuli-responsive emulsions, discussed later. The vertical arrow indicates a catastrophic phase inversion that depends on the volume oil:water; when the ratio of one of the phases is considerably lower than the other phase, then the probability that this becomes the dispersed phase is higher. Note that the catastrophic phase inversion does not affect the particle immersion depth at the interface. From the results in Fig. 4.13 it is clear that the HPs (first column) are apolar, because they are only capable of forming w/o Pickering emulsions for all heptane:water ratios. A transitional emulsion phase inversion from w/o to o/w takes place in the middle of the homologous series, meaning that with the growth of the P(3-TSPM) the JPs become more polar and the affinity of the largest lobe JPs is greater toward water.

Other types of oils differing in their polarity and viscosity can be emulsified in water with JPs. Monomers, fragrant oils, polymers and organic solvents can be emulsified into Pickering emulsions. Further, Honciuc et al. [42] have shown that by changing the polarity of the emulsified oil the interfacial energy of the particles with the oil and water can be estimated. If monomers are used instead as oils, the Pickering emulsion can be subsequently polymerised resulting in solid-state polymers with nanostructured surfaces, see Fig. 4.14, and other advanced materials [42].

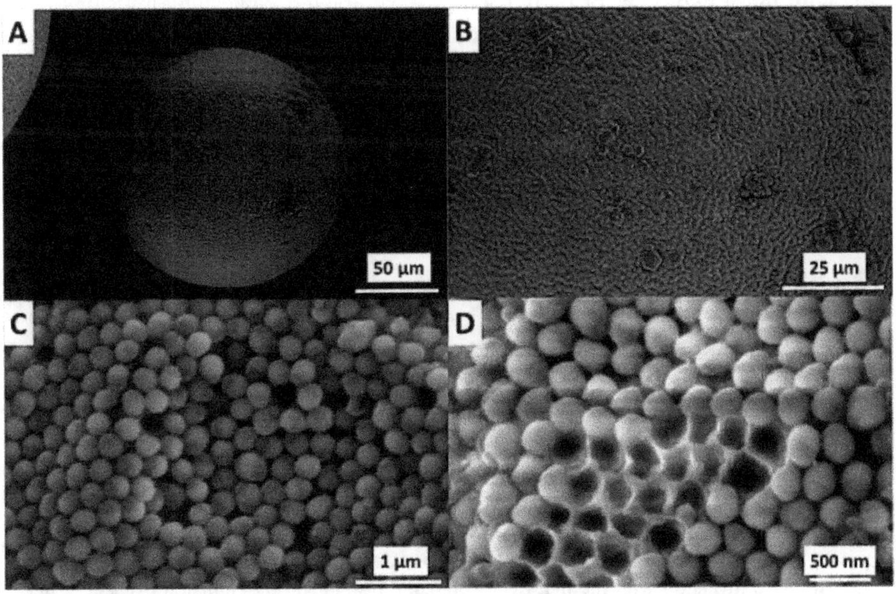

Fig. 4.14 (**a**) Polystyrene/JNP colloidosomes resulting from the polymerisation of a styrene-in-water emulsion obtained with PS/P(3-TSPM) JPs. Reprinted with permission from Ref. [42]. (**b**)-(**d**) zoomed in surface regions of the colloidosome, showing tight packing of the JPs monolayer.

Stimuli-responsive Pickering emulsions can also be designed by using stimuli responsive particles. Upon adsorption of particles at interfaces the emulsion generated acquires the functionality of the populating particles. For example, Tu and Lee [38] have created stimuli-responsive Pickering emulsions from PS/PAA JPs which are capable of phase inversion due to deprotonation at high pH of the -COOH groups and thus polarity of the particle increases and becoming capable of forming o/w emulsions.

In a different example, Pickering emulsions stabilised by PDIPAEMA/P(3-TSPM) JPs it was also possible to induce an emulsion phase inversion by changing in situ the pH value of the water phase below and above the pKa value of the -NR_3 groups at the surface of the PDIPAEMA JP lobe, Fig. 4.15. When the pH is changed in situ and an already formed Pickering emulsion inverses its phase it is called a dynamic transitional emulsion inversion [72] in contrast to static transitional emulsion inversion that assumes the preparation of the emulsion at the given pH. Such pH-responsive Pickering emulsion could be employed in encapsulations and triggered release applications.

The use of Pickering emulsions in phase selective catalysis demonstrates the potential advantages and opportunities offered by the asymmetric architecture of the JPs. A conclusive example is that of Resasco et al. [75] which have produced JPs and loaded them with Pd nanoparticles selectively only on the hydrophobic side to produced Pd/JPs and non-selectively deposited everywhere to produce HPs. Next with these two types of particles they have created Pickering emulsions from decalin and water. The decalin phase contained benzaldehyde that was insoluble in

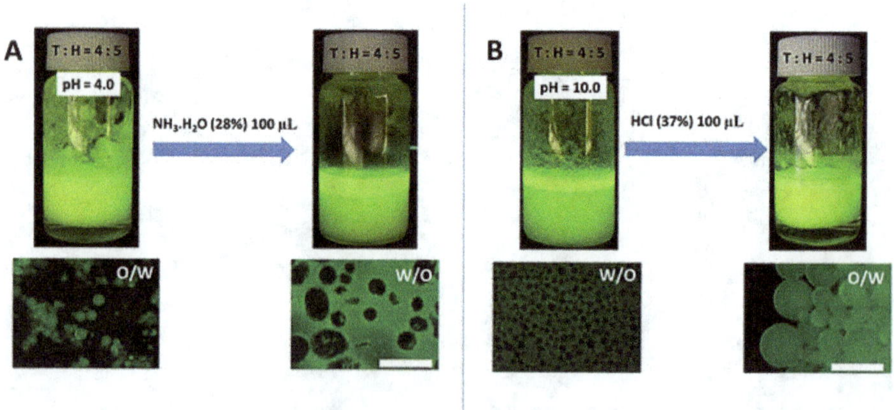

Fig. 4.15 Pickering emulsions stabilised by PDIPAEMA/P(3-TSPM) JPs showing dynamic emulsion phase inversion with the pH: (**a**) as prepared o/w Pickering emulsion (toluene:water = 4:5 ratio, pH = 4.0) changing to w/o after addition of base; (**b**) as prepared w/o Pickering emulsion (toluene:water = 4:5 ratio, pH = 10) changing to o/w after addition of acid. On top photographs of the vials containing the Pickering emulsions, with 0.1% hydrophobic dye and bottom the fluorescence microscopy images showing the corresponding Pickering emulsion type (scale bar = 200 μm). Reprinted with permission from Ref. [62].

water and the water phase contained glutaraldehyde that is insoluble in oil. Next, the two emulsions were hydrogenated and surprising results were obtained: when the catalyst contained Pd on both sides of the Janus particles, high conversion levels were obtained for both reactants, about 80% for glutaraldehyde in the water phase and 100% for benzaldehyde in the oil phase. However, when the catalyst had Pd selectively deposited on the hydrophobic side, the conversion of benzaldehyde was kept at 100%, while the conversion of glutaraldehyde decreased to 2%, demonstrating high phase selectivity, of JP stabilised Pickering emulsions. Similarly, Liu et al. [76] used Au nanoparticle modified SiO_2/PS-PDVB of the JPs as interfacial catalysts for the catalytic reduction of 4-nitroanisole to 4-aminoanisole.

4.7 Self-Assembly of Janus Particles

Similar to molecular surfactants JPs can also self-assemble into suprastructures. The key parameters behind the self-assembly of JPs are the right balance between the repulsive/attractive forces and their geometry that greatly influences the type of suprastructures formed [77, 78]. It is well known that HPs can assemble into colloidal crystals with iridescent appearance and find use in photonics [79, 80], electronics [81, 82], catalysis [83], (bio)sensing [84, 85], etc. In contrast, JPs can give rise to a larger variety of self-assembled suprastructures [86], such as trimmers [87], spherical micelles [88], capsules [89] and crystals [90], which can be of great importance for obtaining novel reconfigurable materials and assemblies at non-equilibrium also referred to as "active matter" that can perform different functions [91]. The different variety of suprastructures that can be formed arise from JPs anisotropy due to geometrical or topological constraints [92, 93], which affect their interaction and packing. JPs may find analogues in nature such as large proteins that exhibit polarity domains on their surface and exhibit anisotropy. Anisotropic proteins form self-assembly structures with precise morphology and specific functional role [94]. The ability of JPs to self-assembly into complex and regular structures that show unusual and reconfigurable properties [91] is a remarkable property and is one of the main driving motivations for further research in this area [93].

The particle–particle interaction drives the self-assembly of JPs in monophasic solvent, but their assembly can also be mediated by a minority liquid in a biphasic system, such as, for example, by placing spherical Au/SiO_2 JPs in a binary system of water/2,6-lutidine binary fluid leading to 2D and 3D clusters or zig-zag chains [95]. Other examples of JPs self-assemblies mediated by liquids in water/n-dodecane mixtures; in this case water was the minority liquid, mediating the self-assembly formation via capillary forces due to the formation of water bridges between particles. By progressively increasing the fraction water, a variety of structures from micelles, worm-like micelles and spherical emulsion droplets of spherical Au/SiO_2 JPs were obtained, Fig. 4.16 [96]. Eventually, when the minority liquid is large enough these transform into colloidosomes of Pickering emulsions, Fig. 4.16g–j;

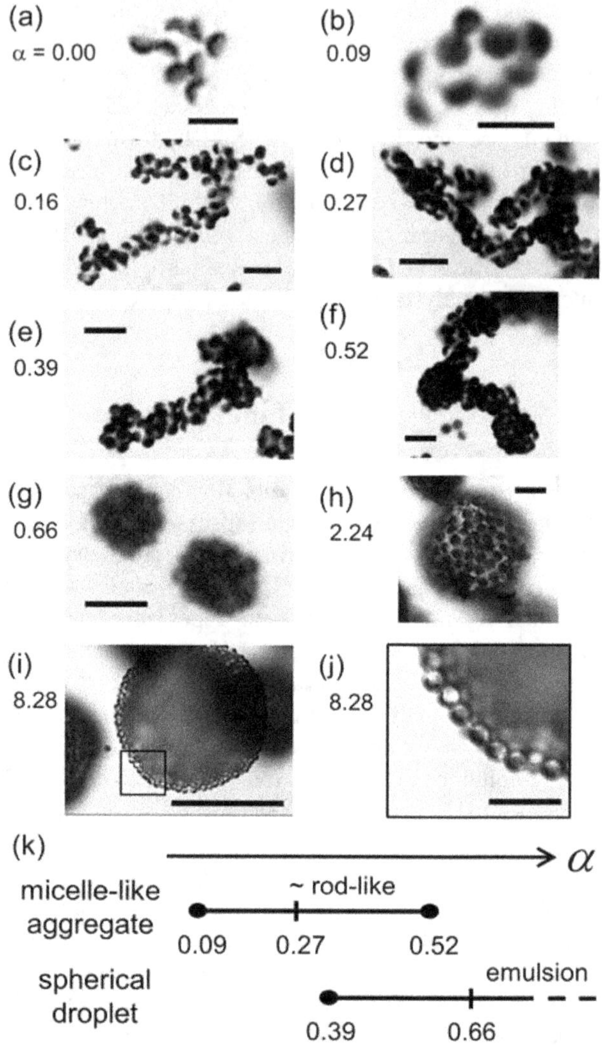

Fig. 4.16 Optical microscope images showing α-dependence of the morphology in self-assembled structures, where α is the fraction of water (minority liquid) added in n-dodecane. (**a–i**) Optical microscope images of typical structures formed at respective α. (**a**) Random aggregate. (**b**) Small micelle-like cluster. (**c–e**) Rod-shaped micelle-like clusters. (**f**) Structure observed at a value of α where rod-shaped micelle-like clusters and spherical droplets coexist. (**g–i**) Spherical droplets in emulsions. (**i**) Hemispherical droplet attached to the bottom of the observation cell. (**j**) Magnified image of the framed region in (**i**). (**k**) Diagram of the α-range of the observed structures. The scale bars are 5 μm in (**a**) and (**b**), 10 μm in (**c**)–(**h**) and (**j**) and 50 μm in (**i**). Reprinted with permission from Ref. [96].

in addition the JPs are oriented with their polar lobe toward water. Hu et al. [97] showed that by further chemical alteration of the metallic side linkable Janus metal-organosilica particles capable of forming dimers and trimers can be created.

Because JPs can adsorb at the air–water interface, they can also act as gas bubble stabilisers and thus air bubbles can act as templates for the self-assembly. Gas bubbles and emulsions have similar properties, e.g. the gas can be thought of as a highly hydrophobic fluid. For example, Fujii et al. [98] obtained large mono-walled vesicles from Au/SiO_2 JPs; the orientation of JPs in the walls could be changed by chemical modification of the Au lobe with different polymers, PS, PPFBEM (Fig. 4.17).

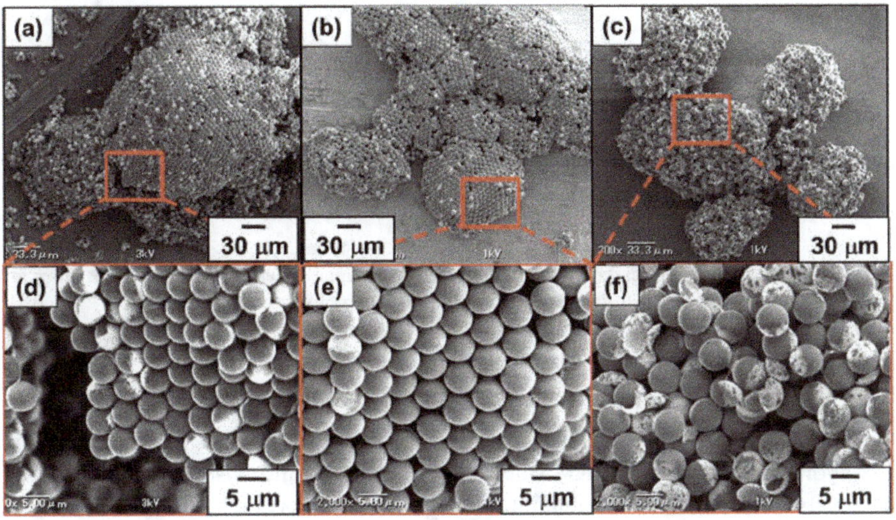

Fig. 4.17 SEM images of bubbles stabilised by the (**a**, **d**) Au-SiO_2, (**b**, **e**) PS-g-Au-SiO_2 and (**c**, **f**) PPFBEM-g-Au-SiO_2 Janus particles. Panels **d–f** are magnifications of panels **a–c**, respectively. Reprinted with permission from Ref. [98].

4.8 JP-Based Nanomotors

Self-propelled, active colloidal systems are of great fundamental interest with potential applications in nanomachineries, nanoscale assembly, catalysis and sensing [99]. Due to their asymmetry JPs possess the right architectures for making nanomotors [100] and the ways to power these externally have been extensively discussed by Shields and Velev [100]. For self-propulsion a motor needs fuel and in order to propel a Janus particle one side of it must be made apt for propelling the particles. One way this can be achieved is by making one lobe of the JPs from a metal that catalyses the decomposition of the H_2O_2 on its surface, such as Pt; the resulting decomposition products O_2 and H_2O act as propelling jet for the JPs. In this way the motion of the particle deviates from that of a pure Brownian motion and enhances the diffusivity of the particle in the bulk solution. The self-propelled JPs are also capable of transporting a cargo. Sanchez et al. [101] coated mesoporous silica nanoparticles on one side by evaporation of 2 nm thin Pt metal. The resulting JPs exhibited an enhanced diffusion coefficient of up to 100% and Rhodamine B could be loaded into the pores of the silica nanoparticles. The presence of H_2O_2 may limit however their application in vivo for drug delivery. Wang et al. [102] used a different fuel/metal system by deposition Ir-metal on one side of silica particles and using hydrazine (well-known monopropellant for rocket motors); N_2, H_2 and NH_3 molecules are generated at the Ir surface as a result the JP motor moves unidirectionally in the direction of the silica face. The group of Joseph Wang also showed that the capabilities of cargo transport of JP self-propelled motors are not

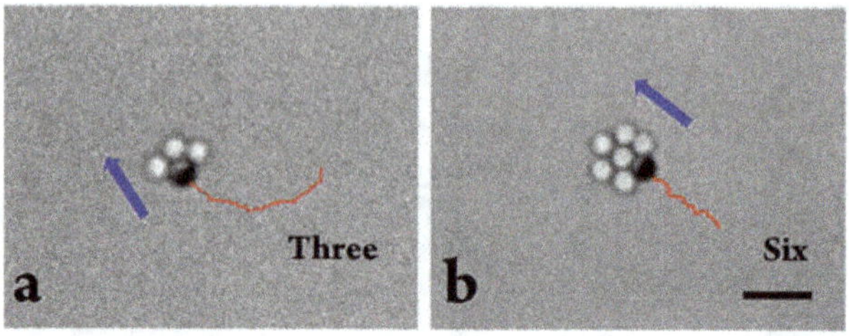

Fig. 4.18 Self-organised cargo loading. Time-lapse images (over 2 s) of the track lines of motor/nonmotor assemblies: transport of 3 (**a**) and 6 (**b**) nonmotor spheres (taken from SI Video 7). Scale bar, 5 μm. Reprinted with permission from Ref. [103].

only limited to small molecules but to other particles; using H_2O_2 fuel and Pt/SiO_2 JPs hydrophobised on the SiO_2 side with OTS, they could observe that JP motors can anchor HPs via hydrophobic interaction then transport them unidirectionally to a different location [103]. Impressively, one JP motor could transport up to six HPs. The mechanism of self-propulsion is well understood [104] and is achieved due to an asymmetric catalytic reaction occurring on the surface of the JPs. The role of the geometric shape in the unidirectional propulsion of JPs was also discussed [105]. By trapping Pt/SiO_2 JPs at the air/water surface Stocco et al. [106] were able to enhance the unidirectional motion of JPs as compared to bulk due to slowing down of the rotational diffusion at the interface and constraining it into a rotational well [36].

JPs' capabilities to self-assemble and unidirectional motion can be combined to achieve very unique operational functions, such as particle cargo applications [102, 103]. In Fig. 4.18a it is depicted the asymmetric cluster formation between a JP and three HPs; the three HPs are all attached to the alike JP lobe. The dark lobe catalyses a chemical decomposition reaction that is able to cargo the other particles and push them unidirectionally to a different location; one JP can transport up to 6 HPs, Fig. 4.18b.

4.9 Conclusions

Janus particles demonstrated a net superiority when compared to HPs in terms of interfacial behaviour and are therefore attractive for use in a plethora of interfacial applications. Stabilisation of Pickering emulsions, stabilisation of gas bubbles and foams are typical applications in which JPs have revealed their versatility. Further, due to their asymmetric architecture JPs open new horizons for particle applications. A few such possibilities have been already demonstrated, such as the self-assembly into reconfigurable suprastructures, paper display applications of bi-coloured JPs or cargo loading operation and transport of up to six particles by a single JP from point A to point B through chemically powered unidirectional motion. The future application potential of the JPs appears limitless and further research may uncover even more extraordinary functions. The strength of these particles lies in the ability to carry different and often contrasting properties and functionalities on each of their lobes. These properties can be bulk-like properties or surface properties. For example, one can combine electric, magnetic and optic properties on each Janus lobes in addition to amphiphilicity, a surface property [6]. By doing this one can couple surface and bulk properties and combine them in surprising new ways. Never before, has the application potential of particles been more exciting than the one opened by the Janus particles.

References

1. D. Wu, J. Chew, A. Honciuc, Polarity reversal in homologous series of surfactant-free Janus nanoparticles: toward the next generation of amphiphiles. Langmuir **32**(25), 6376 (2016)
2. Y. Wang, Y. Wang, D. Breed, V. Manoharan, L. Feng, A. Hollingsworth, M. Weck, D. Pine, Colloids with valence and specific directional bonding. Nature **491**(7422), 51 (2012)
3. E. Passas-Lagos, F. Schüth, Amphiphilic Pickering emulsifiers based on mushroom-type Janus particles. Langmuir **31**(28), 7749 (2015)
4. J. Gong, X. Zu, Y. Li, W. Mu, Y. Deng, Janus particles with tunable coverage of zinc oxide nanowires. J. Mater. Chem. **21**(7), 2067 (2011)
5. A. Walther, X. André, M. Drechsler, V. Abetz, A. Müller, Janus discs. J. Am. Chem. Soc. **129**(19), 6187 (2007)
6. V. Mihali, A. Honciuc, Semiconductive materials with tunable electrical resistance and surface polarity obtained by asymmetric functionalization of Janus nanoparticles. Adv. Mater. Interfaces **4**(23), 1700914 (2017)
7. P. de Gennes, Soft matter. Rev. Mod. Phys. **64**(3), 645 (1992)
8. C. Casagrande, P. Fabre, E. Raphaël, M. Veyssié, "Janus beads": realization and behaviour at water/oil interfaces. Europhys. Lett. **9**(3), 251 (1989)
9. G. Burghard, U. Holtschimdt, G. Koerner, R. Gerd, Particles, modified at their surface by hydrophilic and hydrophobic groups, US patent US4715986A (1984)
10. G. Rossmy, in *Structure, Dynamics and Properties of Disperse Colloidal Systems*, ed. by H. Rehage, G. Peschel (Steinkopff, Darmstadt, 1998), pp. 17–26
11. Y.C. Chen, V. Dimonie, M. El-Aasser, Interfacial phenomena controlling particle morphology of composite latexes. J. Appl. Polym. Sci. **42**(4), 1049 (1991)
12. T. Ondarçuhu, P. Fabre, E. Raphaël, M. Veyssié, Specific properties of amphiphilic particles at fluid interfaces. J. Phys. France **51**(14), 1527 (1990)
13. B. Binks, P. Fletcher, Particles adsorbed at the oil–water interface: a theoretical comparison between spheres of uniform wettability and "Janus" particles. Langmuir **17**(16), 4708 (2001)
14. M. Linder, Hydrophobins: proteins that self assemble at interfaces. Curr. Opin. Colloid Interface Sci. **14**(5), 356 (2009)
15. T. Sarlin, T. Nakari-Setälä, M. Linder, M. Penttilä, A. Haikara, Fungal hydrophobins as predictors of the gushing activity of malt. J. Inst. Brew. **111**(2), 105 (2005)
16. D. Suzuki, H. Kawaguchi, Janus particles with a functional gold surface for control of surface plasmon resonance. Colloid Polym. Sci. **284**(12), 1471 (2006)
17. M. McConnell, M. Kraeutler, S. Yang, R. Composto, Patchy and multiregion Janus particles with tunable optical properties. Nano Lett. **10**(2), 603 (2010)
18. S. Ye, R. Carroll, Design and fabrication of bimetallic colloidal "Janus" particles. ACS Appl. Mater. Interfaces **2**(3), 616 (2010)
19. S. Jiang, Q. Chen, M. Tripathy, E. Luijten, K. Schweizer, S. Granick, Janus particle synthesis and assembly. Adv. Mater. **22**(10), 1060 (2010)

20. L. Hong, S. Jiang, S. Granick, Simple method to produce Janus colloidal particles in large quantity. Langmuir **22**(23), 9495 (2006)
21. D. Suzuki, S. Tsuji, H. Kawaguchi, Janus microgels prepared by surfactant-free Pickering emulsion-based modification and their self-assembly. J. Am. Chem. Soc. **129**(26), 8088 (2007)
22. K. Fujimoto, K. Nakahama, M. Shidara, H. Kawaguchi, Preparation of unsymmetrical microspheres at the interfaces. Langmuir **15**(13), 4630 (1999)
23. T. Skelhon, Y. Chen, S. Bon, Synthesis of "hard-soft" Janus particles by seeded dispersion polymerization. Langmuir **30**(45), 13525 (2014)
24. L. Bradley, K. Stebe, D. Lee, Clickable Janus particles. J. Am. Chem. Soc. **138**(36), 11437 (2016)
25. Y. Sun, F. Liang, X. Qu, Q. Wang, Z. Yang, Robust reactive Janus composite particles of snowman shape. Macromolecules **48**(8), 2715 (2015)
26. M. Hoffmann, Y. Lu, M. Schrinner, M. Ballauff, L. Harnau, Dumbbell-shaped polyelectrolyte brushes studied by depolarized dynamic light scattering. J. Phys. Chem. B **112**(47), 14843 (2008)
27. J.W. Kim, R. Larsen, D. Weitz, Synthesis of nonspherical colloidal particles with anisotropic properties. J. Am. Chem. Soc. **128**(44), 14374 (2006)
28. K. Lee, J. Yoon, J. Lahann, Recent advances with anisotropic particles. Curr. Opin. Colloid Interface Sci. **16**(3), 195 (2011)
29. S. Lone, I. Cheong, Fabrication of polymeric Janus particles by droplet microfluidics. RSC Adv. **4**(26), 13322 (2014)
30. T. Nisisako, T. Torii, T. Takahashi, Y. Takizawa, Synthesis of monodisperse bicolored Janus particles with electrical anisotropy using a microfluidic co-flow system. Adv. Mater. **18**(9), 1152 (2006)
31. M.Á. Fernández-Rodríguez, S. Rahmani, K. Chris, M.Á. Rodríguez-Valverde, M.Á. Cabrerizo-Vílchez, C.A. Michel, J. Lahann, R. Hidalgo-Álvarez, Synthesis and interfacial activity of PMMA/PtBMA Janus and homogeneous nanoparticles at water/oil interfaces. Colloids Surf. A Physicochem. Eng. Asp. **536**, 259 (2018)
32. K.H. Roh, D. Martin, J. Lahann, Triphasic nanocolloids. J. Am. Chem. Soc. **128**(21), 6796 (2006)
33. K.H. Roh, D. Martin, J. Lahann, Biphasic Janus particles with nanoscale anisotropy. Nat. Mater. **4**(10), 759 (2005)
34. R. Shah, J.W. Kim, D. Weitz, Janus supraparticles by induced phase separation of nanoparticles in droplets. Adv. Mater. **21**(19), 1949 (2009)
35. B. Liu, H. Möhwald, D. Wang, Synthesis of Janus particles via kinetic control of phase separation in emulsion droplets. Chem. Commun. **49**(84), 9746 (2013)
36. R. Deng, S. Liu, F. Liang, K. Wang, J. Zhu, Z. Yang, Polymeric Janus particles with hierarchical structures. Macromolecules **47**(11), 3701 (2014)
37. C. Sosa, R. Liu, C. Tang, F. Qu, S. Niu, M. Bazant, R. Prud'homme, R. Priestley, Soft multifaced and patchy colloids by constrained volume self-assembly. Macromolecules **49**(9), 3580 (2016)
38. F. Tu, D. Lee, Shape-changing and amphiphilicity-reversing Janus particles with pH-responsive surfactant properties. J. Am. Chem. Soc. **136**(28), 9999 (2014)
39. S. Jiang, S. Granick, Controlling the geometry (Janus balance) of amphiphilic colloidal particles. Langmuir **24**(6), 2438 (2008)
40. R. Pasquali, M. Taurozzi, C. Bregni, Some considerations about the hydrophilic-lipophilic balance system. Int. J. Pharm. **356**(1–2), 44 (2008)
41. H.M. Gao, Z.Y. Lu, H. Liu, Z.Y. Sun, L.J. An, Orientation and surface activity of Janus particles at fluid-fluid interfaces. J. Chem. Phys. **141**(13), 134907 (2014)
42. D. Wu, B. Binks, A. Honciuc, Modeling the interfacial energy of surfactant-free amphiphilic Janus nanoparticles from phase inversion in Pickering emulsions. Langmuir **34**(3), 1225 (2017)

43. C. Griffin, Calculation of HLB values of non-ionic surfactants. J. Soc. Cosmet. Chem. **5**, 249 (1954)
44. S. Kutuzov, J. He, R. Tangirala, T. Emrick, T. Russell, A. Böker, On the kinetics of nanoparticle self-assembly at liquid/liquid interfaces. Phys. Chem. Chem. Phys. **9**(48), 6351 (2007)
45. N. Glaser, D. Adams, A. Böker, G. Krausch, Janus particles at liquid-liquid interfaces. Langmuir **22**(12), 5227 (2006)
46. M.A. Fernandez-Rodriguez, Y. Song, M.Á. Rodríguez-Valverde, S. Chen, M.A. Cabrerizo-Vilchez, R. Hidalgo-Alvarez, Comparison of the interfacial activity between homogeneous and Janus gold nanoparticles by pendant drop tensiometry. Langmuir **30**(7), 1799 (2014)
47. A. Beloqui Redondo, I. Jordan, I. Ziazadeh, A. Kleibert, J. Giorgi, H. Wörner, S. May, Z. Abbas, M. Brown, Nanoparticle-induced charge redistribution of the air-water interface. J. Phys. Chem. C **119**(5), 2661 (2015)
48. P. Pieranski, Two-dimensional interfacial colloidal crystals. Phys. Rev. Lett. **45**(7), 569 (1980)
49. L. Isa, F. Lucas, R. Wepf, E. Reimhult, Measuring single-nanoparticle wetting properties by freeze-fracture shadow-casting cryo-scanning electron microscopy. Nat. Commun. **2**, 438 (2011)
50. B. Binks, S. Lumsdon, Effects of oil type and aqueous phase composition on oil-water mixtures containing particles of intermediate hydrophobicity. Phys. Chem. Chem. Phys. **2**(13), 2959 (2000)
51. S. Levine, B. Bowen, S. Partridge, Stabilization of emulsions by fine particles II. Capillary and van der Waals forces between particles. Colloids Surf. **38**(2), 345 (1989)
52. N. Vassileva, D. van den Ende, F. Mugele, J. Mellema, Capillary forces between spherical particles floating at a liquid-liquid interface. Langmuir **21**(24), 11190 (2005)
53. P. Kralchevsky, K. Nagayama, Capillary interactions between particles bound to interfaces, liquid films and biomembranes. Adv. Colloid Interface Sci. **85**(2–3), 145 (2000)
54. F. Stillinger, Interfacial solutions of the Poisson-Boltzmann equation. J. Chem. Phys. **35**(5), 1584 (1961)
55. O. Deshmukh, D. van den Ende, M. Stuart, F. Mugele, M. Duits, Hard and soft colloids at fluid interfaces: adsorption, interactions, assembly & rheology. Adv. Colloid Interface Sci. **222**, 215 (2015)
56. Y. Zhang, S. Wang, J. Zhou, R. Zhao, G. Benz, S. Tcheimou, J. Meredith, S. Behrens, Interfacial activity of nonamphiphilic particles in fluid-fluid interfaces. Langmuir **33**(18), 4511 (2017)
57. T. Okubo, Surface tension of structured colloidal suspensions of polystyrene and silica spheres at the air-water interface. J. Colloid Interface Sci. **171**(1), 55 (1995)
58. L. Dong, D. Johnson, The study of the surface tension of charge-stabilized colloidal dispersions. J. Dispers. Sci. Technol. **25**(5), 575 (2005)
59. L. Dong, D. Johnson, Surface tension of charge-stabilized colloidal suspensions at the water-air interface. Langmuir **19**(24), 10205 (2003)
60. K. Casson, D. Johnson, Surface-tension-driven flow due to the adsorption and desorption of colloidal particles. J. Colloid Interface Sci. **242**(2), 279 (2001)
61. X.C. Luu, J. Yu, A. Striolo, Nanoparticles adsorbed at the water/oil interface: coverage and composition effects on structure and diffusion. Langmuir **29**(24), 7221 (2013)
62. D. Wu, A. Honciuc, Design of Janus nanoparticles with pH-triggered switchable amphiphilicity for interfacial applications., ACS Appl. Nano Mater. **1**(1), 471 (2017)
63. B. Peng, L. Zhang, J. Luo, P. Wang, B. Ding, M. Zeng, Z. Cheng, A review of nanomaterials for nanofluid enhanced oil recovery. RSC Adv. **7**(51), 32246 (2017)
64. K. Du, E. Glogowski, T. Emrick, T. Russell, A. Dinsmore, Adsorption energy of nano- and microparticles at liquid-liquid interfaces. Langmuir **26**(15), 12518 (2010)
65. L. Liggieri, F. Ravera, A. Passerone, A diffusion-based approach to mixed adsorption kinetics. Colloids Surf. A Physicochem. Eng. Asp. **114**, 351 (1996)
66. F. Ravera, L. Liggieri, A. Steinchen, Sorption kinetics considered as a renormalized diffusion process. J. Colloid Interface Sci. **156**(1), 109 (1993)

67. V. Dugyala, J. Muthukuru, E. Mani, M. Basavaraj, Role of electrostatic interactions in the adsorption kinetics of nanoparticles at fluid-fluid interfaces. Phys. Chem. Chem. Phys. **18**(7), 5499 (2016)
68. A. Ward, L. Tordai, Time-dependence of boundary tensions of solutions I. The role of diffusion in time-effects. J. Chem. Phys. **14**(7), 453 (1946)
69. Y. Yang, Z. Fang, X. Chen, W. Zhang, Y. Xie, Y. Chen, Z. Liu, W. Yuan, An overview of Pickering emulsions: solid-particle materials, classification, morphology, and applications. Front. Pharmacol. **8**, 287 (2017)
70. P. Finkle, H. Draper, J. Hildebrand, The theory of emulsification. J. Am. Chem. Soc. **45**(12), 2780 (1923)
71. W. Bancroft, *Applied Colloid Chemistry*, 1st edn. (McGraw-Hill Book Co., New York, 1921)
72. A. Kumar, S. Li, C.M. Cheng, D. Lee, Recent developments in phase inversion emulsification. Ind. Eng. Chem. Res. **54**(34), 8375 (2015)
73. R. Aveyard, Can Janus particles give thermodynamically stable Pickering emulsions? Soft Matter **8**(19), 5233 (2012)
74. J. Israelachvili, in *Intermolecular and Surface Forces: Revised* (Academic, Amsterdam, 2011), pp. 415–467
75. J. Faria, M. Ruiz, D. Resasco, Phase-selective catalysis in emulsions stabilized by Janus silica-nanoparticles. Adv. Synth. Catal. **352**, 2359 (2010)
76. Y. Liu, J. Hu, X. Yu, X. Xu, Y. Gao, H. Li, F. Liang, Preparation of Janus-type catalysts and their catalytic performance at emulsion interface. J. Colloid Interface Sci. **490**, 357 (2017)
77. Y. Liu, W. Li, T. Perez, J. Gunton, G. Brett, Self-assembly of Janus ellipsoids. Langmuir **28**(1), 3 (2012)
78. W. Li, J. Gunton, Self-assembly of Janus ellipsoids II: Janus prolate spheroids. Langmuir **29**(27), 8517 (2013)
79. S.H. Kim, S. Lee, S.M. Yang, G.R. Yi, Self-assembled colloidal structures for photonics. NPG Asia Mater. **3**(1), 25 (2011)
80. A. Yethiraj, J. Thijssen, A. Wouterse, A. van Blaaderen, Large-area electric-field-induced colloidal single crystals for photonic applications. Adv. Mater. **16**(7), 596 (2004)
81. J. Anker, W. Hall, O. Lyandres, N. Shah, J. Zhao, R. Van Duyne, Biosensing with plasmonic nanosensors. Nat. Mater. **7**(6), 442 (2008)
82. C. Fenzl, T. Hirsch, O. Wolfbeis, Photonic crystals for chemical sensing and biosensing. Angew. Chem. Int. Ed. **53**(13), 3318 (2014)
83. E. Ortel, S. Sokolov, C. Zielke, I. Lauermann, S. Selve, K. Weh, B. Paul, J. Polte, R. Kraehnert, Supported mesoporous and hierarchical porous Pd/TiO2 catalytic coatings with controlled particle size and pore structure. Chem. Mater. **24**(20), 3828 (2012)
84. J. Anker, W. Hall, O. Lyandres, N. Shah, J. Zhao, R. Van Duyne, Biosensing with plasmonic nanosensors. Nat. Mater. **7**(6), 442 (2008)
85. C. Fenzl, T. Hirsch, O. Wolfbeis, Photonic crystals for chemical sensing and biosensing. Angew. Chem. Int. Ed. **53**(13), 3318 (2014)
86. S. Glotzer, M. Solomon, Anisotropy of building blocks and their assembly into complex structures. Nat. Mater. **6**, 557 (2007)
87. H. Hu, F. Ji, Y. Xu, J. Yu, Q. Liu, L. Chen, Q. Chen, P. Wen, Y. Lifshitz, Y. Wang, et al., Reversible and precise self-assembly of Janus metal-organosilica nanoparticles through a linker-free approach. ACS Nano **10**(8), 7323 (2016)
88. D. Kraft, R. Ni, F. Smallenburg, M. Hermes, K. Yoon, D. Weitz, A. van Blaaderen, J. Groenewold, M. Dijkstra, W. Kegel, Surface roughness directed self-assembly of patchy particles into colloidal micelles. Proc. Natl. Acad. Sci. **109**(27), 10787 (2012)
89. C. Evers, J. Luiken, P. Bolhuis, W. Kegel, Self-assembly of microcapsules via colloidal bond hybridization and anisotropy. Nature **534**(7607), 364 (2016)
90. E. Ducrot, M. He, G.R. Yi, D. Pine, Colloidal alloys with preassembled clusters and spheres. Nat. Mater. **16**, 652 (2017)
91. J. Zhang, S. Granick, Natural selection in the colloid world: active chiral spirals. Faraday Discuss. **191**, 35 (2016)

92. S. Sacanna, M. Korpics, K. Rodriguez, L. Colón-Meléndez, S.H. Kim, D. Pine, G.R. Yi, Shaping colloids for self-assembly. Nat. Commun. **4**, 1688 (2013)
93. V. Manoharan, Colloidal matter: Packing, geometry, and entropy. Science **349**(6251), 1253751 (2015)
94. J. Gunton, A. Shiryayev, D. Pagan, *Protein Condensation: Kinetic Pathways to Crystallization and Disease* (Cambridge University Press, Cambridge, 2007)
95. C. Yu, J. Zhang, S. Granick, Selective Janus particle assembly at tipping points of thermally-switched wetting. Angew. Chem. Int. Ed. **53**(17), 4364 (2014)
96. T. Noguchi, Y. Iwashita, Y. Kimura, Dependence of the internal structure on water/particle volume ratio in an amphiphilic Janus particle-water-oil ternary system: from micelle-like clusters to emulsions of spherical droplets. Langmuir **33**(4), 1030 (2017)
97. H. Hu, F. Ji, Y. Xu, J. Yu, Q. Liu, L. Chen, Q. Chen, P. Wen, Y. Lifshitz, Y. Wang, et al., Reversible and precise self-assembly of Janus metal-organosilica nanoparticles through a linker-free approach. ACS Nano **10**(8), 7323 (2016)
98. S. Fujii, Y. Yokoyama, S. Nakayama, M. Ito, S. Yusa, Y. Nakamura, Gas bubbles stabilized by Janus particles with varying hydrophilic-hydrophobic surface characteristics. Langmuir **34**(3), 933 (2017)
99. K. Dey, F. Wong, A. Altemose, A. Sen, Catalytic motors—quo vadimus? Curr. Opin. Colloid Interface Sci. **21**, 4 (2016)
100. C. Shields, O. Velev, The evolution of active particles: toward externally powered self-propelling and self-reconfiguring particle systems. Chem **3**(4), 539 (2017)
101. X. Ma, K. Hahn, S. Sanchez, Catalytic mesoporous Janus nanomotors for active cargo delivery. J. Am. Chem. Soc. **137**(15), 4976 (2015)
102. W. Gao, A. Pei, R. Dong, J. Wang, Catalytic iridium-based Janus micromotors powered by ultralow levels of chemical fuels. J. Am. Chem. Soc. **136**(6), 2276 (2014)
103. W. Gao, A. Pei, X. Feng, C. Hennessy, J. Wang, Organized self-assembly of Janus micromotors with hydrophobic hemispheres. J. Am. Chem. Soc. **135**(3), 998 (2013)
104. J. Gibbs, Y.P. Zhao, Autonomously motile catalytic nanomotors by bubble propulsion. Appl. Phys. Lett. **94**(16), 163104 (2009)
105. S. Michelin, E. Lauga, Geometric tuning of self-propulsion for Janus catalytic particles. Sci. Rep. **7**, 42264 (2017)
106. X. Wang, M. In, C. Blanc, M. Nobili, A. Stocco, Enhanced active motion of Janus colloids at the water surface. Soft Matter **11**(37), 7376 (2015)

5

Flow and Transport through Porous Media

Matteo Icardi, Gianluca Boccardo, and Marco Dentz

5.1 Introduction

Countless environmental, industrial and biological applications involve fluids flowing through complex media or heterogeneous environments. These can be soil, sand and rocks in aquifers and reservoirs, industrial separation and filtration devices, biological membranes and tissues, composite materials. Although all the fundamental laws and modelling approaches of fluid dynamics still apply, a completely different perspective has to be taken to deal with geometrical and physical complexity and multiscale structure of the underlying media. This is generally done by means of upscaling or averaging techniques, not different than the ones used to deal with the multiscale structure of turbulence. The main difference between flows through porous media and turbulent flows lies in the fact that the latter is an emerging phenomena purely due to the nonlinearity of the Navier–Stokes equations, while the former inherits its multiscale complexity directly from the geometrical and physical properties of the material. This means that, even starting from linearised or simplified flow regimes (e.g., Stokes), interesting emergent macro-scale dynamics can appear due to these properties. The ultimate scientific challenge is to develop a quantitative link between the properties of the media and the upscaled parameters in the macroscopic dynamics. Although a wide range of these emerging dynamics are more easily observed and studied than the turbulent structures (which, by nature, are

M. Icardi (✉)
School of Mathematical Sciences, University of Nottingham, Nottingham, UK
e-mail: matteo.icardi@nottingham.ac.uk

G. Boccardo
DISAT, Politecnico di Torino, Torino, Italy

M. Dentz
IDAEA, CSIC, Barcelona, Spain

hardly reproducible),[1] the high dimensionality of the set of all possible geometrical structures makes a systematic research and a predictive model development (e.g., closures, parameter estimation, etc.) particularly challenging. As turbulence models are naturally first developed and tested on clearly defined scenarios (such as the periodic isotropic turbulence or wall-bounded flows), similarly upscaled porous media models have traditionally been derived from simple granular materials, such as sphere packings. While the former usually assume the existence of a continuum of length scales, as dictated by the classical turbulence theory, the latter typically rely on clearly defined and well-separated scales (usually two or possibly more). These assumptions, however, are only very crude approximations of the actual natural and engineered media, and can lead to significantly misrepresent the overall transport processes.

In this chapter, while presenting the fundamentals of flow and transport through porous media (intended in the classical sense) and some of the specific methodologies and challenges, we take a more general point of view, focused on the underlying upscaling procedures and their assumptions, to help smooth out the still existent barrier between porous media and fluid dynamics research.

5.2 Flow Through Porous and Heterogeneous Media

As already mentioned, although the concept of upscaling and averaging is present in many fluid dynamics problems, particularly relevant to many applications is the understanding of the emerging dynamics of fluid flowing through multiscale (porous) materials. As we will discuss in Sect. 5.2.1, the peculiarity of this problem is due to the presence of large surface areas where no-slip conditions generate, in the first approximation, linear damping in the momentum equation, proportional to an effective parameter known as *permeability* of the media. However, natural porous materials, such as soil and rocks, can have a highly irregular and heterogeneous structure, causing this emerging effect to be significantly space-dependent. Due to the limited a priori knowledge of the exact geology, this space heterogeneity is often modelled as a random field. This gives rise to another important upscaling problem, namely understanding the effect of meso- and macro-scale heterogeneities in the permeability. This is discussed in Sect. 5.2.3.

In both cases, crucial to the upscaling process is the solution of a closure problem, solved on a *representative elementary volume* (REV). This can be understood at different levels. The first simple definition of REV can be based solely based on geometrical information such as the porosity of the material, ϕ, i.e., the volume fraction of void space available for the fluid. This, being a very simple averaging process, can be computed on different length scales ℓ, i.e., $\phi = \phi(\ell)$. For increasing

[1] For example, with the recent development of 3D printing, a wide range of porous media structures can be synthetically recreated and tested.

ℓ, if the media has only finite-size heterogeneities (or well-separated scales) and no fractal structure, this converges to a finite number $0 < \phi_0 < 1$. This implicitly defines a minimal REV of size ℓ_ϕ. Even assuming the existence of this well-defined REV, the actual upscaling of transport and flow processes involves the average of fluctuating quantities (or closure variables), which are solutions of a differential model. This might require a much larger size to converge to a constant. For periodic structures, in the assumption of stationary (fully developed) profiles and local equilibrium (imposed through the pseudo-periodicity of the variables), the periodic cell represents not only the geometric REV but also the right REV for all processes. Relaxing these periodicity assumptions means allowing random "perturbation" in the material that results in a larger geometrical REV, and, possibly, in perturbation in the solution persisting over bigger scales. This means that, to obtain well-defined (e.g., space-independent) macroscopic effective parameters, the existence and size of the REV cannot be known a priori and could be significantly larger than the one obtained from purely geometrical considerations. This consideration applies not only to the upscaling of the flow discussed below (which could indeed need a REV much larger than the geometrical one) but also to the upscaling of transport and reaction processes (discussed in Sect. 5.3), and, more significantly to all non-linear and more complex models (such as multiphase flows).

5.2.1 Darcy's Law

The earliest approaches to the study of flow in porous media were directed to the derivation of simple linear relation between pressure drop and superficial velocity, and implicitly made use of a macroscopic description of a continuous (pseudo-homogeneous) fluid–solid domain. Henry Darcy, who investigated the sand filter system employed in the delivery of freshwater to the city of Dijon, first proposed this relation, now known as *Darcy's law*:

$$-\frac{\delta P}{L} = \frac{\mu}{K} q, \tag{5.1}$$

where δP is the integral pressure drop (or the so-called *pressure head*, including hydrostatic pressure) across the porous medium, L is its length and thus characterises flow in saturated porous media via its permeability, K, and the fluid superficial velocity q. This law, originally derived on purely phenomenological and experimental grounds, can be intuitively extended in three dimensions, as a force balance between pressure gradient and linear wall stresses, neglecting the transient and inertial term in an upscaled form of the Navier–Stokes equation:

$$-\mathbf{K}\nabla P = \mu \mathbf{q}\,, \tag{5.2}$$

where **K** can now be more generally a symmetric tensor, not necessarily isotropic, i.e., pressure gradients in one direction can possibly cause flow to happen in an arbitrary direction, due to non-symmetric porous structures. This result can also be rigorously derived using the tools of homogenisation or volume-averaging [1, 2], upscaling the incompressible Stokes' law to obtain Darcy's law. Equation (5.1), while still being useful in many porous media systems, has been found to have its limitations. The first is related to the relative magnitude of the superficial velocity q. More appropriately, and by analogy with the usual analysis of the laminar–turbulent flow transition, it can be expressed in terms of the Reynolds number where the system's characteristic length is the average grain diameter or pore width. As such, in the vast majority of cases, Darcy's law will find an upper range of validity at Re ranging from one to ten [3]. Other cases, where a more complex equation has to be used, include the already mentioned fractal porous media (where the permeability is no more a constant but a non-local kernel), multiphase flows, non-Newtonian fluids, non-equilibrium flows.

5.2.2 *Extensions of Darcy's Law*

For high Reynolds numbers, the linear relationship expressed in Eq. (5.1) between superficial velocity and the hydraulic gradient ($\delta P/L$) ceases to be valid, making Darcy's law unsuitable for describing the nonlinearities arising under these conditions. Although there has been some controversies [4, 5] about the correct extension of Darcy's law to transitional and turbulent flows, the most commonly used equation that can be used to that end is the *Darcy–Forchheimer equation*:

$$-\frac{\delta P}{L} = \frac{\mu}{K} q + \beta \rho q |q|, \tag{5.3}$$

where β is the so-called inertial flow parameter and, like K, is independent of fluid properties and only depends on the microstructure of the porous medium. Various attempts at an explanation of this phenomenon have been made: the most intuitive of which would be to ascribe this nonlinearity to the onset of turbulence, by immediate analogy with the relationship between head loss and fluid velocity for the flow in pipes, which becomes non-linear right after the transition to the turbulent region corresponding to higher Reynolds numbers. The problem with this approach is that, while for the flow in pipes the laminar and turbulent zones are clearly identifiable, the transition in the case of flow in porous media is much smoother, with no clear separation between the two: this can be related to what is known for flow around spheres, where the same behaviour is found. A number of experiments were conducted in the past to identify the critical Reynolds number associated with the transition to the turbulent zone in porous media, and found it to be several orders of magnitude higher than the Re at which the nonlinearities begin to become apparent [6].

Beyond these difficulties caused by non-trivial changes in the fluid dynamic structure at the pore-scale when transitioning to high Reynolds numbers, there are also a number of other notable extensions, for which, brief pointers follow.

Multiphase and Unsteady Flows

While single-phase flow in porous media (also known as *saturated*) are generally steady, a trivial extension is to add a time derivative term to model unsteadiness caused, for example, by time-dependent pressure boundary conditions. However, when dealing with multiphase flows, the time-dependence naturally appears. While density-driven miscible Darcy's flows are easily obtained, immiscible multiphase extensions of the Darcy's equation rely on much stronger assumptions (see also the discussion in Sect. 5.5). The simplest form of multiphase flow is the *Richards' equation* that describes a water plume travelling through a steady Darcian flow field.

Brinkman

One early and well-known approach to bridge the gap between the free flow and the Darcy's descriptions was put forward by Brinkman, whose eponymous equation adds a viscous term to the Darcy's equation (usually with an effective viscosity which is not necessarily equivalent to the viscosity of the fluid). Rigorous derivations of the Brinkman equation with homogenisation have been proposed [1], with the contemporary presence of the Darcy and Brinkman terms, under a specific scaling for the geometrical properties of the porous structure. Furthermore, the Brinkman equation has found interesting applications as a unified numerical approximation (e.g., penalisation approaches [7]) that, for limiting cases, can recover both Stokes and Darcy equations.

Non-Newtonian

When considering non-Newtonian fluids, the Darcy's law for low fluid velocities is still used, with a modified "porous medium viscosity" comprising the non-Newtonian effects. In the higher velocities ranges, the interplay between shear-thickening effects and turbulent nonlinearities becomes more difficult to understand: both formal attempts of upscaling via volume-averaging [8] and accurate computational pore-scale simulations in reconstructed geometries [9] have been presented.

Knudsen

Finally, the constant assumption of all the theory presented up until this point (and henceforth, excluding this paragraph) has been that of considering the fluid as a continuum and as such, to employ the mentioned *no-slip* condition on the solid matrix boundary. In practice, many real-world systems (e.g., rarefied gases, shale gas) are characterised by Knudsen flow, and are not treatable within the usual framework, leading to non-trivial additions of slip-flow corrections to effective permeability.

5.2.3 Heterogeneous Media

As described in the previous section, the (space) averaged flow behaviour on the scale of a REV is described by the Darcy's equation. Here we consider the transition from the Darcy to larger scales of heterogeneity. For geological porous media, this means an order of metres or hundreds of metres. Here, spatial or ensemble (considering random realisations of geological structures) averaging, or a combination of both, can be used to study emerging macroscopic dynamics. We denote here both averaging operation with the bracket notation $\langle\cdot\rangle$ and we limit our discussion to the steady state Darcy flow equation with heterogeneous medium properties

$$\mathbf{q}(\mathbf{x}) = -\frac{K(\mathbf{x})}{\mu}\nabla P(\mathbf{x}), \qquad \nabla\cdot\mathbf{q}(\mathbf{x}) = 0. \tag{5.4}$$

which is equivalent to a Laplacian equation for the pressure. This description implies that the flow field is helicity-free, i.e., $\mathbf{q}(\mathbf{x})\cdot\nabla\times\mathbf{q}(\mathbf{x}) = 0$. This means that there are no closed streamlines in $d = 2$ dimensional Darcy flow. For $d = 3$ dimensions, zero helicity implies that streamlines are either closed or organised on two-dimensional toruses [10]. These topological properties prohibit chaotic flow and thus have an impact on the stretching of material lines and surfaces.

The systematic upscaling of flow and transport upscaling in heterogeneous porous media has been performed with stochastic approaches in order to model the spatial variability of permeability [11–13]. This is motivated, on the one hand, by the incomplete knowledge of the small scale fluctuations of $K(\mathbf{x})$, and on the other hand by the desire to identify the large scale behaviour due to "typical" spatial random fluctuations and quantify it in terms of only a few geo-statistical characteristics. This requires certain assumptions such as statistical stationarity and ergodicity. In this framework, the log-permeability $f(\mathbf{x}) = \ln[K(\mathbf{x})]$ has been modelled as a multi-Gaussian random field, which implies that $K(\mathbf{x})$ is a multi-lognormal random field. This can be understood as follows. Consider a set of $f(\mathbf{x})$ values evaluated at positions $\mathbf{x}_i$ $(i = 1,\ldots,n)$ in the medium. The set $\{f(\mathbf{x}_i)\}_{i=1}^n$ is modelled now as a spatial stochastic process, which is characterised by a joint Gaussian PDF characterised by the covariance matrix $C_{ij} = C(\mathbf{x}_i - \mathbf{x}_j)$,

$$\mathcal{P}(\{f(\mathbf{x}_i)\}) = \frac{\exp\left[-\sum_{i,j=1}^n f(\mathbf{x}_i)C_{ij}^{-1}f(\mathbf{x}_j)/2\right]}{(2\pi)^n\sqrt{\det(\mathbf{C})}}. \tag{5.5}$$

The variance of $f(\mathbf{x})$ is given by $\sigma_{ff}^2 = C(\mathbf{0})$. The covariance $C(\mathbf{x})$ is typically modelled as a short-ranged function that decays on characteristic length scales, the correlation lengths. For an overview of common covariance models, see Refs. [11–13]. A statistically isotropic medium is characterised by a single correlation scale ℓ. For anisotropic media, the correlation scale depends on the spatial direction.

In this framework flow is upscaled in terms of an effective permeability $\mathbf{K}^e$ tensor [14, 15], which is defined by

$$\langle \mathbf{q}(\mathbf{x}) \rangle = -\frac{\mathbf{K}^e}{\mu} \langle \nabla P(\mathbf{x}) \rangle. \tag{5.6}$$

Here we focus on statistically isotropic media, for which $K^e_{ij} = K^e \delta_{ij}$. Note that K^e is in general not equal to the arithmetic average $\langle K(\mathbf{x}) \rangle$.

In the following, we first report on some exact results for the effective permeability for flow in layered media and two-dimensional multi-Gaussian permeability fields. Then we discuss briefly perturbation theory results and conjectures for three-dimensional media.

Exact Solutions

For layered porous media, permeability is constant along one of the coordinate axis and variable in the other directions. This means the correlation length is infinite along one coordinate axis. For simplicity, we consider the case of 2 spatial dimensions. For a pressure gradient parallel or perpendicular to the layering exact solutions for the effective permeability exist. The direction of the mean pressure gradient in the following is aligned with the x-direction.

For flow aligned with the direction of stratification, the flow problem has an exact solution, which is

$$\mathbf{q}(y) = -K(y) \langle \nabla P(x) \rangle. \tag{5.7}$$

In this case, $K^e = K_A = \langle K(y) \rangle$, the effective permeability is equal to the arithmetic mean permeability. For flow perpendicular to stratification, the exact solution is

$$\mathbf{q}(x) = -\left[\frac{\mu}{L} \int_0^L \frac{dx'}{K(x')} \right]^{-1} \langle \nabla P(x) \rangle. \tag{5.8}$$

where L is the length of the flow domain. Here the effective permeability is given by the harmonic mean $K^e = K_H = 1/\langle K(x)^{-1} \rangle$.

For flow in isotropic two-dimensional multi-lognormal permeability fields with finite correlation length, the effective permeability is exactly given by the geometric mean [16, 17],

$$K^e = K_G \exp\left(-\langle f(\mathbf{x}) \rangle\right). \tag{5.9}$$

This result can be derived based on a duality between stream function and flow potential using the fact that both $K(\mathbf{x})$ and $1/K(\mathbf{x})$ are lognormal distributed.

Perturbation Theory

For three-dimensional heterogeneous porous media, the duality argument invoked for two dimensions does not hold. Thus, the effective permeability has been determined using perturbation theory in the fluctuations of log-permeability about its mean value, $f'(\mathbf{x}) = f(\mathbf{x}) - \langle f(\mathbf{x}) \rangle$, which gives [18–20]

$$K^e = K_G \left(1 + \frac{\sigma_{ff}^2}{6} \right), \tag{5.10}$$

which is strictly valid only for $\sigma_{ff}^2 \ll 1$. For larger values of σ_{ff}^2 and d spatial dimensions, Matheron [18] conjectured the expression

$$K^e = K_G \exp \left[\sigma_{ff}^2 \left(\frac{1}{2} - \frac{1}{d} \right) \right], \tag{5.11}$$

which for $d = 2$ gives the exact result $K^e = K_G$ and in $d = 3$ and small $\sigma_{ff}^2 \ll 1$ is consistent with the perturbation theory result Eq. (5.10). The effective permeability is bounded between the harmonic and arithmetic mean, $K_H \leq K^e \leq K_A$.

5.3 Macroscopic Transport Models

Transport in heterogeneous media can be described by the advection–dispersion equation

$$\frac{\partial c(\mathbf{x}, t)}{\partial t} + \nabla \cdot \mathbf{v}(\mathbf{x}) c(\mathbf{x}, t) - \nabla \cdot \left[\mathbf{D}(\mathbf{x}) \nabla c(\mathbf{x}, t) \right] = 0. \tag{5.12}$$

At the pore-scale, the velocity field $\mathbf{v}(\mathbf{x})$ is obtained from the Stokes equation and the dispersion tensor reduces to $\mathbf{D} = D_m I$, where D_m is the molecular diffusion coefficient and I the identity matrix. At the Darcy scale, a similar equation can be derived by homogenisation or volume-averaging [1, 2], with the flow velocity given by $\mathbf{v}(\mathbf{x}) = \mathbf{q}(\mathbf{x})/\phi$, where ϕ is porosity, which here is assumed to be constant, and the dispersion tensor $\mathbf{D}(\mathbf{x})$ given by the solution of a closure problem. Alternatively, dimensional and phenomenological arguments can lead to the following parameterisation [21, 22]:

$$D_{ij} = \alpha_0 D_m \delta_{ij} + \sum_{k,l=1}^{d} \alpha_{ijkl} \frac{q_k(\mathbf{x}) q_l(\mathbf{x})}{\|\mathbf{q}(\mathbf{x})\|}, \tag{5.13}$$

where $\alpha_0 D_m$ is the effective diffusivity (see Sect. 5.5), and α_{ijkl} are geometrical dispersivities. For an isotropic medium, the α_{ijkl} are given by

$$\alpha_{ijkl} = \alpha_{II}\delta_{ij}\delta_{kl} + \frac{\alpha_I - \alpha_{II}}{2}\left(\delta_{ik}\delta_{jl} + \delta_{il}\delta_{jk}\right). \tag{5.14}$$

This description of dispersion is valid at high Péclet numbers. The Péclet number compares the relative strength of diffusive and advective transport mechanisms and is here defined as $\mathrm{Pe} = VL/D$, where L is a characteristic heterogeneity length scale and V a characteristic velocity.

The advection–dispersion Eq. (5.12) is equivalent to a Ito's stochastic differential equation [23, 24] for the position $\mathbf{x}(t)$ of a solute particle

$$\frac{d\mathbf{x}(t)}{dt} = \mathbf{v}[\mathbf{x}(t)] + \nabla \cdot \mathbf{D}[\mathbf{x}(t)] + \sqrt{2\mathbf{D}[\mathbf{x}(t)]} \cdot \boldsymbol{\xi}(t), \tag{5.15}$$

where $\boldsymbol{\xi}(t)$ is a Gaussian white noise of zero mean, $\langle \boldsymbol{\xi}(t) \rangle = \mathbf{0}$ and covariance $\langle \xi_i(t)\xi_j(t')\rangle = \delta_{ij}\delta(t - t')$. Equation (5.15) is the starting point for random walk particle tracking simulations for the solution of advective–dispersive transport in heterogeneous porous media.

A key issue for transport in heterogeneous media is to quantify the transport behaviours on a scale larger than the characteristic heterogeneity scale. For the transition from pore to Darcy scale, the observation scale is larger than the characteristic pore length, for the transition from Darcy to regional scale, it is larger than the correlation scale of permeability. In the following, we briefly report on upscaling efforts in terms of Fickian transport formulations, the occurrence of anomalous dispersion and modelling approaches to account for non-Fickian transport.

5.3.1 Fickian Dispersion

Before discussing Fickian large scale transport formulations, we briefly summarise some signatures of Fickian transport for one-dimensional transport at constant velocity v_0 and diffusion coefficient D_0. Firstly, for a point-like solute injection, the concentration distribution is Gaussian shaped as

$$c_0(x, t) = \frac{\exp\left[-\frac{(x - v_0 t)^2}{4D_0 t}\right]}{\sqrt{4\pi D_0 t}}. \tag{5.16}$$

The first and second centred moments of $c(x, t)$, denoted by $m(t)$ and $\kappa(t)$ evolve linearly in time as $m(t) = v_0 t$ and $\kappa(t) = 2D_0 t$. Solute *breakthrough*, i.e., the distribution of solute arrival times at a control plane located at the distance x from the plane at which solute is injected, is given by the inverse Gaussian distribution

$$f_0(t,x) = \frac{x \exp\left[-\frac{(x-v_0t)^2}{4D_0t}\right]}{\sqrt{4\pi D_0 t^3}}. \tag{5.17}$$

Hydrodynamic Dispersion

As already mentioned, the upscaling of transport from the pore to the Darcy scale can be approached by stochastic approaches [25], spatial averaging and homogenisation. Under the assumption of local physical equilibrium, these approaches derive for the (homogeneous) Darcy scale the advection–dispersion Eq. (5.12). The hydrodynamic dispersion tensor **D** accounts for the impact of molecular diffusion and pore-scale velocity fluctuations on Darcy-scale solute transport. It is in general a function of the Péclet number. This has been observed both for the longitudinal, i.e., the mean flow direction, and the transverse dispersion coefficients D_L and D_T. For $\mathrm{Pe} \ll 1$, $D_L/D \sim 1$, for $1 < \mathrm{Pe} < \mathrm{Pe}_c$, it behaves as $D_L/D \sim \mathrm{Pe}^{\gamma}$, with $1 < \gamma < 1.5$, and for $\mathrm{Pe} > \mathrm{Pe}_c$ it scales as $D_L/D \sim \mathrm{Pe}$. The critical Péclet number is $\mathrm{Pe}_c \approx 400 - 500$ [26, 27]. These behaviours can be described by the expression [22]

$$D_L = D\alpha + \alpha_I \overline{v} \frac{\mathrm{Pe}}{\mathrm{Pe} + 2 + 4\delta^2}, \tag{5.18}$$

where α accounts for the effect of the tortuous pore geometry on molecular diffusion in the bulk and δ is a parameter that characterises the shape of the pore channels and $\overline{v}$ is the average pore velocity. The second term on the right-hand side of Eq. (5.18) is termed mechanical dispersion. It quantifies solute spreading due to the tortuous streamlines and velocity variability of the pore-scale flow field. Bear [22] proposes to use expression Eq. (5.18) also for the transverse dispersion coefficients D_T. Experimental data suggest that $D_T/D \sim \mathrm{Pe}^{0.95}$ for $1 < \mathrm{Pe} < \mathrm{Pe}_c$ [28].

Macro-Dispersion

For transport upscaling from the Darcy to the regional scale, stochastic perturbation theory gives for macro-scale transport the advection–dispersion equation [20]

$$\phi \frac{\partial c(\mathbf{x},t)}{\partial t} + \langle q \rangle \frac{\partial c(\mathbf{x},t)}{\partial x} - \nabla \cdot \left[\mathbf{D}^* \nabla c(\mathbf{x},t)\right] = 0, \tag{5.19}$$

where the x-axis of the coordinate system is aligned with the mean hydraulic gradient. The Péclet number here is defined as $\mathrm{Pe} = \langle q \rangle \ell / D_L$. For a statistically isotropic heterogeneous porous medium, the macro-dispersion tensor $\mathbf{D}^*$ is diagonal. For $\mathrm{Pe} \gg 1$, the longitudinal macro-dispersion coefficient D_L is given by

$$D_L^* = \sigma_{ff}^2 \ell K_G \langle \nabla P(\mathbf{x}) \rangle + \ldots, \tag{5.20}$$

where the dots denote contributions of the order of D_L and of order σ_{ff}^2. This remarkable result relates the macroscopic dispersion effect due to local scale velocity fluctuations to the statistical medium properties in terms of the variance σ_{ff}^2 and correlation length ℓ of log-permeability and the geometric mean permeability. Perturbation theory in σ_{ff}^2 predicts that D_T^* is of the order of D_T, this means local scale dispersion. While this is exact for two dimensions [29], observations and numerical simulations suggest that it is not valid for three dimensions [30, 31]. In fact, numerical results suggest that $D_T^* \propto \sigma_{ff}^4$ in the advection-dominated limit of $\mathrm{Pe} \to \infty$. The reader is referred to the textbooks by [11–13] for a thorough account of the macro-dispersion approach and stochastic perturbation theory for macro-scale transport in heterogeneous porous media.

5.3.2 Anomalous Dispersion

Fickian dispersion predicts that the first and second centred moments of a solute plume increase linearly with time that the solute distribution is Gaussian shaped, and solute breakthrough can be described inverse Gaussian distributions. Furthermore, within the Fickian dispersion paradigm, mixing is fully characterised by the constant dispersion coefficients. For heterogeneous porous media, and heterogeneous media in general, however, transport does not generally follow Fickian dynamics. Breakthrough curves are characterised by strong tailing, dispersion evolves in general non-linearly in time and spatial plumes do not show Gaussian shapes and are in general characterised by forward or backward tails. Such behaviours are closely related to the notion of incomplete mixing on the support scale. If the support scale is not fully mixed, for example, due to mass transfer between sub-scale mobile and immobile regions, or velocity variability, transport dynamics are history-dependent. Non-Fickian and anomalous transport behaviours have been observed both on the pore [32–34] and Darcy scales [35, 36].

The mechanism that mixes the support scale is diffusion (pore-scale) and hydrodynamic dispersion (Darcy scale). Note that ultimately the mechanism that attenuates concentration contrasts is diffusion. Mechanical dispersion quantifies the spread of a solute distribution due to advective heterogeneity and the formation of filaments, which facilitates the action of diffusion to homogenise concentration, and is discussed further in Sect. 5.3.3. Thus, for high Péclet numbers, the characteristic mixing time scales over the support scale may be significantly larger than the time scale of interest. The prediction of transport in heterogeneous media requires approaches that allow to quantify non-Fickian transport dynamics.

The moment equations and projector formalism approaches [37, 38] are obtained from the stochastic averaging of the local scale heterogeneous transport problem, Eq. (5.12), which yields space- and time-non-local integro-differential equations, whose memory kernels are related to the heterogeneity statistic. Closed-form expressions for the memory kernels are in general difficult to obtain. Fractional advection–dispersion equations [39, 40] are characterised by spatio-temporal kernel function with an asymptotic power-law scaling. This approach can be related to continuous time random walks and Levy flights [41]. In the following, we provide a summary of the continuous time random walk (CTRW) and multi-rate mass transfer (MRMT) frameworks to describe anomalous dispersion in porous media. The CTRW [42–44], and related time domain random walk (TDRW) [45, 46] frameworks, as well as the MRMT approach [47, 48] have been used for transport upscaling in highly heterogeneous porous and fractured media. These approaches account for the heterogeneity-induced distribution of advective and diffusive mass transfer rates and residence times.

Continuous Time Random Walks

The continuous time random walk (CTRW) [49, 50] models particle motion as a random walk in space and time. The concentration distribution, or equivalently, the particle density $c(x, t)$ is given by

$$c(x,t) = \int_{0}^{t} dt' R(x,t') \int_{t-t'}^{\infty} dt'' \psi(t''), \qquad \psi(t) = \int_{-\infty}^{\infty} dx \psi(x,t), \tag{5.21}$$

where $R(x, t)dxdt$ is the average number of times a particle is in $[x, x+dx]\times[t, t+dt]$; $\psi(x, t)$ is the joint PDF of transition length and time. Thus, the right-hand side of Eq. (5.21) denotes the frequency by which a particle arrives at a position x at time t' times the probability that it stays (waits) there for a time smaller that t. $R(x, t)$ satisfies the Chapman–Kolmogorov type equation

$$R(x,t) = c(x,t)\delta(t) + \int_{-\infty}^{\infty} dx' \int_{0}^{t} dt' \psi(x-x', t-t') R(x', t'). \tag{5.22}$$

The first term on the right-hand side denotes the initial particle distribution at time $t = 0$. Combining Eqs. (5.21) and (5.22) gives the generalised master equation [51]

$$\frac{\partial c(x,t)}{\partial t} = \int_{-\infty}^{\infty} dx' \int_{0}^{t} dt' \mathcal{K}(x-x', t-t') \left[c(x',t) - c(x,t)\right], \tag{5.23}$$

where the memory kernel $\mathcal{K}(x, t)$ is defined by its Laplace transform [52] as

$$\mathcal{K}^*(x,\lambda) = \frac{\lambda \psi^*(x,\lambda)}{1 - \psi^*(\lambda)}. \tag{5.24}$$

Laplace transformed quantities are denoted by an asterisk, the Laplace variable is denoted by λ. For short-ranged spatial transitions, Eq. (5.23) can be localised in space such that

$$\frac{\partial c(x,t)}{\partial t} + \int_0^t dt' \left[\kappa_1(t-t') \frac{\partial}{\partial x} - \kappa_2(t-t') \frac{\partial^2}{\partial x^2} \right] c(x,t) = 0, \tag{5.25}$$

where the advection and dispersion kernels are defined by

$$\kappa_1(t) = \int_{-\infty}^{\infty} dx x \mathcal{K}(x,t), \qquad \kappa_2(t) = \frac{1}{2} \int_{-\infty}^{\infty} dx x^2 \mathcal{K}(x,t), \tag{5.26}$$

The fluctuating micro-scale transport dynamics are encoded in the joint PDF $\psi(x,t)$. For purely advective solute transport, for example, the transition length is of the order of the correlation scale ℓ_c of the velocity magnitude v, and the transition time is given kinematically by ℓ_c/v. The distribution $p_s(v)$ of the particle speed sampled equidistantly along a streamline is related to the Eulerian velocity PDF by flux-weighting as [53]

$$p_s(v) = \frac{v p_e(v)}{\langle v_e \rangle}. \tag{5.27}$$

Thus, the joint PDF of transition length and time is

$$\psi(x,t) = \delta(x - \ell_c) \frac{\ell_c^2 p_e(\ell_c/t)}{t^3 \langle v_e \rangle}. \tag{5.28}$$

For transport at an average velocity v_0 over the characteristic length ℓ_0 combined with mass transfer into immobile zones, the transition time distribution is given in Laplace space by Margolin et al. [54]

$$\psi^*(\lambda) = \exp\left(\lambda \tau_0 + \gamma_t \tau_0 \left[1 - p_f^*(\lambda) \right] \right), \tag{5.29}$$

where $\tau_0 = \ell_0/v$ is the advective transition time, γ_t the trapping rate and $p_f(t)$ the distribution of residence times in the immobile regions.

We briefly summarise the transport characteristics for an uncoupled CTRW, this means $\psi(x,t) = \Lambda(x)\psi(t)$, characterised by a power-law long-time scaling of the transition time distribution as $\psi(t) \sim t^{-1-\beta}$ with $0 < \beta < 2$. Such heavy tailed transition distributions imply strong particle retention and thus memory effects. The power-law in $\psi(t)$ directly relates to the solute breakthrough curves. Note that the breakthrough time at a control plane is the sum of n transition times τ_i, where n may be approximated by the average number of spatial steps needed to arrive at the control plane. Thus, the generalised central limit theorem implies that the breakthrough curve scales as $f(t,x) \sim t^{-1-\beta}$. The first and second centred moments of the solute distribution scale asymptotically as $m(t) \sim t^\beta$ and $\kappa_2(t) \sim t^{2\beta}$ for $0 < \beta < 1$ and as $m(t) \sim t$ and $\kappa_2(t) \sim t^{3-\beta}$.

For advective transport upscaling, the CTRW framework has been used together with Markov models for series of velocity magnitudes along streamlines [55–57], which allow for the evolution of the transition time distribution with increasing step number and for the conditioning of transport on initial particle velocities [53]. The CTRW framework has been employed for transport modelling in a wide range of fluctuating environments ranging from the diffusion of charge carriers in impure semi-conductors [50] to diffusion in living cells [58], see also [41, 59].

Multi-Rate Mass Transfer

The multi-rate mass transfer (MRMT) approach [47, 48] separates the support scale into a mobile continuum and a suite of immobile continua, which communicate through linear mass transfer. At each point, the immobile concentration $c_{im}(x, t)$ is related to the mobile concentration $c_m(x, t)$ through the linear relation [48]

$$c_{im}(x, t) = \int_0^t dt' \varphi(t - t') c_m(x, t'). \tag{5.30}$$

The evolution of the mobile concentration is given by the advection–dispersion equation [47]

$$\phi_m \frac{\partial c_m(x, t)}{\partial t} + q \frac{\partial}{\partial x} c_m(x, t) - D\phi_m \frac{\partial^2}{\partial x^2} c_m(x, t) = -\phi_{im} \frac{\partial c_{im}(x, t)}{\partial t}, \tag{5.31}$$

where ϕ_{im} and ϕ_m are the immobile and mobile volume fractions. The memory function $\varphi(t)$ encodes the mass transfer mechanisms between the mobile and immobile continua. For diffusive mass transfer into slab shaped immobile regions, the memory function is defined by its Laplace transform as [48, 60]

$$\varphi^*(\lambda) = \frac{\tanh(\sqrt{\lambda \tau_D})}{\sqrt{\lambda \tau_D}}, \tag{5.32}$$

where τ_D is the characteristic diffusion time across the slab. For spherical inclusions, the memory function is

$$\varphi^*(\lambda) = \frac{3}{\sqrt{\lambda \tau_D}} \left[\coth(\sqrt{\lambda \tau_D}) - \frac{1}{\sqrt{\lambda \tau_D}} \right]. \tag{5.33}$$

For purely diffusive mass transfer the MRMT approach is equivalent to transport under matrix diffusion [61], which describes transport in fractured media under diffusive mass transfer between the fracture and the rock matrix. In general for diffusive mass transfer, the memory function is obtained from the solution of a diffusion problem in a heterogeneous immobile domain [62].

For first-order mass transfer at a single rate ω the memory is given by

$$\varphi(t) = \omega \exp(-\omega t). \tag{5.34}$$

Oftentimes, the mass transfer processes and the geometries of immobile regions are not known in detail. The memory has then been modelled by a superposition of multiple first-order memory functions as [35, 47, 48]

$$\varphi(t) = \int_0^\infty d\omega \mathcal{P}(\omega)\omega \exp(-\omega t), \tag{5.35}$$

where $\mathcal{P}(\omega)$ is the rate distribution, which may be related to the volume fractions of the immobile zones, for example. Other approaches use parametric forms for the memory function, such as truncated power-laws [63].

In this framework, the behaviour of solute breakthrough at asymptotic times follows the time derivative of the memory function

$$f(t, x) = -\frac{x}{q}\frac{\phi_m}{\phi_{im}}\frac{d\varphi(t)}{dt}. \tag{5.36}$$

This means, for a memory function, which asymptotically behaves as a power-law $\varphi(t) \sim t^{-\beta}$, the breakthrough curve scales as $f(t, x) \sim t^{-1-\beta}$ [35, 64]. For matrix diffusion into a semi-infinite slab, the memory function scales as $\varphi(t) \sim t^{-1/2}$ and consequently the breakthrough curve as $f(t, x) \sim t^{-3/2}$, which is a signature of matrix diffusion.

Both the CTRW and MRMT frameworks share similar phenomenology in that they account for memory effects due to a distribution of characteristic mass transfer time scales. The correspondence between the two pictures was discussed in [64–67]. The time behaviour of the spatial moments of the concentration distribution is similar to the ones described by an uncoupled CTRW [64].

5.3.3 *Mixing and Chemical Reactions*

In this section, we are concerned with mixing and reactions in heterogeneous porous media. As chemical reactions are contact processes, mixing and dispersion are key processes for the sound quantification of chemical reactions in heterogeneous media. This refers both to homogeneous, i.e., fluid–fluid, reactions, and to heterogeneous, i.e., fluid–solid, reactions, as outlined in the following. We will first discuss the notions of mixing and dispersion, and specifically the difference between these two processes. Then, we discuss chemical reactions under spatial heterogeneity.

Mixing, Diffusion and Dispersion

In Fickian transport descriptions, the process that leads to the mixing of initially segregated solutes or the mixing of an initially concentrated solute into the ambient fluid is mass transfer due to diffusion or dispersion. From expression Eq. (5.16) we obtain directly that the maximum concentration $c_m(t)$ in one dimension decays as $c_m(t) = 1/\sqrt{4\pi Dt}$. In d spatial dimensions one finds that $c_m(t) = 1/(4\pi Dt)^{d/2}$. Mixing due to molecular diffusion on mesoscopic length scales L is in general slow. The characteristic mixing time is given by $\tau_m = L^2/D$. For a free fluid, stirring or chaotic flow accelerates the mixing process in that it generates laminar structures [68] whose size $l(t)$ increases exponentially fast with time, $l(t) = l_0 \exp(\lambda t)$, where λ here is the Lyapunov exponent. The width of the lamellar structures is limited by stretching and diffusion to the Batchelor scale $s_B = \sqrt{D/\lambda}$ [69]. The number of lamellae in a closed area of size $A = L^2$ increases exponentially fast as $n(t) \sim \ell(t)/L$, while each lamella occupies an area of $A_\ell \sim s_B L$. Complete mixing is achieved when $n(t)A_\ell \sim L^2$. This gives a mixing time $\tau_m = \lambda^{-1} \ln(L^2/\ell_0 s_B)$ which is in general much shorter than the mixing time by diffusion only.

Mixing and Spreading in Porous Media

Here we are concerned with mixing in flows through heterogeneous porous media. We have seen above that solute transport has been quantified in terms of hydrodynamic dispersion (pore to Darcy scale) and macro-dispersion (Darcy to regional scale), which simulates that the support scale is well-mixed. This Fickian paradigm, however, breaks down for transport in heterogeneous media, for which anomalous or non-Fickian transport behaviours are observed. These involve history-dependence, which implies that the support scale cannot be considered well-mixed. Unlike for stirring in a free fluid, for porous media flows, the "stirring" is done by the medium itself, whose structure leads to tortuous path-lines and velocity heterogeneity. Chaotic flow patterns are in general prohibited topologically for steady two-dimensional flows. In three dimensions, steady pore-scale flow is chaotic [70], which may lead to similar mixing as in chaotic flow in a free fluid [71]. The existence of low velocity regions, stagnant zones in the wake of solute grains, velocity variability between pores and intra-grain mass transfer, however, leads to incomplete mixing on the REV scale and thus history-dependent transport [72, 73].

The topological properties of Darcy-scale flow through porous media prohibit chaotic flow [74], see also Sect. 5.2.3. Thus, while the action of the spatially variable flow velocity leads to the creation of lamellar structures [75], their lengths cannot increase exponentially fast [76]. In fact, the extension of a solute distribution due to such advection-induced spreading can be measured by the concept of macro-dispersion. When lamellae form along the mean flow direction, the separation distance between them is given by the characteristic transverse heterogeneity length scale ℓ. Thus, the time scale to mix the heterogeneous concentration distribution is given by the time for transverse dispersion over the distance ℓ, $\tau_D = \ell^2/D_T$. These mechanisms are accounted for by effective dispersion coefficients [77, 78]. Unlike macro-dispersion, the concept of effective dispersion does not account for the effect of purely advective spreading [79, 80], but quantifies the combined effect of local scale dispersion and advective heterogeneity, which eventually leads to mixing.

Scalar Dissipation and Concentration Statistics

In order to illustrate the relative role of local scale dispersion and spreading as quantified by macro-dispersion in solute mixing, we consider the evolution of the variance of the concentration fluctuation about its ensemble mean value, $c'(\mathbf{x}, t) = c(\mathbf{x}, t) - \langle c(\mathbf{x}, t)\rangle$, defined here by

$$\sigma_c^2(t) = \int d\mathbf{x}\langle c'(\mathbf{x}, t)^2\rangle. \tag{5.37}$$

From the Darcy-scale advection–dispersion Eq. (5.12) one can derive [81]

$$\begin{aligned}\frac{d\sigma_c^2(t)}{dt} &= -\frac{2}{\phi}\int d\mathbf{x}\langle\nabla\tilde{c}(\mathbf{x}, t)\cdot\mathbf{D}\nabla\tilde{c}(\mathbf{x}, t)\rangle+ \\ &+\frac{2}{\phi}\int d\mathbf{x}\nabla\langle c(\mathbf{x}, t)\rangle\cdot\mathbf{D}^*\nabla\langle c(\mathbf{x}, t)\rangle.\end{aligned} \tag{5.38}$$

The first term on the right-hand side is denoted as scalar dissipation rate. It quantifies the destruction of concentration variance due to local dispersive mass transfer. The second term on the right-hand side quantifies the creation of concentration variance due to spreading as quantified by macro-dispersion.

The mixing process can also be described in terms of the evolution of the PDF of concentration values $c(\mathbf{x}, t)$ in the heterogeneous mixture, which can be defined by

$$p(c; \mathbf{x}, t) = \langle\delta[c - c(\mathbf{x}, t)]\rangle. \tag{5.39}$$

Based on the advection–dispersion Eq. (5.12) one can derive the following evolution equation for the PDF [82]:

$$\begin{aligned}\phi\frac{\partial p(c; \mathbf{x}, t)}{\partial t} &+ \langle\mathbf{q}\rangle\cdot\nabla p(c; \mathbf{x}, t) - \nabla\cdot\mathbf{D}^*\nabla p(c; \mathbf{x}, t) = \\ &= -\frac{\partial}{\partial c}\langle\nabla\cdot\mathbf{D}\nabla c(\mathbf{x}, t)|c\rangle p(c; \mathbf{x}, t).\end{aligned} \tag{5.40}$$

The term on the right-hand side is the average over the local dispersive flux terms conditional to concentration, which represents a closure problem. The celebrated interaction by exchange with the mean (IEM) closure [83] approximates this expression as

$$\langle\nabla\cdot\mathbf{D}\nabla c(\mathbf{x}, t)|c\rangle = \frac{\gamma_{IEM}}{2}\left(c - \langle c(\mathbf{x}, t)\rangle\right), \tag{5.41}$$

where γ_{IEM} is a rate constant that may be related to local scale dispersion and the local dissipation scales. This closure implies for the scalar dissipation rate

$$\frac{2}{\phi}\int d\mathbf{x}\langle\nabla c'(\mathbf{x},t)\cdot\mathbf{D}\nabla c'(\mathbf{x},t)\rangle = \gamma_{IEM}\sigma_c^2(t). \tag{5.42}$$

This closure approximation has been applied to predict mixing in heterogeneous porous media, but is not able to match the numerically observed evolution of the scalar dissipation rate [84]. The IEM closure has several shortcomings for porous media mixing. Firstly it implicitly assumes that the concentration PDF is Gaussian shaped or approximately Gaussian, while in porous media they are typically highly non-Gaussian [75]. Secondly, it assumes a constant local mixing scale, while the mixing scale in porous media evolves with time [85]. Alternative approaches employ parametric forms for the concentration PDF, such as beta-distributions, which can be parameterised by the concentration mean and variance [86], mapping approaches [87] and stochastic mixing models for the evolution of concentration [88].

Lamellar Mixing

Recently, the problem of mixing in porous media has been addressed using a lamellar mixing approach [75, 89]. The mixing process can be roughly separated in two regimes. In an early time regime, the initial solute distribution spreads out and advective heterogeneity generates a lamellar organisation of the concentration field. In the late time regime, the lamellar organisation is destroyed due to coalescence of adjacent lamellae. In both regimes, the concentration PDF can be constructed in terms of the concentration contents of individual lamellae and their interactions.

In the early time regime, lamellae are non-interacting and the evolution of the concentration content of the mixture can be understood by the superposition of the concentration contents of isolated lamellae, which is fully determined by fluid stretching and local scale dispersion [69, 90]. The concentration across a stretched lamella is Gaussian shaped and given by [75]

$$c(z,t) = \frac{c_0 \exp\left(\frac{z^2\ell(t)^2}{s_0^2[1+4\theta(t)]}\right)}{\sqrt{1+4\theta(t)}}, \qquad \theta(t) = \frac{D}{\phi s_0^2}\int_0^t dt'\ell(t')^2, \tag{5.43}$$

where z is the coordinate across the lamella, $\ell(t) = l(t)/l(0)$ the relative strip elongation and s_0 the initial strip width. Note that the concentration here depends on the elongation $\ell(t)$. The PDF of concentration values across a strip is then given by

$$p(c|c_m) = \frac{1}{2c\sqrt{\ln(c_m/\epsilon)\ln(c_m/c)}}, \tag{5.44}$$

where ϵ is a lower concentration cut-off and $c_m(t) = c(z = 0, t)$ is the maximum strip concentration, which again depends on elongation $\ell(t)$. For heterogeneous media, the strip elongation $\ell(t)$ is the result of the random deformations a strip experiences as it is transported through the medium. For heterogeneous porous media, elongation is dominated by intermittent shear events along a trajectory [76]. The mean elongation may follow power-law behaviours $\langle \ell(t) \rangle \sim t^{\alpha}$ with $1/2 < \alpha < 2$. The maximum concentration can be approximated by $c_m(t) \approx 1/\ell(t)\sqrt{Dt}$ [75], because it decays at the same rate as the area of the lamella increases. Thus, the PDF of elongation can be mapped onto the PDF $p_m(c_m, t)$ of maximum concentrations, and the global concentration PDF is obtained through superposition of the local laminar PDFs as

$$p(c, t) = \int dc_m \, p_m(c_m, t) p(c|c_m). \tag{5.45}$$

The decisive step here is to recognise that the concentration field at early times is organised in a lamellar structure and that the concentration content of a lamella depends explicitly on the strip elongation. This allows obtaining the concentration PDF by mapping from the PDF of strip elongations.

With increasing time, the length of the lamellae becomes larger than the mixing support, which increases slower than the lamella elongation ($\sim \sqrt{t}$ for dispersive growth), or is constant in the case of a confined domain. Thus, the lamellae need to fold back to each other, which in the late time regime leads to diffusive overlap and the formation of lamella aggregates through a random aggregation process. The PDF of maximum concentrations of lamella aggregates after n aggregations is given by the gamma distribution [69]

$$p_m(c_m, t) = \frac{\left(c_m/\langle c_m(t) \rangle\right)^{n-1} \exp\left(-c_m/\langle c_m(t) \rangle\right)}{\Gamma(n) \langle c_m(t) \rangle}. \tag{5.46}$$

In the centre of the plume, the number $n(t)$ of lamellae in the aggregate is related to the average maximum concentration as $n(t)\langle c_m(t) \rangle \sim 1/\sqrt{\kappa^*(t)}$, where $\kappa^*(t)$ is the spatial variance of the average solute concentration, which can be described by macro-dispersion. This sets the concentration PDF at the plume centre as a result of random aggregation of lamellae. The evolution equation for the PDF $p(c, t)$ of the concentration content in the mixture as a result of random aggregation is discussed in [69].

Chemical Reactions

In this section, we will briefly discuss the upscaling of chemical reactions in heterogeneous porous media and the influence of mixing on the reaction efficiency. Chemical reactions are contact processes and thus depend on the availability of reacting species and on the mechanisms that bring them into contact. In a well-mixed reactor, in which stirring-induced mixing is exponential as described above, mass transfer is not a limiting process. For a heterogeneous porous medium, in which mixing is much slower, reactions may be limited by the transport rate, i.e., the efficiency by which they are brought into contact. In the former case, the chemical reaction is rate limited, in the latter transport limited. These situations are distinguished by the Damköhler number

$$\mathrm{Da} = k_r \tau_m, \tag{5.47}$$

where k_r is a reaction rate and τ_m a characteristic mass transfer time scale. For $\mathrm{Da} < 1$, the chemical reaction is rate limited, for $\mathrm{Da} \gg 1$ it is mixing, or transport limited.

Incomplete Mixing

A key issue for reaction upscaling in porous media is the notion of a well-mixed support scale. We have seen in the previous sections that mixing in porous media is slow because the "stirring" by the porous medium is much less efficient than stirring-induced chaotic advection in a free fluid. Spatial heterogeneity and consequently slow mass transfer between different compartments of a heterogeneous porous medium leads to reactant segregation and thus to a reduction of the reaction rate compared to a well-mixed system. We have seen in Sect. 5.3.2 that incomplete mixing on the support scale leads to non-Fickian transport and history-dependent transport phenomena. The same mechanisms lead to reaction behaviours that are different from the ones measured in well-mixed laboratory environments [91, 92]. However, traditional Darcy-scale reactive transport modelling is based on the advection–dispersion equation for the species concentration c_i combined with a kinetic rate law determined for a well-mixed environment,

$$\phi \frac{\partial c_i(\mathbf{x}, t)}{\partial t} + \mathbf{q} \cdot \nabla c(\mathbf{x}, t) - \nabla \cdot \left[\mathbf{D} \nabla c(\mathbf{x}, t)\right] = \sum_j \nu_{ij} r_j[\mathbf{c}(\mathbf{x}, t)]), \tag{5.48}$$

where ν_{ij} are stoichiometric coefficient, r_j is the reaction rate of the jth reaction and $\mathbf{c}(\mathbf{x}, t)$ is the vector of concentrations of the reacting species. This formulation assumes that the support scale is a well-mixed environment. This means that the time scale by which the concentration on the support scale homogenises after a concentration perturbation due to mass transfer is smaller than the reaction time scale. Only under these conditions can the reaction rate on the right-hand side be identified with the one obtained in a well-mixed environment. The upscaling of reactive transport from the pore to the Darcy scale can be formally studied, for example, using volume-averaging [93, 94] or homogenisation theory [95]. The validity of Darcy-scale reaction–dispersion models such as Eq. (5.48) have been investigated in detail by Kechagia et al. [96] and Meile and Tuncay [97]. These studies systematically show discrepancies between the average reaction rates and reaction rates predicted by the advection–dispersion reaction equation.

Similar observations have been made for the upscaling from Darcy to regional scale [98–101]. Incomplete mixing on the support scale leads in general to the reduction of the mixing efficiency compared to equivalent Fickian large scale models. The segregation of reactants on the support scale can be addressed using multi-continuum approaches [102, 103], which resolve concentration variability on the support scale due to subregions of slow and fast mass transfer.

An illustrative example for the impact of chemical heterogeneity on reactivity is diffusion in a medium characterised by a spatial distribution of specific reactive surfaces $\sigma_r(\mathbf{x})$ at which species A reacts to C. The concentration c_A of A evolves according to the reaction–diffusion equation

$$\frac{\partial c_A(\mathbf{x},t)}{\partial t} - D\nabla^2 c_A(\mathbf{x},t) = -k\sigma_r(\mathbf{x})c_A(\mathbf{x},t). \tag{5.49}$$

We consider $\sigma_r(\mathbf{x}) = 0, 1$ randomly distributed in space with a characteristic distance ℓ_c and $\mathrm{Da} = kL^2/D \gg 1$, this means fast chemical reactions. Thus one would assume that the effective reaction rate is given by the mean diffusion time between reactive spots, $k^e = 1/\tau_D$, where $\tau_D = \ell_c^2/D$ is the average diffusion time between reaction spots. However, the total mass of A decays at long times as a stretched exponential [104],

$$c_A \sim \exp\left[-\beta\left(t/\tau_D\right)^{d/(d+2)}\right], \tag{5.50}$$

with β being a constant. It decays slower than the exponential decay predicted by τ_D. This is due to the fact that the space between reaction spots has a finite probability to be arbitrarily large. Thus, segregation due to spatial heterogeneity leads to a slower decay than what would be predicted by mean field theory.

Mixing-Limited Reactions

For mixing-limited chemical reactions, i.e., for high Da, the "stirring" action of the heterogeneous porous medium enhances the mixing efficiency compared to purely diffusive mass transfer and may lead to the formation of localised mixing and reaction hotspots [105, 106]. At high Da, the reaction rate is directly proportional to the mixing rate and the reactive transport problem can be mapped onto a conservative transport problem plus a speciation relation [107]. We illustrate this briefly for a fast reversible bimolecular reaction $A + B \rightleftharpoons C \downarrow$ on the Darcy scale. Chemical equilibrium is described by the mass action law

$$c_A c_B = K, \tag{5.51}$$

where c_A and c_B are the concentrations of species A and B and K is the equilibrium constant. The reactive transport problem is described by Eq. (5.48) for each species ($i = A, B$), which for a heterogeneous medium reads as

$$\phi \frac{\partial c_i(\mathbf{x},t)}{\partial t} + \nabla \cdot \mathbf{q}(\mathbf{x}) c_i(\mathbf{x},t) - \nabla \cdot \mathbf{D}(\mathbf{x}) \nabla c_i(\mathbf{x},t) = -r(\mathbf{x},t), \tag{5.52}$$

where we have assumed the same dispersivity for both species, and where $r(\mathbf{x}, t)$ denotes the "equilibrium" reaction rate. It is determined by observing that $\xi = c_A - c_B$ is a conserved variable (sometimes called mixture fraction) and obeys the conservative advection–dispersion equation

$$\phi \frac{\partial \xi(\mathbf{x},t)}{\partial t} + \nabla \cdot \mathbf{q}(\mathbf{x}) \xi(\mathbf{x},t) - \nabla \cdot \mathbf{D}(\mathbf{x}) \nabla \xi(\mathbf{x},t) = 0. \tag{5.53}$$

Secondly, both c_A and c_B depend only on u through the mass action law, Eq. (5.51),

$$c_A(u) = \frac{\xi}{2} + \sqrt{\frac{\xi^2}{4} + K}, \tag{5.54}$$

and analogously for c_B. Using this relation Eq. (5.52) together with Eq. (5.53) gives the following equation for the reaction rate:

$$r(\mathbf{x},t) = \frac{1}{\phi} \frac{d^2 c_A(\xi)}{d\xi^2} \left[\nabla \xi(\mathbf{x},t) \cdot \mathbf{D}(\mathbf{x}) \nabla \xi(\mathbf{x},t) \right], \tag{5.55}$$

The expression in the square brackets is identical to the scalar dissipation rate, see Eq. (5.38), which measures the rate of mixing of dissolved substances. This expression illustrates the direct dependence of the reaction rate at high Da with the mixing rate as quantified by the scalar dissipation rate. The impact of medium heterogeneity on the reaction rate has been investigated using the PDF of the conserved components $\xi(\mathbf{x}, t)$, which can be mapped directly on the PDF of the species concentrations via Eq. (5.54) [108, 109].

5.4 Multiphase and Surface Processes

In the previous sections, we presented the challenges related to the dispersion and the upscaling of bulk reactions in the fluid. These, when the Fickian assumption is not valid, can be conveniently approximated with multi-continuum models. This is not dissimilar from what is obtained in proper multiphase systems. For example, *conjugate heat transfer* problems, that would require the coupled solution of heat transfer in the fluid and solid region, can be averaged to obtain two-phase formulations with appropriate transfer terms. However, these transfer terms are local (in time and space) and linear only under local equilibrium conditions in both phases. An alternative approach is to perform the upscaling in two steps. First the phase which relaxes faster to equilibrium is approximated macroscopically with an equivalent heat transfer coefficient (possibly non-constant) at the solid boundary.

This first upscaling reduces the dimensionality of the fast dynamics in a sub-domain, effectively representing it as a surface process, leaving us with the task of averaging transport in one phase only with complex boundary conditions.

Among these surface processes we can identify the following three categories:

- *Dynamic conditions:* One of the most general surface process is the one encountered in adsorption–desorption processes that can be modelled with the following boundary conditions:

$$\mathbf{j}\cdot\mathbf{n} = (\mathbf{u} - D_m\nabla)\, c\cdot\mathbf{n} = b\,(c, s)\,,$$

  where $\mathbf{j}$ is the flux, b represents the adsorption/desorption/transfer processes, r is a source/sink describing chemical reactions and $s = s(\mathbf{x}, t)$ is the adsorbed concentration on the surface. This has been solved separately with a surface ordinary differential equation

$$\frac{\partial s}{\partial t} = -b\,(c, s)\,.$$

- *Mixed conditions:* When the time scale defined by b is fast enough, one can explicitly find $s = s(c)$ such that $b(c, s) = 0$. This means that the above conditions simplify to a, possibly non-linear, mixed boundary condition of the type

$$\mathbf{j}\cdot\mathbf{n} = (\mathbf{u} - D_m\nabla)\, c\cdot\mathbf{n} = b\left(c, s\,(c)\right) = f(c).$$

  Linearising f we obtain a mixed (Robin) boundary condition

$$\mathbf{j}\cdot\mathbf{n} = f_0 + f'c. \tag{5.56}$$

- *Simple conditions:* Assuming infinitely slow adsorption/reaction process f, the condition above reduces to a fixed flux (Neumann) condition

$$\mathbf{j}\cdot\mathbf{n} = f_0,$$

  while, in the opposite case of infinitely fast surface process, we can retrieve a simple fixed concentration (Dirichlet) condition.

Clearly, the validity of these increasingly simplifying assumptions has to be verified case by case, and we consider this as part of the upscaling process. In all three cases, however, standard upscaling techniques (such as volume-averaging and homogenisation) can be applied only when the surface process is slow compared to advection and diffusion. When instead this is not the case, more advanced techniques have to be used [110, 111]. This is similar to the case of diffusion-limited bulk reactions (see above) that generally results in upscaled dispersion and velocity coefficients significantly different from the non-reactive case.

5.4.1 *Mass and Heat Transfer*

The transport and deposition of particles in porous media are fundamental multiscale phenomena present in a number of natural and engineered processes. Although we refer here to the transport of physical particles, the discussion below conceptually applies, with minor modifications, also to heat transfer mechanisms.

The classical theoretical framework typically used is the *colloid filtration theory* but, more in general, advances in this field generally belong to the so-called *soft matter* physics. Several complex physical mechanism, in fact, can arise due to the complex particle–particle and particle–wall interaction. In this chapter, however, we focus on a simple advection–diffusion description, in the dilute limit, with negligible Stokes and Reynolds numbers. Therefore most hydrodynamical interactions (sometimes called hydrodynamic retardation effects) between the particles and the surface of the solid grains, and the DLVO interactions happen at a very small scale and are therefore taken into account only at the boundaries, by modified boundary conditions. This is known as the Smoluchowski–Levich approximation that results in the molecular diffusion coefficient D_m being constant, obtained, for example, via the Stokes–Einstein relation for diffusion of spheres in liquids. For larger particles instead, one should consider many other effects that act possibly also far from the wall, such as modified suspension viscosity, lift forces arising in small Reynolds number flows, the Faxen correction, due to the perturbed flow around the particles, and possibly also particle rotation, particle collisions, etc. Although some of these additional physics can easily be included in the upscaling, we limit ourselves to the effect of a surface process in the upscaling, namely in the equivalent macroscopic dispersion and reaction coefficients.

In the dimensional analysis of mass transfer phenomena, the most used dimensionless quantity is the Sherwood number, describing the ratio between convective mass transfer and diffusive transport, which is the analogue of the Nusselt number used in heat transfer. It is defined as:

$$\mathrm{Sh} = \frac{hL}{D_m}, \tag{5.57}$$

where L is a characteristic length (in porous media application generally taken to be equal to the pore or grain size), and D_m is the molecular diffusion coefficient. The mass transfer coefficient h is defined as the molar flux through the surface per unit surface, normalised by ΔC the concentration driving force. This representation implicitly considers the mass transfer as an equivalent diffusion process through the surface. If this is more often true at the micro-scale, after the upscaling, this effective transfer coefficient scaled by the diffusion strongly depends on the other transport mechanisms, such as advection, gravity and all the other external forces.

In the earlier studies of filtration, the most common upscaling approach was to determine a single parameter describing the filter effectiveness: this is obtained from its features and the operating conditions under investigation, and is defined the *collector efficiency* η_D [112]. This efficiency coefficient is then assumed to be the product of two separated effects: the so-called attachment efficiency α, describing the probability of a particle that has reached the solid grain to be adsorbed, and the purely fluid-dynamical term η_0 to model the transport the bulk of the fluid to the surface of the grains. The latter is the often decomposed as a sum of different contributions due to, for example, Brownian diffusion, steric interception and inertial (and gravitational) effects. Some early works [113] analytically obtained, for Sh in idealised geometries, expressions such as

$$\mathrm{Sh} = As^{\frac{1}{3}}\mathrm{Pe}^{\frac{1}{3}}\,, \quad \eta = 4As^{\frac{1}{3}}\mathrm{Pe}^{-\frac{2}{3}},$$

where As is a parameter depending on the porous medium porosity ϕ. Many other such relationships are available connecting the system features, in terms of geometrical features and fluid dynamic conditions, to an approximate particle deposition efficiency η_D [114]. There are a number of issues facing these models; first of all they are most often based on a single idealised geometrical model representing the porous medium, thus failing to grasp the pore-scale complexities and heterogeneity and its effect on particle filtration. Another conceptual hurdle in the application of these models is the difficult translation of the obtained efficiency parameter η_D into an effective macro-scale reaction term employable in a macroscopic transport equation [115], and to understand its dependence on the flow parameters and its inseparable connection with the effective dispersion and velocity.

From Surface Processes to Averaged Reaction Rates

We follow here [115], showing how to obtain a stationary effective reaction rate for a periodic geometry and its dependence on the (reactive) boundary condition and flow parameters, for arbitrary reaction/deposition regimes. To this aim, we consider Eq. (5.12) and we assume that the detailed surface processes can be averaged in a small boundary layer around the solid matrix and approximated with a generic effective linearised mixed boundary condition (see Eq. (5.56)) of the type:

$$-D_m \nabla c \cdot \mathbf{n} = -r\frac{\alpha}{\alpha - 1}c + r_0 \quad \text{on}\,\Gamma, \tag{5.58}$$

where α is the deposition/attachment efficiency, Γ is the porous matrix surface area, r is the surface transfer coefficient and r_0 is a constant surface flux. When $\alpha = 1$, Eq. (5.58) is equivalent to a Dirichlet conditions $c = 0$ on the solid grains (perfect sink). In a particle-based Lagrangian framework, such as the one considered when performing random walk simulations, this efficiency α can be interpreted as related to the probability of a single colloidal particle of attaching to the collector surface upon collision [116].

Applying a simple volume-average over a fixed REV Ω, defined as $\bar{\cdot} = \frac{1}{V}\int \cdot\, dv$ (with V being the total volume), the divergence theorem and the boundary condition, Eq. (5.58), we obtain

$$\frac{\partial \overline{c}}{\partial t} + \frac{1}{V}\int_{\partial\Omega} (c\,\mathbf{u}\cdot\mathbf{n} - D_m \nabla c\cdot\mathbf{n})\,\mathrm{d}s = -\frac{1}{V}\int_{\Gamma} r\frac{\alpha}{\alpha-1} c\,\mathrm{d}s + r_0. \tag{5.59}$$

Considering a box with periodic boundary conditions on $y-$ and $z-$directions, and no accumulation (stationary, local equilibrium hypothesis), it is possible to identify the second surface integral on the LHS of Eq. (5.59) with the total flux F through the $x-$boundaries of the domain (inlet and outlet in this case). Being the starting equation linear, it is reasonable to assume the above quantity (the average mass flux) to be a linear function of the average concentration, with a macroscopic *effective reaction rate* R defined as:

$$R = \frac{F - r_0}{\overline{c}V}. \tag{5.60}$$

This quantity is simply computable from a micro-scale simulation on the periodic cell, simply looking at inlet–outlet fluxes and averaged volume concentration, even in the case of perfect sink ($r \to \infty$) condition. Assuming a Fickian macroscopic dispersion (see sections above), we can therefore postulate a closed-form for the one-dimensional macroscopic advection–diffusion–reaction equation for $c = (\overline{c}/c_\infty)$, in dimensionless form

$$\frac{\partial C}{\partial t_{\mathrm{diff}}} + \varepsilon \mathrm{Pe}\frac{\partial C}{\partial X} - \frac{\mathbf{D}}{D_m}\frac{\partial^2 C}{\partial X^2} = -\mathrm{Da}DC + \mathrm{Da}_0, \tag{5.61}$$

where X represents the (dimensionless) macroscopic space variable and we have defined $t_{\mathrm{diff}} = \frac{tD_m}{L^2}$, and the Damköhler numbers as $\mathrm{Da} = \frac{R_s L^2}{D_m}$, and $\mathrm{Da}_0 = \frac{r_0 L}{q}$.

For a periodic FCC packing [115] we obtain the following qualitative upscaling law for the long-time effective macroscopic reaction rate, as a function of the microscopic Damköhler number $\mathrm{Da}_m = \frac{rL^2}{D}$:

$$\mathrm{Da} = \begin{cases} K_1(\mathrm{Da}_m) & \text{for} \quad \mathrm{Pe} \lesssim 10 \\ K_2(\mathrm{Da}_m)\mathrm{Pe}^{0.15} & \text{for} \quad \mathrm{Pe} \gtrsim 10, \mathrm{Da}_m \gtrsim 1 \\ K_3(\mathrm{Da}_m) & \text{for} \quad \mathrm{Da}_m \lesssim 1 \end{cases}$$

with constants K_i. This qualitative behaviour is universal, although the exponent 0.15 and the constants could possibly depend on the specific geometry. The dependence of Da with respect to Da_m, on the other hand, is linear (independent of Pe) for slow surface processes while, for infinitely fast processes, it saturates to a constant (that depends on Pe). This is the typical behaviour for reactions happening on a localised lower-dimensional manifold where mixing can totally control the reaction.

5.5 Conclusions

In the previous sections we highlighted the most important physical models and macroscopic equations that can be relevant in porous and heterogeneous materials. We identified several assumptions and limitations of the upscaling processes.

Non-Equilibrium and Lack of Scale Separation

When no upscaling is available that can decouple the physical scales or when transition out-of-equilibrium effects are important at the micro-scale, alternative techniques can be used such as numerical multiscale approaches such as multiscale FEM, variational, heterogeneous or hybrid multiscale methods. These methods allow for a generic system to be solved efficiently explicitly accounting for some micro-scale information. They usually rely on a pre-processing offline step (similarly to the cell-problems in classical upscaling) or on an online dual-resolution computational approach.

Suspensions and Interfacial Flows

Multiphase flows such as suspensions and interfacial flows can be upscaled with the approaches described above, only under local equilibrium, and when the forces acting on each phase are relatively small. The presence of complex momentum transfer (in the case of suspensions) and strong localised forces, such as surface tension (in the case of interfacial flows), makes the standard upscaling inadequate. An intuitive interpretation of this inadequacy is the fact that volumetric/ensemble averages cannot properly represent interfacial forces and configuration-dependent forces. Time-dependent, non-linear, non-local and memory effects can arise at the macro-scale.

Appendix A: Homogenisation and Two-Scale Expansions

In this section we briefly sketch the main ideas and steps in the formal derivation of macroscopic equations using the method of periodic homogenisation with two-scale asymptotic expansion. The method has been initially proposed in theoretical mechanics for the study of composite materials and subsequently extensively studied in mathematical analysis, for elliptic (diffusion) operators and variational problems [117–121]. Quite interestingly, the first results in the area were developed in parallel both for periodic and random (stationary ergodic) media. Being the underlying mathematical techniques significantly different, the latter is also called stochastic homogenisation and has lately seen important developments [122–124], making it, in some cases, a more realistic and conceptually deeper alternative to the former. While the formal steps of periodic homogenisation are easily accessible with a basic background of asymptotic methods and PDEs, the mathematical proofs concerning the existence of a macroscopic limit (and the convergence towards it, [125]) require extensive knowledge of functional analysis and PDE theory and a much more complex notation, clearly beyond the scope of this chapter. For porous media applications we refer to [1] for a comprehensive, yet accessible, treatise.

We can identify the following steps that are usually needed to perform the two-scale expansion and derive a homogenised equation:

1. Define the scale separation parameter $\epsilon = \frac{\ell}{L}$, where L is a characteristic macroscopic length (e.g., the total domain size) and ℓ is a microscopic length (e.g., the size of the pores). While mathematical homogenisation theory deals with the existence, uniqueness, form and convergence rate of the problem when $\epsilon \to 0$, the more applied approach is to guess such a limit exist and constructively find its form and validity (measured in terms of corrections of the order of ϵ).
2. Put the (microscopic) equations in dimensionless form, to highlight and recognise the *scaling*, which is the crucial assumption for the validity of what follows. This is not strictly necessary but one of the advantages of homogenisation theory is, in fact, the clear identification of validity regimes of the macroscopic equation.
3. Write down spatial (and possibly temporal) coordinates, derivatives and fields as an asymptotic expansion in terms of ϵ. Neglecting time, given a spatial coordinate $\mathbf{x} \in \Omega \in [0, L]^d$ and a field $c = c(\mathbf{x})$, a *two-scale* expansion is performed as follows:

$$\mathbf{x} = \mathbf{x}_0 + \epsilon \mathbf{x}_1 \tag{5.62a}$$

$$c(\mathbf{x}) \approx c_0(\mathbf{x}) + \epsilon c_1(\mathbf{x}) + \epsilon^2 c_2(\mathbf{x}) = c_0(\mathbf{x}_0, \mathbf{x}_1) + \epsilon c_1(\mathbf{x}_0, \mathbf{x}_1) + \epsilon^2 c_2(\mathbf{x}_0, \mathbf{x}_1) \tag{5.62b}$$

$$\nabla c = \left(\nabla_{\mathbf{x}_0} + \frac{1}{\epsilon}\nabla_{\mathbf{x}_1}\right) c \approx \frac{1}{\epsilon}\nabla_{\mathbf{x}_1} c_0 + \left(\nabla_{\mathbf{x}_0} c_0 + \nabla_{\mathbf{x}_1} c_1\right) + \epsilon\left(\nabla_{\mathbf{x}_0} c_1 + \nabla_{\mathbf{x}_1} c_2\right), \tag{5.62c}$$

   where $\mathbf{x}_0 \in \Omega \in [0, L]^d$ defines the macro-scale and $\mathbf{x}_1$ the (periodic with period $[0, \ell]^d$) micro-scale. While this expansion is formally always valid, its usefulness, as we will further explain below, relies on a wide *separation of scales* in the domain which is not always satisfied in real porous media.
4. Starting from the lowest power of ϵ, hierarchically define and solve (when a solution is trivially found) the cascade of equations for $c_0, c_1, \ldots$. Usually it is enough to solve for the first two terms[2] to get a closed-form and computable parameters of the macroscopic limit (c_0). However, to achieve that, additional assumptions have to be made to close the system of equations. In particular, a simplified dependence of c_1 on the macro-scale c_0 has to be assumed (when it is not obtained formally), guessing a separable structure of the type:

[2]With a total of three equations for three lowest powers of ϵ, in the case second-order PDEs.

$$c_1(\mathbf{x}_0, \mathbf{x}_1) = \mathbf{w} \cdot \nabla_{\mathbf{x}_0} c_0 + \overline{c_1}, \tag{5.63}$$

where the vectorial field $\mathbf{w} = \mathbf{w}(\mathbf{x}_1)$, periodic and with zero average, $\overline{\mathbf{w}} = \mathbf{0}$, is also called *corrector* and can be found by solving the so-called *cell-problem*, usually derived, after some manipulation, from the first correction equation for c_1. This decomposition c_1 well represents the scale-separation hypothesis.

5. Depending on the problem, also an average on c_0 (or, when c_0 is constant on the micro-scale, on $c_0 + \epsilon c_1$) might be defined to get rid of the micro-scale dependence, defining a new macroscopic field

$$\overline{c_0}(\mathbf{x}_0) = \overline{c_0(\mathbf{x}_0, \mathbf{x}_1)} = \int_{[0,\ell]^d} c_0(\mathbf{x}_0, \mathbf{x}_1) d\mathbf{x}_1$$

and a macroscopic evolution equation for $\overline{c_0}$ (or $\overline{c_0 + \epsilon c_1}$) can be derived from the terms of order ϵ^0.

Example: Reaction–Diffusion in a Perforated Domain

It is important to notice that the steps outlined above are a *constructive formal* approach for homogenisation that is attractive (at least as a first approach) for new problems but does not cover the large number of possible problems for which a homogenised limit exists. This is much less general than, for example, variational approaches in which a *given* macroscopic limit is guessed and proven to be the correct limit. As an example, we illustrate the steps above for the simple case of pure diffusion with slow superficial reaction (or equivalently, heat transfer with a prescribed heat flux at the boundary), i.e.,

$$\nabla \cdot \big(D \nabla c(\mathbf{x})\big) = f(\mathbf{x}) \tag{5.64}$$

with constant diffusion[3] coefficient D, and a generic space-dependent source/sink term f. The equation is defined on a perforated (porous) domain $\Omega \in [0, L]^d$ in d-dimensions, with periodic microstructure (cell) $\Omega_\epsilon \in [0, \ell]^d$. For simplicity we assume also periodicity in Ω but this can be easily replaced with any other simple (macroscopic) boundary conditions. On the internal boundaries (the porous matrix) we impose a linear mixed (normal flux) condition:

$$D \nabla_{\mathbf{n}} c = kc + r, \tag{5.65}$$

[3]Traditionally, homogenisation is performed on continuous domains with space-dependent oscillating diffusion coefficient. However, the case studied here, more relevant to porous media applications, can be interpreted as a limiting case in which the diffusion coefficient tends to a patch-wise constant. This is what is done practically when solving porous media problems with immersed boundaries, penalisation or diffuse domain methods.

where k is a superficial reaction term (or heat transfer coefficient), and r is a constant superficial source (or sink) term. Despite its simplicity, this includes already important applications such as linear isotherm adsorption or heat transfer. The extension to the mass or heat transfer through an interface within two domain or two phases, instead of a boundary condition, is a possible extension which is discussed in Sect. 5.4.

Inserting the expansions Eq. (5.62) into Eq. (5.64), yields

$$\nabla_{\mathbf{x}_0} \cdot \left[D \left(\nabla_{\mathbf{x}_0} + \frac{1}{\epsilon} \nabla_{\mathbf{x}_1} \right) (c_0 + \epsilon c_1) \right] +$$
$$+ \frac{1}{\epsilon} \nabla_{\mathbf{x}_1} \cdot \left[D \left(\nabla_{\mathbf{x}_0} + \frac{1}{\epsilon} \nabla_{\mathbf{x}_1} \right) (c_0 + \epsilon c_1) \right] =$$
$$= f_0(\mathbf{x}_0, \mathbf{x}_1) + \epsilon f_1(\mathbf{x}_0, \mathbf{x}_1),$$

while the expansion of the boundary condition can be rewritten as

$$D \left(\mathbf{n} \cdot \nabla_{\mathbf{x}_0} + \frac{1}{\epsilon} \mathbf{n} \cdot \nabla_{\mathbf{x}_1} \right) (c_0 + \epsilon c_1) = k c_0 + \epsilon k c_1 + r.$$

At this point, since we have not put the equation in dimensionless form, it is important to identify the regime of interest. We will focus here on the simplest regime for which the homogenisation approach outlined above works seamlessly. This is the case when all coefficients (D, f) are of the same order, namely of order one, and the boundary coefficients $(k = \epsilon k_1, r = \epsilon r_1)$ are of order ϵ^1.

Collecting now terms with equal power of ϵ, and taking into account the assumptions on the coefficients, the only term of order ϵ^{-2} leads to the linear homogeneous equation

$$\nabla_{\mathbf{x}_1} \cdot \left(D \nabla_{\mathbf{x}_1} c_0 \right) = 0$$

that turns out to be a simple (linear homogeneous) equation for the variable c_0 at the micro-scale, i.e., in each periodic cell Ω_ϵ (since only derivatives with respect to $\mathbf{x}_1$ appear). The largest terms in the boundary condition, i.e., a simple no-flux (homogeneous Neumann) condition, are $\mathbf{n} \cdot \nabla_{\mathbf{x}_1} c_0 = 0$ on the internal solid boundaries, with periodic boundary conditions on the external boundaries. This equation is trivially satisfied by a constant, i.e., a function $c_0 = c_0(\mathbf{x}_0)$ which is a function of $\mathbf{x}_0$ only.

For the terms of order ϵ^1, we obtain the following equation:

$$\nabla_{\mathbf{x}_0} \cdot \left(D \nabla_{\mathbf{x}_1} c_0 \right) + \nabla_{\mathbf{x}_1} \cdot \left(D \nabla_{\mathbf{x}_0} c_0 \right) + \nabla_{\mathbf{x}_1} \cdot \left(D \nabla_{\mathbf{x}_1} c_1 \right) = 0.$$

Given the conclusion above, i.e., c_0 constant at the micro-scale, the first term disappears and, replacing the assumption Eq. (5.63),[4] we obtain the equation

$$\nabla_{\mathbf{x}_1} \cdot \left(D\nabla_{\mathbf{x}_1} \left(\mathbf{w} \cdot \nabla_{\mathbf{x}_0} c_0 \right) + D\nabla_{\mathbf{x}_0} c_0 \right) = 0$$

or, since this has to hold for an arbitrary $\nabla_{\mathbf{x}_0} c_0$, equivalently written as a vectorial *cell-problem* for the corrector $\mathbf{w}$

$$\nabla_{\mathbf{x}_1} \cdot D \left(\nabla_{\mathbf{x}_1} \mathbf{w} + I \right) = \mathbf{0} \tag{5.66}$$

with I being the identity matrix, with boundary condition

$$D\mathbf{n} \cdot \left(\nabla_{\mathbf{x}_1} \left(\mathbf{w} \cdot \nabla_{\mathbf{x}_0} c_0 \right) + \nabla_{\mathbf{x}_0} c_0 \right) = D\mathbf{n} \cdot \left(\nabla_{\mathbf{x}_1} \mathbf{w} + I \right) \cdot \nabla_{\mathbf{x}_0} c_0 = 0$$

or, equivalently, in vectorial form,

$$D\mathbf{n} \cdot \left(\nabla_{\mathbf{x}_1} \mathbf{w} + I \right) = \mathbf{0}. \tag{5.67}$$

The next scale, ϵ^0, reads

$$\nabla_{\mathbf{x}_0} \cdot \left(D \left(\nabla_{\mathbf{x}_0} c_0 + \nabla_{\mathbf{x}_1} c_1 \right) \right) + \nabla_{\mathbf{x}_1} \cdot \left(D \left(\nabla_{\mathbf{x}_0} c_1 + \nabla_{\mathbf{x}_1} c_2 \right) \right) = f_0(\mathbf{x}_0, \mathbf{x}_1)$$

that, by using Eq. (5.63), the conclusions obtained above and the boundary conditions

$$D\mathbf{n} \cdot \left(\nabla_{\mathbf{x}_0} c_1 + \nabla_{\mathbf{x}_1} c_2 \right) = k_1 c_0 + r_1$$

can be averaged over the periodic cell to obtain the *effective* equation

$$\nabla_{\mathbf{x}_0} \cdot \left(\mathbf{D} \nabla_{\mathbf{x}_0} \overline{c_0} \right) = \overline{f_0(\mathbf{x}_0, \mathbf{x}_1)} - \frac{\alpha}{\phi} \left(k_1 \overline{c_0}^{\Gamma} + r_1 \right), \tag{5.68}$$

where α is the specific surface area, ϕ is the porosity of the porous material, $\mathbf{D}$ is the (tensorial anisotropic) effective diffusion coefficient $\mathbf{D} = \overline{DI + \nabla_{\mathbf{x}_1} \mathbf{w}}$ that is computed from the cell-problem defined above. This gives us a macroscopic governing equation for $\overline{c_0 + \epsilon c_1}$ (which, in this case, is equivalent to c_0, being it constant over $\mathbf{x}_1$, and c_1 with mean zero), where surface terms integrated over the surface ($\overline{c_0}^{\Gamma}$ is, in fact, a surface average which, in this case, is again equivalent to c_0) appear in the right-hand side, together with f_0, as bulk reaction terms.

[4]Which, in this case, is a unique and exact decomposition since this equation is defined up to an additive constant, $\overline{c_1}$, and a multiplicative constant, $\nabla_{\mathbf{x}_0} c_0$.

Physical Interpretation and Limitations

With respect to other upscaling techniques, the steps performed above do not to have a direct physical motivation and, therefore, it might be hard to understand and assess the validity of the underlying assumptions. However there are a few considerations that can be made:

- Despite the apparent complexity, the cell-problem Eqs. (5.66) and (5.67) has a clear physical interpretation, being each component of $\mathbf{w}$ a harmonic function (i.e., a solution of the Laplace equation, while in other cases it would be the solution of the underlying microscopic physics), with periodic boundary conditions and a fixed gradient at the boundary.[5] These periodic cell-problems can often be shown to be equivalent to more intuitive closure problem, similar to the ones obtained in volume-averaging, where a concentration gradient is imposed at the external boundaries, respectively, in the x-, y- or z-direction. The periodicity, however, enforces also that a fully developed concentration profile (and not just a constant) is obtained while, as we will discuss in the next section, more general closure problems can be obtained by volume-averaging, with arbitrary boundary conditions.
- Two often overlooked assumptions for homogenisation, possibly more important than the separation of scales or the periodicity of the geometry (which can be simply assumed as a first-order approximation) are the *local equilibrium* and *stationarity* conditions, implicitly assumed when imposing periodicity of all fields. This is particularly relevant for processes with long pre-asymptotic transition times (lengths) to equilibrium. In those cases, homogenisation is only able to retrieve the asymptotic stationary macroscopic model which might be significantly inaccurate for studying the initial times of processes like mixing. Even if, in principle, homogenisation can include explicitly the time-dependence both in the macro- and microscales, the periodic assumption implicitly defines steady *closure* (cell) problems. Generalisations of the homogenisation techniques to overcome these assumptions have been recently proposed [126, 127].
- Another limitation is the particular *scaling* chosen which, on the one hand, clearly states the validity of the model, but, on the other hand, can be very restrictive. This aspect is also often overlooked when, for example, non-linear models, multiphase or fast reactions are considered. In those cases, additional terms arise in the leading orders that, only in some cases, can be tamed by ad hoc homogenisation techniques.
- Homogenisation should be thought of as a rigorous and robust analysis of the multiscale features of the model (convergence properties to the macroscopic solution, identification of validity regimes and approximation errors) and not

---

[5] Also a source term on the right-hand side would appear to counterbalance advection or faster reaction and obtain a periodic solution.

only as generic way of discovering emerging macroscopic equations. For more complex models, it is usually advisable to first tackle the problem with more phenomenological macroscopic models based, for example, on mixture theory and variation principles, using conservation laws and the second law of thermodynamics. Complementarily, volume-averaged computational results can be used to compute the parameters of these models, while homogenisation gives directly explicit formulas for upscaled parameters. From the practical point of view, both approaches require the solution a microscopic *cell/closure* problem.

Appendix B: Volume/Ensemble Averaging

An upscaling approach, developed extensively for porous media in the last 30 years [2, 128–130], is the theory of volume-averaging with its many variants. This is conceptually equivalent to the techniques more commonly used in fluid dynamics, turbulence, combustion and multiphase flows [82, 131, 132], such as the large eddy simulation. Closer instead to the Reynolds averaging in the RANS equations is instead the concept of ensemble averaging used predominantly in a surprisingly abundant amount of theoretical work in stochastic hydrology [11, 12, 133, 134]. These approaches have been separately developed by the porous media and fluid dynamics communities without significant connections. While it is out of the scope of this chapter to explain any of these approaches in detail, we offer here a very brief introduction and some general comments about the applicability, similarities, differences and meaning of these methodologies. The interested reader is referred to the previously cited works and to the review [135].

Both spatial averaging and ensemble/perturbation methods rely on the following steps:

- Write the heterogeneous coefficients of the governing equation (e.g., velocity field, diffusion/dispersion coefficient) with a simple decomposition into a mean and fluctuating term, which for a generic quantity g reads

$$g(\mathbf{x}) = \int_{\Omega} w(\mathbf{y}) g(\mathbf{x}+\mathbf{y}) \mathrm{d}\mathbf{y} + \tilde{g}(\mathbf{x}) = \overline{g}(\mathbf{x}) + \tilde{g}(\mathbf{x})$$

for volume-averaging (better denoted, in this formulation, as generalised volume-averaging or *space convolution filter*), where $w(\mathbf{x})$ is function with a compact and localised support such that $\int_{\Omega} w(\mathbf{y})\mathrm{d}\mathbf{y} = 1$. The size of the support, ℓ, which defines the averaging length scale, should be related to the REV size (when the REV exists). Alternatively, for ensemble averaging, assuming that now g is a random field, we can similarly define

$$g(\mathbf{x}, \omega) = \int_{\omega} g(\mathbf{x}, \omega) \mathrm{d}\mu(\mathbf{x}, \omega) + g'(\mathbf{x}) = \langle g \rangle (\mathbf{x}) + g'(\mathbf{x}),$$

where $\mu(\mathbf{x}, \omega)$ is the probability associated with the random field measure, ω is a random event. While the first averaging is well-defined both for pore-scale (i.e., perforated) and macro-scale (i.e., continuous) domains, the second one strictly applies only to macro-scale (e.g., permeability) heterogeneities, although extensions are possible.

- Applying the averaging operator to a generic transport equation for the field c, an unclosed equation for the mean (which we denote here generally with $\overline{c}$ for both averages) is found, where the unclosed terms are either non-linear terms (for both volume and ensemble averages) or boundary terms (for volume-averaging). In fact, while in standard single-phase turbulence modelling, the averaging volume only contains fluid, in porous media the so-called *spatial averaging theorem* applies

$$\overline{\nabla c} = \nabla \overline{c} + \frac{\alpha}{\varepsilon} \overline{c\mathbf{n}}^{\Gamma},$$

where, as discussed in the homogenisation section, $\overline{\cdot}^{\Gamma}$ denotes an average over the fluid–solid interface, and $\mathbf{n}$ the normal to the surface. A generalisation is possible when dealing with generic multiphase systems whose interfaces between phases can be mobile. In that case, a more complex treatment has been used to derive upscaled equations for all phases [2, 131, 136].

- To obtain a first-order upscaled model, the unclosed equation for $\overline{c}$ is closed with a combination of geometrical[6] and physical/phenomenological arguments, and by the solution of the *closure* problem for the fluctuations $\tilde{g}$. The latter can be easily obtained by subtracting the equation for $\overline{g}$ from the initial equation. This, however, often contains other unclosed terms that could be written in terms of higher order moments of coefficients, e.g., g, and variables, e.g., c.
- Higher order approximations can be obtained either by defining new problems for the higher order moments or, exploiting the linearity (or linearisation) of the transport equation, in defining a new sequence of problems to find an approximation for c of the type:

$$c = c_0 + c_1 + c_2 + \dots,$$

where, for example, c_0 is obtained by neglecting all unclosed terms in the equation for $\overline{c}$.

[6]Here is where the hypotheses on the porous media structure are introduced, through estimates of the (tensorial) spatial moments $\overline{\mathbf{n}\mathbf{y}^j}^{\Gamma}$ with $j = 0, 1, 2, \dots$ denoting the order of the tensorial product and $\mathbf{y}$ being the spatial coordinate.

References

1. U. Hornung, *Homogenization and Porous Media*, vol. 6 (Springer Science & Business Media, Berlin, 2012)
2. S. Whitaker, *The Method of Volume Averaging*, vol. 13 (Springer Science & Business Media, Berlin, 1998)
3. S.M. Hassanizadeh, W. Gray, High velocity flow in porous media. Transp. Porous Media **2**(6), 521 (1987)
4. D. Lasseux, A.A. Abbasian Arani, A. Ahmadi, On the stationary macroscopic inertial effects for one phase flow in ordered and disordered porous media. Phys. Fluids **23**(7), 73103 (2011)
5. E. Skjetne, J.L. Auriault, High-velocity laminar and turbulent flow in porous media, Transp. Porous Media **36**(2), 131 (1999)
6. C.R. Dudgeon, An experimental study of the flow of water through coarse granular media. La Houille Blanche **7**, 785 (1966)
7. P. Angot, C.H. Bruneau, P. Fabrie, A penalization method to take into account obstacles in incompressible viscous flows. Numer. Math. **81**(4), 497 (1999)
8. R.E. Hayes, A. Afacan, B. Boulanger, A.V. Shenoy, Modelling the flow of power law fluids in a packed bed using a volume-averaged equation of motion. Transp. Porous Media **23**(2), 175 (1996)
9. T. Tosco, D. Marchisio, F. Lince, R. Sethi, Extension of the Darcy-Forchheimer law for shear-thinning fluids and validation via pore-scale flow simulations. Transp. Porous Media **96**(1), 1 (2013)
10. V.I. Arnol'd, On the topology of three-dimensional steady flows of an ideal fluid. J. Appl. Math. Mech. **30**, 223 (1966)
11. L.W. Gelhar, *Stochastic Subsurface Hydrology* (Prentice-Hall, Upper Saddle River, 1993)
12. G. Dagan, *Flow and Transport in Porous Formations* (Springer Science & Business Media, Berlin, 2012)
13. Y. Rubin, *Applied Stochastic Hydrogeology* (Oxford University Press, New York, 2003)
14. P. Renard, G. de Marsily, Calculating equivalent permeability: a review. Adv. Water Resour. **20**, 253 (1997)
15. X. Sanchez-Vila, A. Guadagnini, J. Carrera, Representative hydraulic conductivities in saturated groundwater flows. Rev. Geophys. **44**, RG3002 (2006)
16. J.B. Keller, A theorem on the conductivity of a composite medium. J. Math. Phys. **5**, 548 (1964)
17. D.S. Dean, I.T. Drummond, R.R. Horgan, Effective transport properties for diffusion in random media. J. Stat. Mech. **7**, P07013 (2007)
18. G. Matheron, Composition des perméabilités en milieu poreux héterogène. Méthode de Schwydler et règles de pondération, Rev. l'Institute Français du Pet. **Mars**, 443 (1967)
19. A.L. Gutjahr, L.W. Gelhar, A.A. Bakr, J.R. MacMillan, Stochastic analysis of spatial variability in subsurface flows 2. Evaluation and applications. Water Resour. Res. **14**, 953 (1978)
20. L.W. Gelhar, C.L. Axness, Three-dimensional stochastic analysis of macrodispersion in aquifers. Water Resour. Res. **19**(1), 161 (1983)
21. A.E. Scheidegger, General theory of dispersion in porous media. J. Geophys. Res. **66**, 3273 (1961)
22. J. Bear, *Dynamics of Fluids in Porous Media* (American Elsevier, New York, 1972)
23. H. Risken, *The Fokker-Planck Equation* (Springer, Heidelberg, 1996)
24. B. Noetinger, D. Roubinet, A. Russian, T. Le Borgne, F. Delay, M. Dentz, J.R. De Dreuzy, P. Gouze, Random walk methods for modeling hydrodynamic transport in porous and fractured media from pore to reservoir scale. Transp. Porous Media, 1–41 (2016)
25. P.G. Saffman, A theory of dispersion in a porous medium. J. Fluid Mech. **6**(03), 321 (1959)
26. B. Bijeljic, M.J. Blunt, Pore-scale modeling and continuous time random walk analysis of dispersion in porous media. Water Resour. Res. **42**, W01202 (2006)

27. M. Icardi, G. Boccardo, D.L. Marchisio, T. Tosco, R. Sethi, Pore-scale simulation of fluid flow and solute dispersion in three-dimensional porous media. Phys. Rev. E **90**(1), 13032 (2014)
28. B. Bijeljic, M.J. Blunt, Pore-scale modeling of transverse dispersion in porous media. Water Resour. Res. **43**, W12S11 (2007)
29. S. Attinger, M. Dentz, W. Kinzelbach, Exact transverse macro dispersion coefficient for transport in heterogeneous media. Stoch. Env. Res. Risk A. **18**, 9 (2004)
30. L.W. Gelhar, C. Welty, K.R. Rehfeldt, A critical review of data on field-scale dispersion in aquifers. Water Resour. Res. **28**(7), 1955 (1992)
31. A. Beaudoin, J.R. Dreuzy, Numerical assessment of 3-D macrodispersion in heterogeneous porous media. Water Resour. Res. **49**, 2489 (2013)
32. B. Bijeljic, P. Mostaghimi, M.J. Blunt, Signature of non-Fickian solute transport in complex heterogeneous porous media. Phys. Rev. Lett. **107**(20), 204502 (2011)
33. V.L. Morales, M. Dentz, M. Willmann, M. Holzner, Stochastic dynamics of intermittent pore-scale particle motion in three-dimensional porous media: Experiments and theory. Geophys. Res. Lett. **44**, 9361 (2017)
34. E. Crevacore, T. Tosco, R. Sethi, G. Boccardo, D.L. Marchisio, Recirculation zones induce non-Fickian transport in three-dimensional periodic porous media. Phys. Rev. E **94**(5) (2016)
35. R. Haggerty, S.A. McKenna, L.C. Meigs, On the late time behavior of tracer test breakthrough curves. Water Resour. Res. **36**(12), 3467 (2000)
36. P.K. Kang, T. Le Borgne, M. Dentz, O. Bour, R. Juanes, Impact of velocity correlation and distribution on transport in fractured media: field evidence and theoretical model. Water Resour. Res. **51**, 940 (2015)
37. S.P. Neuman, Eulerian-Lagrangian theory of transport in space-time nonstationary velocity fields: exact nonlocal formalism by conditional moments and weak approximation. Water Resour. Res. **29**(3), 633 (1993)
38. J.H. Cushman, X. Hu, T.R. Ginn, Nonequilibrium statistical mechanics of preasymptotic dispersion. J. Stat. Phys. **75**(5/6), 859 (1994)
39. D.A. Benson, S.W. Wheatcrat, M.M. Meerschaert, Application of a fractional advection-dispersion equation. Water Resour. Res. **36**, 1403 (2000)
40. J.H. Cushman, T.R. Ginn, Fractional advection-dispersion equation: A classical mass balance with convolution–Fickian flux. Water Resour. Res. **36**, 3763 (2000)
41. R. Metzler, J. Klafter, The random walk's guide to anomalous diffusion: a fractional dynamics approach. Phys. Rep. **339**(1), 1 (2000)
42. B. Berkowitz, H. Scher, Anomalous transport in random fracture networks. Phys. Rev. Lett. **79**(20), 4038 (1997)
43. M. Dentz, A. Cortis, H. Scher, B. Berkowitz, Time behavior of solute transport in heterogeneous media: transition from anomalous to normal transport. Adv. Water Resour. **27**(2), 155 (2004)
44. B. Berkowitz, A. Cortis, M. Dentz, H. Scher, Modeling non-Fickian transport in geological formations as a continuous time random walk. Rev. Geophys. **44**, RG2003 (2006)
45. V. Cvetkovic, H. Cheng, X.H. Wen, Analysis of nonlinear effects on tracer migration in heterogeneous aquifers using Lagrangian travel time statistics. Water Resour. Res. **32**(6), 1671 (1996)
46. F. Delay, J. Bodin, Time domain random walk method to simulate transport by advection-diffusion and matrix diffusion in fracture networks. Geophys. Res. Lett. **28**, 4051 (2001)
47. R. Haggerty, S.M. Gorelick, Multiple-rate mass transfer for modeling diffusion and surface reactions in media with pore-scale heterogeneity. Water Resour. Res. **31**(10), 2383 (1995)
48. J. Carrera, X. Sánchez-Vila, I. Benet, A. Medina, G. Galarza, J. Guimerà, On matrix diffusion: formulations, solution methods, and qualitative effects. Hydrogeol. J. **6**, 178 (1998)
49. E.W. Montroll, G.H. Weiss, Random walks on lattices, 2. J. Math. Phys. **6**(2), 167 (1965)
50. H. Scher, M. Lax, Stochastic transport in a disordered solid. I. Theory. Phys. Rev. B **7**(1), 4491 (1973)
51. V.M. Kenkre, E.W. Montroll, M.F. Shlesinger, Generalized master equations for continuous-time random walks. J. Stat. Phys. **9**(1), 45 (1973)

52. M. Abramowitz, I.A. Stegun, *Handbook of Mathematical Functions* (Dover, Mineola, 1972)
53. M. Dentz, P.K. Kang, A. Comolli, T. Le Borgne, D.R. Lester, Continuous time random walks for the evolution of Lagrangian velocities. Phys. Rev. Fluids **1**, 74004 (2016)
54. G. Margolin, M. Dentz, B. Berkowitz, Continuous time random walk and multirate mass transfer modeling of sorption. Chem. Phys. **295**, 71 (2003)
55. T. Le Borgne, M. Dentz, J. Carrera, Spatial Markov processes for modeling Lagrangian particle dynamics in heterogeneous porous media. Phys. Rev. E **78**, 41110 (2008)
56. P.K. Kang, M. Dentz, T. Le Borgne, R. Juanes, Spatial Markov model of anomalous transport through random lattice networks. Phys. Rev. Lett. **107**, 180602 (2011)
57. P. De Anna, T. Le Borgne, M. Dentz, A.M. Tartakovsky, D. Bolster, P. Davy, Flow intermittency, dispersion, and correlated continuous time random walks in porous media. Phys. Rev. Lett. **110**(18), 184502 (2013)
58. E. Barkai, Y. Garini, R. Metzler, Strange kinetics of single molecules in living cells. Phys. Today **8**, 29 (2012)
59. J. Klafter, I. Sokolov, Anomalous diffusion spreads its wings. Phys. World **18**(8), 29 (2005)
60. C.F. Harvey, S.M. Gorelick, Temporal moment-generating equations: Modeling transport and mass transfer in heterogeneous aquifers. Water Resour. Res. **31**(8), 1895 (1995)
61. P. Maloszewski, A. Zuber, On the theory of tracer experiments in fissured rocks with a porous matrix. J. Hydrol. **79**, 333 (1985)
62. P. Gouze, Z. Melean, T. Le Borgne, M. Dentz, J. Carrera, Non-Fickian dispersion in porous media explained by heterogeneous microscale matrix diffusion. Water Resour. Res. **44**, W11416 (2008)
63. M. Willmann, J. Carrera, X. Sanchez-Vila, Transport upscaling in heterogeneous aquifers: What physical parameters control memory functions? Water Resour. Res. **44**, W12437 (2008)
64. M. Dentz, B. Berkowitz, Transport behavior of a passive solute in continuous time random walks and multirate mass transfer. Water Resour. Res. **39**(5), 1111 (2003)
65. R. Schumer, D.A. Benson, M.M. Meerschaert, B. Bauemer, Fractal mobile/immobile solute transport. Water Resour. Res. **39**(10), 1296 (2003)
66. D.A. Benson, M.M. Meerschaert, A simple and efficient random walk solution of multi-rate mobile/immobile mass transport equations. Adv. Water Resour. **32**(4), 532 (2009)
67. A. Comolli, J.J. Hidalgo, C. Moussey, M. Dentz, Non-Fickian transport under heterogeneous advection and mobile-immobile mass transfer. Transp. Porous Media **115**(2), 265 (2016)
68. E. Villermaux, J. Duplat, Mixing as an aggregation process. Phys. Rev. Lett. **91**, 18 (2003)
69. J. Duplat, E. Villermaux, Mixing by random stirring in confined mixtures. J. Fluid Mech. **617**, 51 (2008)
70. D.R. Lester, G. Metcalfe, M.G. Trefry, Is chaotic advection inherent to porous media flow? Phys. Rev. Lett. **111**(17), 174101 (2013)
71. M. Kree, E. Villermaux, Scalar mixtures in porous media. Phys. Rev. Fluids **2**, 104502 (2017)
72. F. Gjetvaj, A. Russian, P. Gouze, M. Dentz, Dual control of flow field heterogeneity and immobile porosity on non-Fickian transport in Berea sandstone. Water Resour. Res. **51**, 8273 (2015)
73. M. Dentz, M. Icardi, J.J. Hidalgo, Mechanisms of dispersion in a porous medium. J. Fluid Mech. **841**, 851 (2018)
74. G. Sposito, Steady groundwater flow as a dynamical system. Water Resour. Res. **30**(8), 2395 (1994)
75. T. Le Borgne, M. Dentz, E. Villermaux, The lamellar description of mixing in porous media. J. Fluid Mech. **770**, 458 (2015)
76. M. Dentz, D.R. Lester, T.L. Borgne, F.P.J. de Barros, Coupled continuous time random walks for fluid stretching in two-dimensional heterogeneous media. Phys. Rev. E **94**(6-1), 061102 (2016)
77. M. Dentz, H. Kinzelbach, S. Attinger, W. Kinzelbach, Temporal behavior of a solute cloud in a heterogeneous porous medium, 1, Point-like injection. Water Resour. Res. **36**(12), 3591 (2000)

78. M. Dentz, F.P.J. de Barros, Mixing-scale dependent dispersion for transport in heterogeneous flows. J. Fluid Mech. **777**, 178 (2015)
79. P.K. Kitanidis, Prediction by the method of moments of transport in heterogeneous formations. J. Hydrol. **102**, 453 (1988)
80. G. Dagan, Transport in heterogeneous porous formations: spatial moments, ergodicity, and effective dispersion. Water Resour. Res. **26**, 1287 (1990)
81. V. Kapoor, P.K. Kitanidis, Concentration fluctuations and dilution in aquifers. Water Resour. Res. **34**, 1181 (1998)
82. S.B. Pope. Turbulent flows. Meas. Sci. Technol. **12**(11) (2001)
83. J. Villermaux, J.C. Devillon, in *Proceedings of the 2nd International Symposium on Chemical Reaction Engineering* (Elsevier, New York, 1972)
84. J.R. De Dreuzy, J. Carrera, M. Dentz, T. Le Borgne, Time evolution of mixing in heterogeneous porous media. Water Resour. Res. **48**, W06511 (2012)
85. T. Le Borgne, M. Dentz, P. Davy, D. Bolster, J. Carrera, J.R. de Dreuzy, O. Bour, Persistence of incomplete mixing: A key to anomalous transport. Phys. Rev. E **84**, 015301(R) (2011)
86. E. Caroni, V. Fiorotto, Analysis of concentration as sampled in natural aquifers. Transp. Porous Media **59**(1), 19 (2005)
87. D.M. Tartakovsky, P.C. Lichtner, R.J. Pawar, PDF methods for reactive transport in porous media. Acta Univ. Carol. Geol. **46**, 113 (2002)
88. A. Bellin, D. Tonina, Probability density function of non-reactive solute concentration in heterogeneous porous formations. J. Contam. Hydrol. **94**, 109 (2007)
89. E. Villermaux, Mixing by porous media. C. R. Mécanique **340**, 933 (2012)
90. W.E. Ranz, Application of a stretch model to mixing, diffusion and reaction in laminar and turbulent flows. AIChE J. **25**(1), 41 (1979)
91. C.I. Steefel, D.J. DePaolo, P.C. Lichtner, Reactive transport modeling: An essential tool and a new research approach for the Earth sciences. Earth Planet. Sci. Lett. **240**, 539 (2005)
92. M. Dentz, T. LeBorgne, A. Englert, B. Bijeljic, Mixing, spreading and reaction in heterogeneous media: A brief review. J. Contam. Hydrol. **120–121**, 1 (2011)
93. D.A. Edwards, M. Shapiro, H. Brenner, Dispersion and reaction in two-dimensional model porous media. Phys. Fluids A **5**, 837 (1993)
94. M. Quintard, S. Whitaker, Convection, dispersion and interfacial transport of contaminants: Homogeneous media. Adv. Water Resour. **17**, 221 (1994)
95. A. Mikelic, V. Devigne, C.J. Van Duijn, Rigorous upscaling of the reactive flow through a pore, under dominant Peclet and Damkohler number. Siam J. Math. Anal. **38**, 1262 (2006)
96. P.E. Kechagia, I.N. Tsimpanogiannis, Y.C. Yortsos, P.C. Lichtner, On the upscaling of reaction-transport processes in porous media with fast or finite kinetics. Chem. Eng. Sci. **57**(13), 2565 (2002)
97. C. Meile, K. Tuncay, Scale dependence of reaction rates in porous media. Adv. Water Resour. **29**, 62 (2006)
98. F.J. Molz, M.A. Widdowson, Internal inconsistencies in dispersion-dominated models that incorporate chemical and microbial kinetics. Water Resour. Res. **24**(4), 615 (1988)
99. T.R. Ginn, C.S. Simmons, B.D. Wood, Stochastic-convective transport with nonlinear reaction: biodegradation with microbial growth. Water Resour. Res. **31**(11), 2689 (1995)
100. V. Kapoor, L. Gelhar, F. Miralles-Wilhelm, Bimolecular second-order reactions in spatially varying flows: segregation induced scale-dependent transformation rates. Water Resour. Res. **33**, 527 (1997)
101. C.M. Gramling, C.F. Harvey, L.C. Meigs, Reactive transport in porous media: A comparison of model prediction with laboratory visualization. Environ. Sci. Technol. **36**, 2508 (2002)
102. P.C. Lichtner, Q. Kang, Upscaling pore-scale reactive transport equations using a multiscale continuum formulation. Water Resour. Res. **43**, W12S15 (2007)
103. M. Dentz, P. Gouze, J. Carrera, Effective non-local reaction kinetics for transport in physically and chemically heterogeneous media. J. Contam. Hydrol. **120–121**, 222 (2011)
104. D. Ben-Avraham, S. Havlin, *Diffusion and Reactions in Fractals and Disordered Systems* (Cambridge University Press, Cambridge, 2004)

105. J.J. Hidalgo, M. Dentz, Y. Cabeza, J. Carrera, Dissolution patterns and mixing dynamics in unstable reactive flow. Geophys. Res. Lett. **42**, 6357 (2015)
106. M. Pool, M. Dentz, Effects of heterogeneity, connectivity and density variations on mixing and chemical reactions under temporally fluctuating flow conditions and the formation of reaction patterns. Water Resour. Res. **54** (2018). https://doi.org/10.1002/2017WR021820
107. M. De Simoni, J. Carrera, X. Sánchez-Vila, A. Guadagnini, A procedure for the solution of multicomponent reactive transport problems. Water Resour. Res. **41**(11) (2005). https://doi.org/10.1029/2005WR004056
108. O.A. Cirpka, R.L. Schwede, J. Luo, D. M., Concentration statistics for mixing-controlled reactive transport in random heterogeneous media. J. Contam. Hydrol. **98**, 61 (2008)
109. A. Bellin, G. Severino, A. Fiori, On the local concentration probability density function of solutes reacting upon mixing. Water Resour. Res. **47**, W01514 (2010)
110. G. Allaire, A.L. Raphael, Homogenization of a convection–diffusion model with reaction in a porous medium. Comptes Rendus Math. **344**(8), 523 (2007)
111. R. Mauri, Dispersion, convection, and reaction in porous media. Phys. Fluids A Fluid Dyn. **3**(5), 743 (1991)
112. K.M. Yao, *Influence of Suspended Particle Size on the Transport Aspect of Water Filtration*. Ph.D. thesis, University of North Carolina (Chapel Hill, North Carolina, 1968)
113. R. Pfeffer, J. Happel, An analytical study of heat and mass transfer in multiparticle systems at low Reynolds numbers. AIChE J. **10**(5), 605 (1964)
114. G. Boccardo, R. Sethi, D.L. Marchisio, Fine and ultrafine particle deposition in packed-bed catalytic reactors. Chem. Eng. Sci. **198**, 290–304 (2019)
115. G. Boccardo, E. Crevacore, R. Sethi, M. Icardi, A robust upscaling of the effective particle deposition rate in porous media. J. Contam. Hydrol. **212**, 3–13 (2017)
116. G. Boccardo, I.M. Sokolov, A. Paster, An improved scheme for a robin boundary condition in discrete-time random walk algorithms. J. Comput. Phys. **374**, 1152 (2018)
117. I. Babuška, in *Proceedings of the Third Symposium on the Numerical Solution of Partial Differential Equations–III* (Elsevier, Amsterdam, 1976), pp. 89–116
118. S.M. Kozlov, Averaging of random operators. Mat. Sb. **151**(2), 188 (1979)
119. E. Sánchez-Palencia, Non-homogeneous media and vibration theory. Lect. Notes Phys. **127** (1980)
120. L. Tartar, in *North-Holland Mathematics Studies*, vol. 30 (Elsevier, Amsterdam, 1978), pp. 472–484
121. J.L. Lions, G. Papanicolaou, A. Bensoussan, *Asymptotic Analysis for Periodic Structures* (North-Holland, Amsterdam, 1978)
122. A. Gloria, F. Otto, in *ESAIM Proc., CEMRACS 2013 – Modelling and Simulation of Complex Systems: Stochastic and Deterministic Approaches*, vol. 48 (EDP Sciences, Les Ulis, 2015), pp. 80–97
123. X. Blanc, C. Le Bris, F. Legoll, Some variance reduction methods for numerical stochastic homogenization. Phil. Trans. R. Soc. A **374**(2066), 20150168 (2016)
124. S. Armstrong, T. Kuusi, J.C. Mourrat, Quantitative stochastic homogenization and large-scale regularity, arXiv Prepr. arXiv1705.05300 (2017)
125. G. Allaire, Homogenization and two-scale convergence. SIAM J. Math. Anal. **23**(6), 1482 (1992)
126. P. Haynes, J. Vanneste, Dispersion in the large-deviation regime. Part 1: shear flows and periodic flows. J. Fluid Mech. **745**, 321 (2014)
127. M. Icardi, Multiscale model reduction for advection diffusion problems in periodic media. Internal report
128. B.D. Wood, F. Cherblanc, M. Quintard, S. Whitaker, Volume averaging for determining the effective dispersion tensor: closure using periodic unit cells and comparison with ensemble averaging. Water Resour. Res. **39**(8) (2003)
129. M. Quintard, S. Whitaker, Transport in ordered and disordered porous media II: generalized volume averaging. Transp. Porous Media **14**(2), 179 (1994)

130. W.G. Gray, C.T. Miller, Thermodynamically constrained averaging theory approach for modeling flow and transport phenomena in porous medium systems: 1. Motivation and overview. Adv. Water Resour. **28**(2), 161 (2005)
131. D.A. Drew, S.L. Passman, *Theory of Multicomponent Fluids*, vol. 135 (Springer Science & Business Media, Berlin, 2006)
132. R.O. Fox, *Computational Models for Turbulent Reacting Flows* (Cambridge University Press, Cambridge, 2003)
133. S.P. Neuman, Eulerian-Lagrangian theory of transport in space-time nonstationary velocity fields: exact nonlocal formalism by conditional moments and weak approximation. Water Resour. Res. **29**(3), 633 (1993)
134. D.M. Tartakovsky, S.P. Neuman, Transient flow in bounded randomly heterogeneous domains: 1. Exact conditional moment equations and recursive approximations. Water Resour. Res. **34**(1), 1 (1998)
135. J.H. Cushman, L.S. Bennethum, B.X. Hu, A primer on upscaling tools for porous media. Adv. Water Resour. **25**(8–12), 1043 (2002)
136. W.G. Gray, P.C.Y. Lee, On the theorems for local volume averaging of multiphase systems. Int. J. Multiphas. Flow **3**(4), 333 (1977)

6

Turbulence Research: Advances in Particle Tracking Velocimetry

Nathanaël Machicoane, Peter D. Huck, Alicia Clark, Alberto Aliseda, Romain Volk, and Mickaël Bourgoin

6.1 Introduction

Flow velocity measurements based on the analysis of the motion of particles imaged with digital cameras have become the most commonly used measurement technique in contemporary fluid mechanics research [1, 2]. *Particle image velocimetry* (PIV) and *particle tracking velocimetry* (PTV) are two widely used methods that enable the characterisation of a flow based on the motion of particles, from Eulerian (PIV) or Lagrangian (PTV) points of view. Several aspects influence the accuracy and reliability of the measurements obtained with these techniques [2]: resolution (temporal and spatial), dynamical range, the capacity to measure 2D or 3D components of velocity in a 2D or 3D fluid domain, statistical convergence, etc. These imaging and analysis considerations depend on the hardware (camera resolution, repetition rate, on-board memory, optical system, etc.) but also on the software (optical calibration relating real-world coordinates to pixel coordinates, particle identification and tracking algorithms, image correlation, dynamical post-processing, etc.) used in the measurements. In this context, particle tracking velocimetry can provide highly resolved, spatially and temporally, measurements of the flow velocity (if the particles are flow tracers) or of particle velocities (if the particles immersed in the flow have their own dynamics) in experimental

N. Machicoane · A. Clark · A. Aliseda
University of Washington, Department of Mechanical Engineering, Seattle, WA, USA

P. D. Huck
University of Washington, Department of Mechanical Engineering, Seattle, WA, USA

Laboratoire de Physique, ENS de Lyon, CNRS and Université de Lyon, Lyon, France

R. Volk · M. Bourgoin (✉)
Laboratoire de Physique, ENS de Lyon, CNRS and Université de Lyon, Lyon, France
e-mail: mickael.bourgoin@ens-lyon.fr

fluid mechanics research and applications [2–5]. A frequent implementation of this method in the laboratory is based on taking a pair of images (with double exposure cameras, typical of PIV) in rapid succession followed by a larger time interval before the next pair of images. A second common implementation of this method starts with the capture of a long sequence of images, all equally separated by a small time interval (with high-speed cameras). In the first case, the particle tracking velocimetry technique provides a single vector per particle in a pair of consecutive images, with subsequent velocity measurements in other image pairs being uncorrelated. The high-speed image sequence, on the contrary, provides the opportunity to track the same particle over multiple (n) images and provides several (n-1) correlated velocity (or n-2 acceleration) measurements, at different locations but along the same particle trajectory.

There are three recent contributions implemented by the authors and summarised in this chapter that apply equally to both versions of the particle tracking velocimetry technique: each one advances important aspects in one of the stages of the measurement of velocity from particle images. The first contribution (Sect. 6.2) provides an optical-model-free calibration technique for multi-camera particle tracking velocimetry and potentially also for particle image velocimetry. This method is simpler to apply and provides equal or better results than the pinhole camera model originally proposed by Tsai in 1987 [6]. In the context of particle tracking with applications in fluid mechanics, particle centre detection and tracking algorithms have been the focus of more studies [7, 8] than optical calibration and 3D position determination. Although many strategies with various degrees of complexity have been developed for camera calibration [9–13], most existing experimental implementations of multi-camera particle tracking use Tsai pinhole camera model as the basis for calibration. Using plane-by-plane transformations, it defines an interpolant that connects each point in the camera sensor to the actual light beam across the measurement volume. As it does not rely on any a priori model, the method easily handles potential complexity and non-linearity in an optical setup while remaining computationally efficient in stereo-matching 3D data. In opposition, Tsai approach, sketched in Fig. 6.1, is based on the development on a physical model for the cameras arrangement with several parameters (the number depending on the complexity). The model assumes that all ray of light received on the camera sensor pass through an optical centre (pinhole) for each camera. The quality of the inferred transformation will therefore be sensitive to variations of the setup leading to calibration data which may no longer match the model due to optical distortions, for instance. Besides, Tsai model requires non-linear elements to account for each aspect of the optical path. In practice, realistic experimental setups are either complex and time-consuming to model via individual optical elements in the Tsai method or over-simplified by ignoring certain elements such as windows, or compound lenses, with loss of accuracy.

The second contribution (Sect. 6.3) addresses the reconstruction of trajectories from the set of particle positions detected in the image sequence, an important aspect of particle tracking velocimetry [8, 14–17]. It describes the practical implementation of two recent developments: shadow particle velocimetry using parallel light

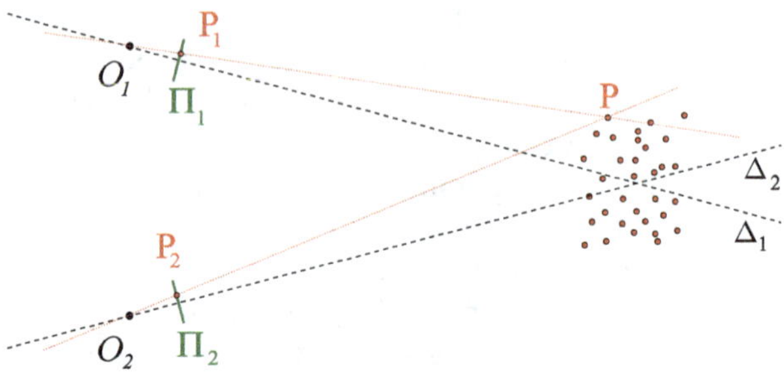

Fig. 6.1 Sketch of Tsai pinhole camera model and stereo-matching: the position of a particle in the real world corresponds to the intersection of 2 lines Δ_1 and Δ_2, each emitted by the camera centres O_1 and O_2 and passing through the position of particles P_1 and P_2 detected on each camera plane Π_1 and Π_2

combined with pattern tracking [18, 19] and trajectory reconstruction based on an extension of the four-frame best estimate (4BE) method. While the former was developed originally to access the size, orientation, or shape of the tracked particles, the latter is an extension of previous tracking algorithms [17] (which also extended previous algorithms) and which can be easily implemented as an add-on to an existing tracking code.

Finally, Sect. 6.4 describes a method to estimate noiseless velocity and acceleration statistics from particle tracking velocimetry tracks. This is a crucial step because imaging techniques may introduce noise into the detection of particle centres, which is then amplified when computing successive temporal or spatial derivatives. The position signal is then usually time-filtered prior to differentiation [5, 20], a procedure that increases the signal-to-noise ratio at the cost of signal alteration. The method described here, inspired by work in this area [21, 22], is based on computing the statistics of the particles displacements with increasing time lag, does not require any kind of filter, and allows for the estimation of noiseless statistical quantities both in the Lagrangian framework (velocity and acceleration time correlation functions) and in the Eulerian framework (statistics of spatial velocity increments) [23, 24].

Note that this chapter does not intend to review all the possible extensions of particle tracking velocimetry and has been limited to some recent developments from the authors' groups, which we believe can be useful and easily implemented to improve the accuracy of already operational PTV systems in other groups or which may help users developing new PTV experiments. Many other interesting advances have been developed over the past decade. We can, for instance, mention the use of inverse-problem digital holography [25–27], which allows to track particles in 3D with one single camera, new algorithms allowing to track particles in highly seeded flows such as the shake the box method [28] or the tracking of particles with rotational dynamics [29, 30], which allows to investigate simultaneously the translation and rotation of large objects transported in a flow.

6.2 A Model-Free Calibration Method

6.2.1 Principle

3D particle imaging methods require an appropriate calibration method to perform the stereo-matching between the 2D positions of particles in the *pixel coordinate system* for each camera and their absolute 3D positions in the *real-world coordinate system*. The accuracy of the calibration method directly impacts the accuracy of the 3D positioning of the particles in real-world coordinates.

The calibration method proposed here (further discussed in [31]) is based on the simple idea that no matter how distorted a recorded image is, each bright point on the pixel array is associated with the ray of light that produced it. As such, the corresponding light source (typically a scatterer particle) can lie anywhere on this ray of light. An appropriate calibration method should be able to directly attribute to a given doublet (x_p, y_p) of pixel coordinates its corresponding ray path. If the index of refraction in the measurement volume of interest is uniform (so that light propagates along a straight line inside the measurement volume) each doublet (x_p, y_p) can be associated with a straight line d (defined by 6 parameters in 3D: a position vector $\mathbf{O}_\Delta(x_p, y_p)$ and a displacement vector $\mathbf{V}_\Delta(x_p, y_p)$), regardless of the path outside the volume of interest, which can be very complex as material interfaces and lenses are traversed. The calibration method described here builds a *pixel-to-line* interpolant $\mathcal{I}$ that implements this correspondence between pixel coordinates and each of the 6 parameters of the ray of light: $(x_p, y_p) \xrightarrow{\mathcal{I}} (\mathbf{O}_\Delta, \mathbf{V}_\Delta)$. While this method may seem similar to Tsai approach which also designates a ray of light for each doublet (x_p, y_p), there is a significant difference in that Tsai approach assumes a camera model and is sensitive to deviations in the actual setup from this idealised optical model. The proposed approach does not rely on any *a priori* model and is only based on empirical interpolations from the actual calibration data. Thus, the new method implicitly takes into account optical imperfections, media inhomogeneities (outside the measurement volume) or complex lens arrangements. Additionally, the generalisation of the method to cases where light does not propagate in a straight line is straightforward: it is sufficient to build the interpolant with the parameters required to describe the expected curved path of light in the medium of interest (for instance, a parabola in the case of linear stratification).

6.2.2 Practical Implementation

An implementation of the method proposed is used to build the interpolant $\mathcal{I}$ from experimental images of a calibration target with known patterns at known positions. The process described here concerns only one camera for clarity. In general, in a realistic multi-camera system, the protocol has to be repeated for each camera independently.

A calibration target, consisting of a grid of equally separated dots, is translated perpendicularly to its plane (along the OZ axis) using a micropositioning stage, and is imaged at several known Z positions by every camera simultaneously. In total, N_Z images are taken by each camera: I_j is the calibration image when the plane is at position Z_j (with $j \in [1, N_Z]$). For an example highlighting the quality of the calibration method, $N_Z = 13$ planes were collected across the measurement volume. The calibration protocol, sketched in Fig. 6.2, then proceeds as follows:

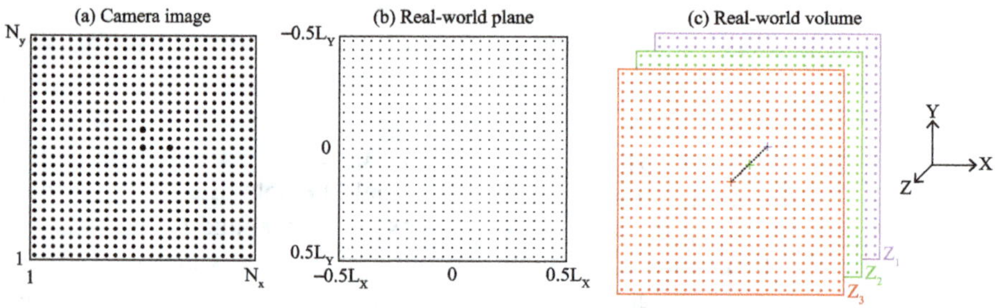

Fig. 6.2 Sketch of the calibration method. (**a**) Image I_j of the calibration target (over $N_x \times N_y$ pixels) located at one position Z_j (in real-world coordinates). From this image the centre of the images of the calibration dots in pixel coordinates $((x_j^k, y_j^k)_{k\in[1;N_j]})$ is determined. (**b**) Corresponding known location of the centre of the calibration points in real-world coordinates $((X_j^k, X_j^k)_{k\in[1;N_j]})$. From $(x_j^k, y_j^k)_{k\in[1;N_j]}$ and $(X_j^k, X_j^k)_{k\in[1;N_j]}$, the coefficients of the transformation $\mathcal{T}_j$ connecting pixel and real-world coordinates of the target located at Z_j are evaluated (the procedure is repeated for several target positions $Z_{j\in[1,N_z]}$) using least squares methods [32]. From a practical point of view the transformations $\mathcal{T}_j$ can be easily determined using ready to use algorithms, such as the `fitgeotrans` function in Matlab®. Note that for the simplicity of the illustration of the method, we show here a situation with no optical distortion and no perspective deformation, where the plane-by-plane transformation $\mathcal{T}_j$ is just given by a magnification factor M_j between pixel and real-world coordinates. In an actual experiment, perspective effects would require at least a linear projective transformation, defined by a 2×2 matrix $M_j^{\alpha\beta}$ with at least 4 coefficients to be estimated for each plane position Z_j. More realistic situations would require higher order polynomial transformations including a larger number of coefficients [32]; a third polynomial transformation embeds, for instance, 10 coefficients per plane). (**c**) Stacks of calibration planes at 3 different positions $(Z_{j=1,2,3})$ in 3D real-world coordinates (for simplicity, only 3 planes are illustrated, although in an actual calibration more planes may be used for better accuracy). The 3 coloured crosses illustrate the 3 projections (one on each of the 3 planes, the colour of the points corresponds to the colour of the plane onto which it is projected) in real-world coordinates $((X, Y, Z)_{j=1,2,3})$ of an arbitrary point (x, y) in pixel coordinates to which the 3 transformations $\mathcal{T}_{j=1,2,3}$ have been applied. These projections are distributed along a path of light corresponding to the line in real-world coordinates that projects onto the point (x, y) in the camera pixel coordinates. Since in a homogeneous medium light propagates in straight lines, the path of light is simply determined by a linear fit (dashed line), in 3D real-world coordinates, of the three points $((X, Y, Z)_{j=1,2,3})$. Using more calibration planes leads to more points for the linear fit and hence to a better accuracy. This procedure then directly connects the pixel coordinate (x, y) into the corresponding ray of light that produces it. Note that the fit is only done within the calibration volume where the target is translated along the N_z planes and does not extend to the cameras

1. **Dot centres detection.** For each calibration image I_j the dot centres are detected, giving a set $(x_j^k, y_j^k)_{k\in[1;N_j]}$ of pixel coordinates. N_j is the number of dots actually detected on each image I_j. *Real-world* coordinates of the dots $(X_j^k, Y_j^k, Z_j^k)_{k\in[1;N_j]}$ are known. Lowercase coordinates represent *pixel coordinates*, while uppercase coordinates represent *absolute real-world coordinates.*
2. **2D Plane-by-plane transformations.** For each position Z_j of the calibration target, the measured 2D pixel coordinates $(x_j^k, y_j^k)_{k\in[1;N_j]}$ and the known 2D real-world coordinates $(X_j^k, Y_j^k)_{k\in[1;N_j]}$ are used to infer a spatial transformation $\mathcal{T}_j$ projecting 2D pixel coordinates onto 2D real-world coordinates in the plane XOY at position Z_j. Different type of transformations can be inferred, from a simple linear projective transformation, to high order polynomial transformations if non-linear optical aberrations need to be corrected (common optical aberrations are adequately captured by a third-order polynomial transformation). This is a standard planar calibration procedure, where an estimate of the accuracy of the 2D plane-by-plane transformation can be obtained from the distance, in pixel coordinates, between $(x_j^k, y_j^k)_{k\in[1;N_j]}$ and $\mathcal{T}_j^{-1}\left(X_j^k, Y_j^k, Z_j\right)_{k\in[1;N_j]}$. The maximum error for the images used here is less than 2 pixels, corresponding in the present case to a maximum error of about 1/10th of the diameter of the dots in the calibration image.
3. **Building the pixel-line interpolant and stereo-matching.** The key step in the calibration method is building the pixel-to-line transformation. For a given pixel coordinate (for instance, corresponding to the centre of a detected particle), this is simply done by applying the successive inverse plane-by-plane transformations $\mathcal{T}_j^{-1}$ to project the pixel position to real space at each plane. This builds a set of points (one per plane) which define the line of sight corresponding to the considered pixel coordinate. The line is then determined by a linear fit of these points. For practical purposes, instead of repeating this procedure every time for every detected particle, we rather chose to build a pixel-line interpolant, $\mathcal{I}$, which directly connects pixels coordinates to a ray path. To achieve this, a grid of $N_\mathcal{I}$ interpolating points in pixel coordinates $(x_l^\mathcal{I}, y_l^\mathcal{I})_{l\in[1,N_\mathcal{I}]}$ is defined, for which the ray paths have to be computed. The inverse transformations $\mathcal{T}_j^{-1}$ are then used to project each point of this set back onto the real-world planes (X, Y, Z_j), for each of the N_Z positions Z_j. Each interpolating point $(x_l^\mathcal{I}, y_l^\mathcal{I})$ is therefore associated with a set of N_Z points in real world $(X_l^\mathcal{I}, Y_l^\mathcal{I}, Z_j)$. Conversely, these points in real world can be seen as a discrete sampling of the ray path which impacts the sensor of the camera at $(x_l^\mathcal{I}, y_l^\mathcal{I})$. If light propagates along a straight line, the N_Z points $(X_l^\mathcal{I}, Y_l^\mathcal{I}, Z_j)$ should be aligned. By a simple linear fit of these points, each interpolating point $(x_l^\mathcal{I}, y_l^\mathcal{I})$ is related to a line Δ_l, defined by a point $\mathbf{O}_{\Delta_l} = (X_l^0, Y_l^0, Z_l^0)$ and a vector $\mathbf{V}_{\Delta_l} = (Vx_l, Vy_l, Vz_l)$ (hence 6 parameters for each interpolating point). Each of these rays from the $N_\mathcal{I}$ interpolation points is used to compute the interpolant $\mathcal{I}$, which allows any pixel coordinate (x, y) in the camera to be connected to its ray path $(O_\Delta, \mathbf{V}_\Delta)$ corresponding to all possible positions of light sources that could produce a bright spot in (x, y).

Stereo-matching, or finding the 3D position of a point (or particle), is performed by finding a set of rays from each camera that cross (or almost cross) in the vicinity of the same spot in the volume of interest. The most probable 3D location of the corresponding particle is then taken as the 3D position that minimises the total distance to all those rays. The interpolant described in the method is created using every pixel in the cameras, as this step is done only once, but the method can be applied with a subset of the pixel array. For a setup with moderate optical distortion, a loose interpolating grid with a few hundreds points (typically, 20×20) is largely sufficient. As a matter of fact, using the interpolant is not mandatory, as all the calibration information is embedded in the plane-by-plane transformations. Third-order polynomial plane-by-plane transformation embeds 10 parameters each (5 polynomial coefficients for each of the X and Y transformations). If, instead, 7 calibration planes are used, the calibration information embeds about 70 parameters in total. Using the interpolant approach is above all a practical solution, while the interpolation information embeds a massive number of hidden parameters (6 per interpolation point) and is therefore expected to be highly redundant. Therefore, it is generally unnecessary to build the interpolant on a too refined grid (however, the added computational cost is minimal as the interpolant is only built once per calibration procedure, and can be stored in a small file for later use). This may happen for systems with important small-scale and heterogeneous optical distortions, in which case higher order plane-by-plane transformations (hence embedding more parameters) would also be necessary.

6.2.3 *Results: Comparison with Tsai Model*

The calibration procedure proposed by Tsai [6] has been widely used to recover the optical characteristics of an imaging system to reconstruct the 3D position of an object. The accuracy of the proposed imaging calibration procedure is assessed by comparing it with a simple implementation of Tsai model. A camera model accounting only for radial distortion is used. While improved optical elements in Tsai model could increase the accuracy, they come at an increased operator workload.

Our stereoscopic optical arrangement (see Refs. [31, 33] for more details), typical of PTV in a 1 cm thick laser sheet, focuses on the geometrical centre of a water flow inside an icosahedron, with both cameras objectives mounted in a Scheimpflug configuration. A plate mounted parallel to the laser sheet with 2 mm dots, attached to a micrometric traverse (with 10 μm accuracy), is used as a target. Both calibration methods use 13 target images, 1 mm apart from each other along the Z axis.

The calibration method uses the 2D positions of the target dots, and provides a series of positions that cannot exactly match the 3D real coordinates because, in both methods, the model parameters are obtained by solving an over-constrained linear system in the least-square sense. The calibration error, i.e., the absolute difference between the (known) real coordinates and the transformed ones, is computed to evaluate the calibration accuracy. This error can be estimated along each direction or as a norm: $d = \sqrt{d_X^2 + d_Y^2 + d_Z^2}$ (Table 6.1). Figure 6.3 plots the total 3D error averaged over the 13 planes used, for both the proposed method and Tsai model.

The accuracy of the proposed calibration is superior to that of the Tsai method (in its simplest implementation). The error is at least 300% smaller (depending on which component is considered) and is reduced to barely 0.5 pixel. It is important to note that the error map obtained with the Tsai method (Fig. 6.3b) seems to display a large bias along Y that could be due to the use of Scheimpflug mounts, which are typically not included in this Tsai calibration, and to the angle between the cameras and the tank windows. This hypothesis was verified by comparing the two calibrations procedures in more conventional conditions, where they give similar results with a very small error.

For the present optical arrangement and the new calibration method, the error in the Y positioning is the smallest. Indeed, due to the shape of the experiment (an icosahedron), the y axis of the camera sensor is almost aligned with the Y direction so that this coordinate is fully redundant between the cameras, while the x axes of each camera sensor form an angle $\alpha \simeq \pi/3$ with the X direction so that the precision on X positioning is lower. This directly impacts the precision on the Z positioning, whose error is almost equal to the X positioning error.

Table 6.1 Spatial average of the absolute deviation from the expected position of the targets

	d_X (μm)	d_Y (μm)	d_Z (μm)	d (μm)
Proposed calibration	32.7	12.6	39.2	59
Tsai model	121	171.1	112.7	266.6

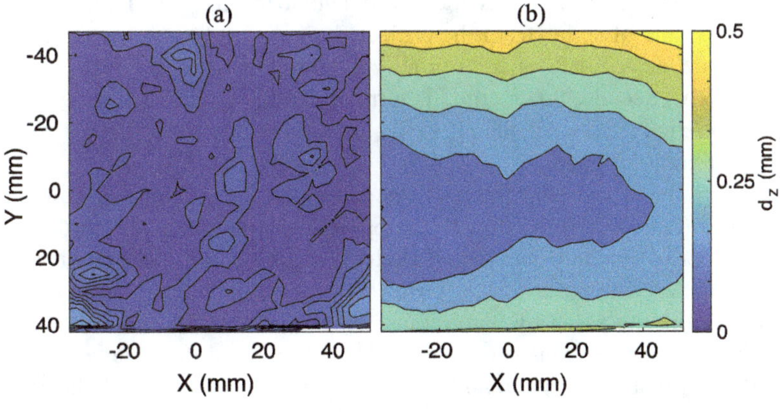

Fig. 6.3 Calibration error averaged along Z using the proposed calibration method (**a**) or Tsai model (**b**)

6.2.4 Discussion

Up to 13 planes were used to build the operator that yields the camera calibration. While two planes are the minimum required for the method, a larger number of planes imaged provide better accuracy. In this case study, the major sources of optical distortion were the Scheimpflug mounts, the imperfect lenses, and the non-perpendicular interfaces. 7 planes provided an optimal trade-off between high accuracy and simplicity, with an error only 2% larger than the 13 planes setup, while using only 3 planes yields an 10% larger error. The fact that few planes are sufficient to obtain a good accuracy of the calibration is likely related to the fact that the third-order polynomial plane-by-plane transformations are sufficient to handle most of the distortions, including those originating from the optics, from the tilt and shift system and from the refraction at the air–water interface, so that the projection of a pixel position to real space is accurately aligned along a line which defines the corresponding line of sight. Few points are then needed to accurately fit the line parameters (using more points essentially ensures a more robust fit with respect to small errors in the plane-by-plane transformations). When dealing with a more complex experiment, i.e., with a refraction index gradient, increasing the number of planes in the calibration would improve the results allowing to accurately capture the curvature of the light rays.

The proposed calibration method has several advantages that make it worth implementing in a multi-camera particle imaging setup. First, it requires no model or assumption about the properties of the optical path followed by the light in the different media outside the volume of interest. It only requires light to propagate in straight line. The method simply computes the equation for propagation of light in space. This ray line equation is fully determined by the physical location of the calibration dots located at known positions in space. Note that the present calibration method is versatile enough so that the linear propagation constraint can be easily relaxed. This can be useful, for instance, to calibrate stratified flows, with spatial variations of optical index. It is then sufficient to change the linear fit used to determine the line of sight (from the projected pixel coordinates to the planes), by an appropriate curved path of light (a polynomial fit may often be a good enough approximation). Second, this method is turnkey for any typical optical system. The implementation of the new method is easily done and can be used retroactively using previous calibration images.

Let us briefly discuss the improved accuracy of the calibration, compared to the model of Tsai. The reason for the improved accuracy is mainly hidden in the higher number of (hidden) parameters actually defining both calibration methods. As pointed out earlier, in the new proposed calibration all the calibration parameters are embedded in the plane-by-plane transformations, with 10 parameters for each third-order polynomial transformation. Using 13 calibration planes ends up with 130 hidden calibrating parameters. These reduce to 70 when using 7 planes. In any case this is much larger than the number of parameters embedded in the Tsai model (which has typically 6 external parameters defining the position and the orientation of the equivalent pinhole camera) and several internal parameters (focal length, pixel aspect ratio, optical distortion parameters, etc.), typically of order 10. It is therefore not surprising that the present method gives better accuracy. Note also that the present comparison may be unfair to the Tsai model, as we have not considered more sophisticated pinhole camera models, properly accounting, for instance, for tilt and shift corrections, and which would naturally embed a larger number of parameters and an increased accuracy. Such extension of the pinhole approach is based on sophisticated physical and geometrical models, with algorithms that tend to be tedious to implement. A big advantage of the present calibration is its versatility and ease of algorithmic implementation, which remains identical whatever the complexity of the optical path. Finally, note that while the proposed method has a larger number of parameter, they only come from empirical determination and are obtained automatically through the calibration process, and there is no need to prescribe *a priori* a set of parameters tightened to a specific model requiring choices from the user. This makes the method not only more accurate but also adaptable and objective.

To conclude, the model-free calibration method proposed can be easily implemented with both the calibration image acquisition and spatial detection of target points currently standard in the field. The calibration algorithm and the operator calculation to convert pixel locations to physical locations, with minimal errors, can easily be programmed in any language available to experimentalists (the reader can contact the authors for source codes to implement the calibration algorithms). The new method is at least equally, and frequently more, accurate than the commonly used Tsai model, and it can be used more easily and in a wider range of optical configurations. As experimental setups become more complicated with more optical and light refraction elements, this method should prove simpler to implement and more accurate than the model-based Tsai one.

6.3 Particle Tracking Algorithms

Section 6.3.1 describes the implementation of particle tracking velocimetry in a von Kármán flow using parallel light beams and two cameras forming an angle of 90°. As described below, the originality of this implementation of PTV is in the combination of parallel illumination and of pattern tracking (rather than particle tracking), which makes the calibration and the matching particularly simple and accurate. It is well suited to the tracking of small objects in a large volume using only two standard LEDs as light sources. In this setup, tracking is performed independently on the 2 views using a nearest neighbour algorithm prior to stereo-matching 2D tracks. Section 6.3.2 describes recent improvements of the tracking algorithms which use more than two consecutive frames in order to increase track lengths.

6.3.1 *Shadow Particle Tracking Velocimetry*

Experimental Setup

Particle tracking has been performed in a tank with a 15 cm × 15 cm square cross-section, where a von Kármán flow is created between two bladed discs, of radius $R = 7.1$ cm and separated by 20 cm, counter-rotating at constant frequency Ω (Fig. 6.4a). The flow has a strong mean spatial structure arising from the counter-rotation of the discs. The azimuthal component resulting from this forcing is of order $2\pi R\Omega$ near the discs' edge and zero in the mid-plane ($z = 0$), creating a strong axial gradient (Fig. 6.4a). The discs also act as centrifugal pumps ejecting fluid radially outward in their vicinity, resulting in a large-scale poloidal recirculation with a stagnation point in the geometrical centre of the cylinder (Fig. 6.4b). Using water to dilute an industrial lubricant, Ucon^{TM}, a mixture with a viscosity $\nu = 8.2\ 10^{-6}\,\text{m}^2\,\text{s}^{-1}$ and a density of $\rho = 1000\,\text{kg}\,\text{m}^{-3}$ allows for the production of an intense turbulence with a Taylor-based Reynolds number $R_\lambda = 200$ and a dissipative length scale $\eta = 130$ microns (see Table 6.2 for more details on

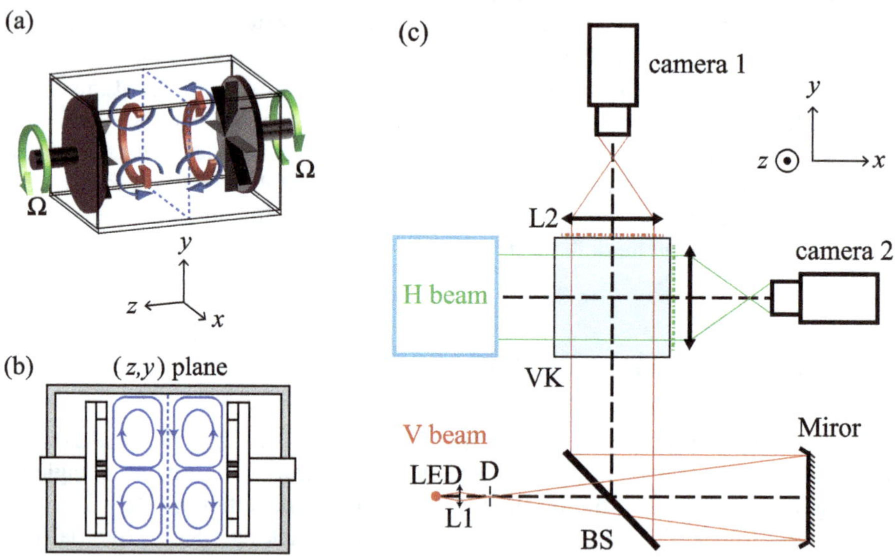

Fig. 6.4 (**a**) Sketch of the counter-rotating von Kármán flow. Arrows indicate the topology of the mean flow, the dashed line indicates the mid-plane of the vessel. (**b**) Schematic cut of the vessel along the (z, x) or (z, y) plane. (**c**) Optical setup for S-PTV with 2 identical optical arrangements forming an angle $\theta = 90$ degrees (only the vertical arm is described). The 1W LED source is imaged in the focus of a parabolic mirror to form a large collimated beam. A converging lens and a diaphragm are used to make the LED a better point-like source of light. Light propagates through the flow volume passing through a beam splitter (BS) before being collected using a 15 cm large lens that redirects the collimated light into the camera objective. The optical system [L_2+objective] is focused on the camera sides of the vessel, marked with a dashed-dotted line

Table 6.2 Parameters of the flow

Ω	v'_x	v'_y	v'_z	v'	τ_η	η	ϵ	R_λ	Re
Hz	$mmss^{-1}$	$mmss^{-1}$	$mmss^{-1}$	$mmss^{-1}$	ms	μm	$W.kg^{-1}$	–	–
4.2	0.39	0.37	0.24	0.34	2.9	154	1.0	155	16,200
5.5	0.50	0.49	0.33	0.45	2.0	128	2.1	190	21,200
6.9	0.62	0.62	0.41	0.56	1.5	111	3.6	225	26,700

Ω, rotation rate of the discs; the dissipative time scale is estimated from the zero-crossing ($t_0 = (t_{0x} + t_{0y} + t_{0z})$) of the acceleration auto-correlation functions: $t_0 \simeq 2.2\tau_\eta$ [36], the dissipation rate ε is estimated as $\varepsilon = \nu/\tau_\eta^2$, and the dissipative length scale is $\eta = (\nu^3/\varepsilon)^{1/4}$. The rms velocities are obtained at the geometrical centre of the flow using data points situated in a ball with a 1 cm radius. The Taylor-based Reynolds number is estimated as $Re_\lambda = \sqrt{15v'^4/\nu\varepsilon}$ with $v' = \sqrt{(v_x'^2 + v_y'^2 + v_z'^2)/3}$. The large-scale Reynolds number is $Re = 2\pi R^2\Omega/\nu$. The kinematic viscosity of the water-Ucon™ mixture is $\nu = 8.2\ 10^{-6}\,m^2s^{-1}$ with a density $\rho = 1000\,kg\,m^{-3}$

the flow parameters). This setup allows for the tracking of Lagrangian tracers (250 μm polystyrene particles with density $\rho_p = 1060\,kg\,m^{-3}$) in a large volume $6 \times 6 \times 5.5\,cm^3$ centred around the geometrical centre of the flow ($(x, y, z) = (0, 0, 0)$). Two high-speed video cameras (Phantom V.12, Vision Research, Wayne, NJ.) with a resolution of 800×768 pixels, and a frame rate up to $f_s = 12\,kHz$ are used. This sampling frequency is sufficient to resolve particle accelerations, calculated by taking the second derivative of the trajectories.

The camera setup uses a classical ombroscopy configuration [34], with parallel illumination. We have recently used such a setup (depicted in Fig. 6.4c) for Lagrangian studies of turbulence [35]; we will use the data from this experiment to illustrate the present section. It consists of 2 identical optical configurations with a small LED located at the focal point of a large parabolic mirror (15 cm diameter, 50 cm focal length) forming 2 collimated beams which are perpendicular to each other in the measurement volume. A converging lens and a diaphragm are used to make the LED a better point-like source of light. This large parallel ray of light then reflects on a beam splitter and intersects the flow volume before being collected by the camera sensor using a doublet consisting of a large lens (15 cm in diameter, 50 cm focal length) and a 85 mm macro camera objective. All optical elements are aligned using large (homemade) reticles, which also precisely measure the magnification in each arrangement. When placing an object in the field of view, it appears as a black shadow on a white background, corresponding to the parallel projection of the object on the sensor. Thanks to the parallel illumination, the system has telecentric properties. The particle size and shape do not depend then on the object-to-camera distance, as opposed to classical lighting schemes where due to perspective the apparent object size changes with the object-to-camera distance. The telecentricity also makes the calibration of each camera trivial as there is a simple, unique, and homogeneous magnification factor relating the (x, y) pixel coordinates to the (X, Z) real-world coordinates for one camera and to (Y, Z) real-world coordinates for the other camera. In addition, the optical arrangement is rigorously implemented so that the Z real-world coordinate is exactly redundant between the 2 cameras. This makes the matching step (detailed below) both simple and accurate. When particles are tracked, camera 1 will provide their (x_1, z_1) 2D positions, while camera 2 will measure their (y_2, z_2) positions. As the z coordinate is redundant, a simple equation $z_2 = az_1 + b$ accounts for slight differences in the magnification and centring between both arrangements.

The Trajectory Stereo-Matching Approach

Given the magnification of the setup (1/4, 1 px equals 90 μm), the depth of field of the optical arrangement is larger than the experiment. As both beams do not overlap in the entire flow domain, particles situated in one light beam but outside the common measurement volume can give a well-contrasted image on one camera while not being seen by the other. Such a situation could lead to an incorrect stereo-matching event when many particles are present. This is illustrated in Fig. 6.5a,

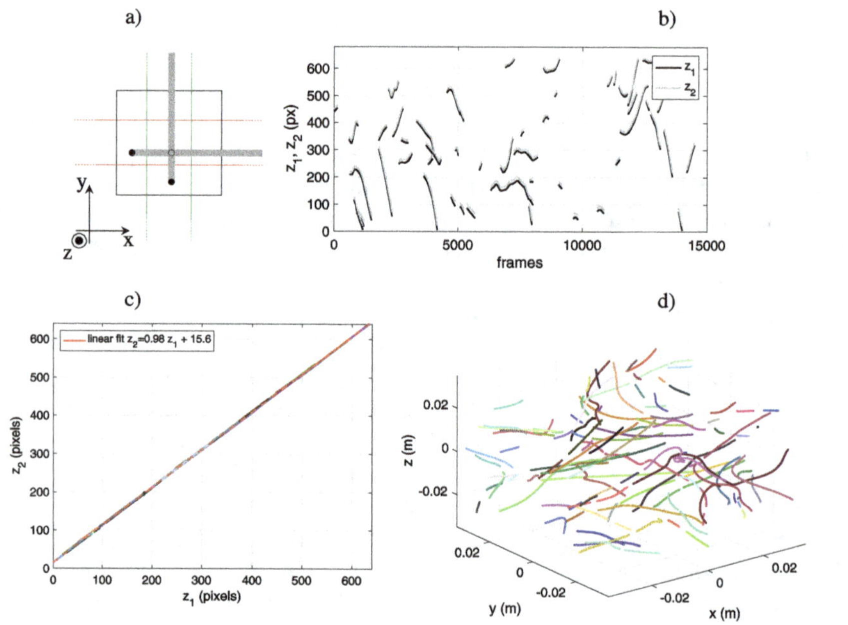

Fig. 6.5 (**a**) Scheme of the intersecting parallel light beams showing individual particle stereo-matching is not reliable. The black dots are two particles at the same z position outside of the beams overlapping region and the dashed circle is a particle at the same z position within the region (both situations being measured identically by the cameras). (**b**) Time evolution of the raw z (redundant) coordinate of the same particles as obtained with 2D tracking with camera 1 and camera 2. Only 38 matched trajectories are plotted. (**c**) Affine relation between $z_2 = az_2 + b$ ($a = 0.98$, $b = 15.6$ px) measured with 1900 trajectories corresponding to 6×10^5 data points. (**d**) A random sample of 150 trajectories in the vessel obtained from the same movie

where the shadows left by two particles situated at the same z position but outside of the beams overlap (black dots) could be interpreted as one "ghost" particle within the overlapping region (dashed circle). To mitigate these errors, we construct 2D trajectories for each camera using the (x_1, z_1) and (y_2, z_2) coordinates separately. Once tracked in time, these trajectories, instead of individual particle positions, may be stereo-matched. This approach is similar to the "pattern matching" originally proposed by Guezennec et al. [16], in contrast with the particle-matching strategy, used in many recent studies, which perform stereo-matching on individual particles before tracking. The advantage of this method, in particular when it is combined with telecentric illumination, is that neither stereo-matching nor tracking errors are made, as will be detailed below. However, one must track many more 2D trajectories that are stereo-matched. Another drawback is the projection of 3D positions into a plane, which strongly decreases the inter-particle distance making the apparent particle overlap an issue when the particle diameter becomes large with respect to the effective measurement volume. However, the presence of redundancy in the z coordinate may be used to overcome such indetermination when the apparent proximity results only from the projection.

We implement a 2D tracking scheme using a simple method inspired from previous works [8, 17, 20]. This tracking procedure searches for particles in frame $n + 1$ whose distance from particles in frame n is smaller than a specified value. If only one particle is found in the vicinity of the last point of a track, this track is continued. When multiple candidates are found, the track is stopped and new tracks are initiated with these new particles. Particles in frame n+1 which do not match with any of the existing tracks in frame n initiate new trajectories. This procedure, whose improvement is described in the next subsection, results in a collection of 2D trajectories with various lengths.

Stereo-matching is then performed by identifying trajectories with $z_1(t) \simeq z_2(t)$ using the relation $z_2 = az_1 + b$ as shown in Fig. 6.5b. This calibration relation is determined recursively using a dilute ensemble of particles for which the initial identification of a single pair of 2D trajectories gives a first estimate of the relationship between z_2 and z_1. As more trajectories are found, the affine relationship is refined until the maximum possible amount of trajectories for a single experiment is obtained. In this recursive manner, the tracking algorithm is self-calibrating. Here, the parameters are $a = 0.98$, $b = 15.6$ px estimated from 1900 matched trajectories, corresponding to $6\ 10^6$ data points as shown in Fig. 6.5c. Together with the pixel-to-mm conversion from one of the cameras, this method provides all relevant information about particle positions in world coordinates. Note that the temporal support for the 2D tracks $z_1(t)$ and $z_2(t)$ for a given particle may not be identical (the track may be longer on one camera than on the other or may start and end at slightly different times). When it comes to analysing 3D Lagrangian statistics, only the portions of trajectories over a common temporal interval are kept. In addition, only trajectories with sufficient temporal overlap (typically 70 time-steps, i.e., approximately $2.5\tau_\eta$) are matched, in order to prevent anomalous trajectories due to possible ambiguities when matching short patterns. Such an occurrence becomes increasingly unlikely as the trajectory duration threshold is increased. A false trajectory can only occur when the relationship $z_2 = az_1 + b$ becomes undetermined, which may happen, for instance, when two particles are close to colliding and the matching of the two nearby particles becomes ambiguous. Such a situation remains however an extraordinarily rare event in dilute situations. After tracking and stereo-matching, each pair of movies gives an ensemble of trajectories from which all single particle statistics can be computed as shown in Fig. 6.5d.

Flow Measurements

Measurements are performed in a volume ($6 \times 6 \times 5.5\,\mathrm{cm}^3$) larger than one integral scale ($L_v = \mathrm{v}'^3\epsilon^{-1} \simeq 4.8\,\mathrm{cm}$) of an inhomogeneous flow. As the statistics are subsampled spatially and temporally, a large number of trajectories are then needed to achieve a good statistical convergence. We record 200 sets of movies with a duration of 1.3 s at 12 kHz and obtain $\mathcal{O}(1000)$ tracer trajectories per set. A statistical ensemble of $\mathcal{O}(10^5)$ trajectories with mean durations $\langle t \rangle \sim 0.25/\Omega$ permits the spatial convergence of both Eulerian and Lagrangian statistics. The flow properties are obtained from the PTV data and are given in Table 6.2 together with the energy dissipation ε. The latter quantity is estimated by calculating the zero-crossing time τ_0 of the acceleration auto-correlation curves which is empirically known to be related to the Kolmogorov time scale τ_η ($\tau_0 \simeq 2.2\tau_\eta$) [36] and thus, energy dissipation. The fluctuating velocity of the flow is found to be proportional to the propeller frequency Ω (Table 6.2) due to inertial steering at the bladed discs which forces the turbulence that becomes full-developed, provided $Re = \frac{2\pi R^2\Omega}{\nu} > 3300$ [37]. In what follows, we focus our analysis on the case $\Omega = 5.5$ Hz. of trajectories, each containing the temporal evolution of the Lagrangian velocity at the particle position. Based on this ensemble of trajectories, one may reconstruct the mean velocity field in 3D,

$$\langle \mathbf{v} \rangle (x, y, z) = (\langle v_x \rangle, \langle v_y \rangle, \langle v_z \rangle),$$

and the rms fluctuations of each velocity component (v'_x, v'_y, v'_z). This is achieved by an Eulerian averaging of the Lagrangian dataset on a Cartesian grid of size 12^3, which corresponds to a spatial resolution of 5 mm in each direction. The choice of the grid size must fulfil several criteria: it must be small compared to the typical scale of the mean flow properties (here, $L_v \sim 4.8\,\mathrm{cm}$), but large enough so that statistical convergence is achieved. Here, the grid size was chosen so that there are at least $\mathcal{O}(1000)$ trajectories in each bin, enough to converge both mean and rms values.

Figure 6.6a, b displays two cross-sections of the reconstructed mean flow in two perpendicular planes, the mid-plane $\Pi_{xy} = (x, y, z = 0)$ and $\Pi_{yz} = (x = 0, y, z)$, a horizontal plane containing the axis of rotation of the discs. We observe a mean flow structure that is close to the schematic view of Fig. 6.4a. The flow is almost radial and convergent with $\langle v_z \rangle \sim 0$ in Π_{xy}, with a z component which reverses under the transformation $z \to -z$ (Fig. 6.6b). We also observe a strong y-component of the velocity in Π_{yz} which reverses under the transformation $y \to -y$ and corresponds to the differential rotation imposed by the discs. These cross-sections also reveal that the flow has the topology of a stagnation point at the geometric centre (0, 0, 0), as was shown in another von Kármán flow with a circular section [38]. With a 3D measurement of the mean flow, it is possible to compute spatial derivatives along all directions. This leads to $\partial_x \langle v_x \rangle \sim \partial_y \langle v_y \rangle \simeq -1.5\ \Omega$ for the stable directions, and $\partial_z \langle v_z \rangle \sim 3.0\ \Omega$ for the unstable direction. Note that the sum of these terms must be zero because this quantity is the divergence of the mean flow. This condition is found to be well satisfied although the velocity components were computed independently without any constraint. The verification that the flow is divergence-free is then an *a posteriori* test that the reconstruction of the mean flow is physically sound. Figure 6.6c, d displays rms values of velocity fluctuations

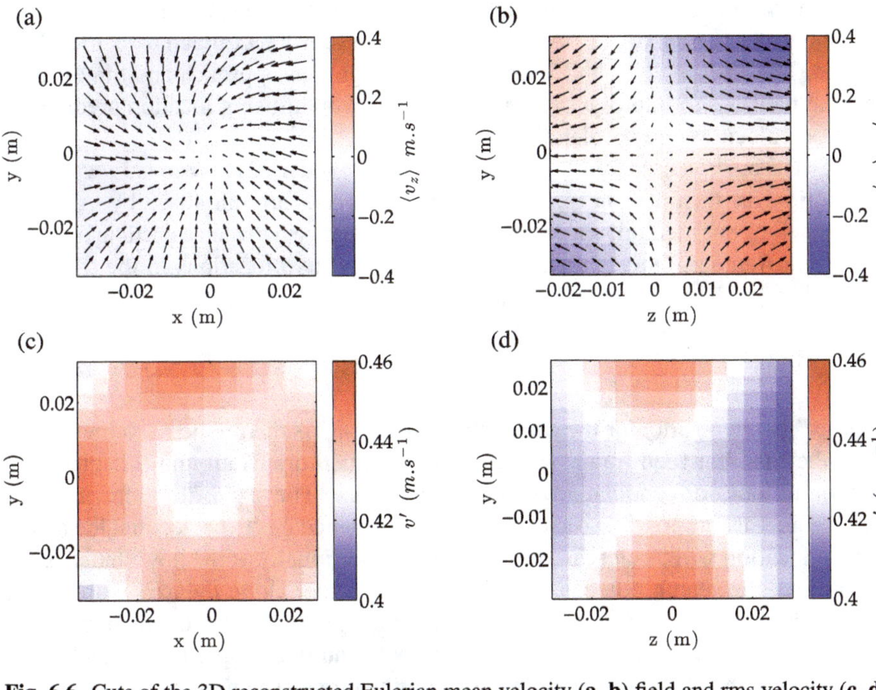

Fig. 6.6 Cuts of the 3D reconstructed Eulerian mean velocity (**a, b**) field and rms velocity (**c, d**). The reconstruction is achieved by computing the mean $\langle \mathbf{v} \rangle$ and rms values (v'_x, v'_y, v'_z) of the velocity in each bin of a Cartesian grid of size 12^3. (**a**) $\Pi_{xy} = (x, y, z = 0)$ plane. Arrows are $(\langle v_x \rangle, \langle v_y \rangle)$, the colour coding for the $\langle v_z \rangle$. (**b**) $\Pi_{yz} = (x = 0, y, z)$ plane. Arrows are $(\langle v_y \rangle, \langle v_z \rangle)$, the colour coding for the $\langle v_x \rangle$. (**c**) rms value of velocity fluctuations $v' = \sqrt{(v'^2_x + v'^2_y + v'^2_z)/3}$ in the $\Pi_{xy} = (x, y, z = 0)$ plane. (**d**) rms value of velocity fluctuations in the $\Pi_{yz} = (x = 0, y, z)$ plane

$v' = \sqrt{(v_x'^2 + v_y'^2 + v_z'^2)/3}$ in $\Pi_{xy} = (x, y, z = 0)$ and $\Pi_{yz} = (x = 0, y, z)$. These maps reveal that the flow properties are anisotropic and inhomogeneous at large scales, as previously observed in similar setups [39].

6.3.2 Improved Four-Frame Best Estimate

As mentioned in the previous section, using only two frames and a nearest neighbour criterion may lead to multiple candidates for a given track or wrong matches when increasing the number of particles in the field of view. To overcome such limitation, four-frame tracking methods were developed, as, for instance, the "four-frame minimal acceleration method" (4MA), developed by Maas et al. [14], which minimises the change in acceleration along the track, or the further extension by Ouellette et al., known as "four-frame best estimate" particle tracking method (4BE) which minimises the distance between the prediction of particle position two time-step forward in time and all the particles detected at that time [17]. The 4BE method was shown [17] to have an improved tracking accuracy compared to the 4MA method. The 4BE method builds on a nearest neighbour approach and three-frame tracking methods to improve tracking performance by utilising location predictions based on velocities and accelerations.

The 4BE method uses four frames ($n-1, n, n+1$, and $n+2$) to reconstruct particle trajectories, as illustrated in Fig. 6.7a. Individual tracks are initialised by using the nearest neighbour method, which minimises the distance between a particle in frame $n-1$ and frame n. Once a track is started, the first two locations in the track are used to predict the position $\tilde{x}_i^{n+1}$ of the particle in frame $n+1$:

$$\tilde{x}_i^{n+1} = x_i^n + \tilde{v}_i^n \Delta t, \tag{6.1}$$

where x_i^n is the position of the particle in frame n, $\tilde{v}_i^n$ is the predicted velocity, and Δt is the time between frames. A search box is then drawn around this predicted location to look for candidates to continue the track. The size of the search box is set to be as small as possible (usually a few pixels) since it is expected that the actual particle location will be close to the prediction. Additionally, if the flow statistics are anisotropic, the search box can be adjusted to be larger along the axis with higher velocity fluctuations and smaller in the directions with smaller fluctuations. This decreases computational costs because it limits the number of particles found in the initial search, thus limiting possible track continuations. The particles found within this bounding box can then be used to predict a set of positions $\tilde{x}_i^{n+2}$ in frame $n+2$:

$$\tilde{x}_i^{n+2} = x_i^n + \tilde{v}_i^n (2\Delta t)^2 + \frac{1}{2}\tilde{a}_i^n (2\Delta t)^2, \tag{6.2}$$

where x_i^n, $\tilde{v}_i^n$, and Δt are the same as above, and $\tilde{a}_i^n$ is the predicted acceleration. As in the previous frame, $n+1$, a search box is drawn around each of the $\tilde{x}_i^{n+2}$ predicted locations. Each of these bounding boxes is then interrogated for particles. Using these particle locations, the track is determined by minimising the cost function ϕ_{ij}^n:

$$\phi_{ij}^n = ||x_j^{n+2} - \tilde{x}_i^{n+2}||. \tag{6.3}$$

Equation (6.3) minimises the distance between the actual (x_j^{n+2}) and predicted ($\tilde{x}_i^{n+2}$) particle locations, thus minimising changes in acceleration for a given track. An optional upper threshold, typically half the length of the search box, can be set on the cost function to help limit tracking error. The particle, and, therefore, the track that minimises this cost function and falls within the threshold is then defined as the correct track and all other possible tracks are discarded. It is also important to note that a track is discarded if at any point it does not contain any particles in the search box in frames $n+1$ or $n+2$.

While 4BE with nearest neighbour initialisation (4BE-NN) is a very good compromise between tracking accuracy and efficiency (low computational cost), there are certain cases where it starts to fail. For instance, it is not suitable for situations where the particle displacement starts to be comparable to the inter-particle distance. Therefore, we have developed a modified initialisation (MI) method for 4BE (4BE-MI) that is more effective at detecting tracks than the current nearest neighbour initialisation [40]. Figure 6.7b shows the modified 4BE algorithm. This method uses a search box based on the estimated maximum particle

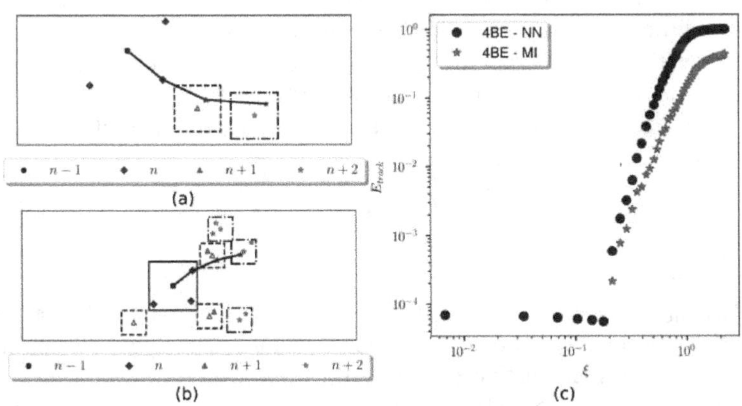

Fig. 6.7 **(a)** 4BE-NN. Particle locations are denoted with filled symbols, whereas predicted particle locations are denoted with hollow symbols. The boxes represent the bounding boxes used in the algorithm. The predicted path is overlaid in the figure. **(b)** 4BE-MI. The initial bounding box (now shown in the figure) allows for more potential tracks to be examined when searching for the correct track. **(c)** Comparison of tracking performance for 4BE-NN and 4BE-MI methods. At values of $\xi < 0.2$, the tracking error is zero for 4BE-MI

displacement between two frames to initialise tracks. The size of this search box is determined based on the flow characteristics (instantaneous spatial-averaged velocity, velocity fluctuations in all three directions, etc.), but it is always larger than the size of the search box used for track continuation (which is only aimed at accounting for the error in evaluating the next position in the track). This allows the algorithm to explore multiple possible trajectories for each particle and eliminates the assumption that the closest particle in the next frame is the only option when starting a track. It also enables a track to be constructed based on knowledge of the flow physics as a feature of the initialisation.

The performance of the 4BE algorithm both with and without the modified initialisation scheme was analysed using direct numerical simulation (DNS) data of a turbulent channel available through the Johns Hopkins University Turbulence Databases [41]. The DNS was performed in a $8\pi \times 2 \times 3\pi$ domain using periodic boundary conditions. The Reynolds number was $Re = \frac{U_c h}{\nu} = 2.2625 \times 10^4$, where U_c and h are, respectively, the channel centre-line velocity and height. The flow was initially seeded with tracer particles throughout the entire volume. The particles were then advected through the channel for each time-step based on the resolved DNS flow field. The trajectories were cut in a subdomain of the channel, creating an ersatz of particle entering and leaving the measurement volume as is typical in experiments. The trajectories generated were then used to benchmark the tracking scheme by comparing tracking results to the known trajectories.

Several datasets were generated by varying the distances that the particles moved between frames. This generated data over a wide range of ξ, defined as the ratio of the average distance each particle moves between frames to the average separation between particles in a frame. When ξ is small, tracking is easy because the particles move very little between frames and there are not many particles to consider for track continuation. However, as this ratio increases, tracking becomes more difficult because the particles move a large amount between frames and there are many particles per frame. Figure 6.7c shows the tracking error E_{track} plotted against ξ. The tracking error is defined as:

$$E_{\text{track}} = \frac{N_{\text{imperfect}}}{N_{\text{total}}}, \tag{6.4}$$

where $N_{\text{imperfect}}$ is the total number of imperfect tracks and N_{total} is the total number of tracks in the dataset generated. A perfect track must start at the same point as the actual track and must contain no spurious locations.

Figure 6.7c shows how the tracking error E_{track} is decreased when using the modified initialisation scheme. E_{track} is equal to zero, meaning there are no erroneous tracks computed, up to approximately $\xi = 0.2$ for the modified initialisation scheme. Additionally, at all values of ξ, the modified initialisation scheme performs better than the nearest neighbour initialisation scheme. This shows the advantage of the modified initialisation scheme in creating trajectories in flow with large particle displacements or high particle density.

6.4 Noise Reduction in Post-Processing Statistical Analysis

Particle tracking velocimetry leads to a collection of tracks, $(\mathbf{x}_j(t))_{j\in[1,N]}$, from which turbulent statistics, such as the mean flow and velocity fluctuations, may be computed. Most of the desired quantities have in common that they require taking the derivative of the particle positions, which inevitably leads to noise amplification. In the Lagrangian framework, single particle (two-time) statistics such as velocity or acceleration auto-correlation functions are of great interest; they will be considered in Sect. 6.4.1. In the Eulerian framework, moments of velocity differences separated by a distance $\mathbf{r}$ (structure functions) are of great importance; these two particle statistics will be addressed in Sect. 6.4.2.

The method presented below seeks to obtain unbiased one- and two-point statistics of experimental signal derivatives without introducing any filtering. It is valid for any measured signal whose typical correlation scale is much larger than the noise correlation scale. While one aims to obtain the real signal $\hat{x}$, the presence of noise b implies that one actually measures $x(t) = \hat{x} + b$. For simplicity, we consider the case of a temporal signal $x(t)$ that is centred, i.e., $\langle x\rangle = 0$, and is obtained by considering $x(t) - \langle x\rangle$, where $\langle\cdot\rangle$ is an ensemble average.

The method is based on the temporal increment dx of the signal x over a time dt that we express as $dx = x(t+dt) - x(t) = d\hat{x} + db$. Assuming that the increments of position and noise are uncorrelated, the position increment variance is written as $\left\langle (dx)^2\right\rangle = \left\langle (d\hat{x})^2\right\rangle + \left\langle (db)^2\right\rangle$. Introducing the velocity $\hat{v}$ and acceleration $\hat{a}$ through a second-order Taylor expansion $\hat{x}(t+dt) = \hat{x}(t) + \hat{v}\, dt + \hat{a}\, dt^2/2 + o(dt^2)$, one obtains:

$$\left\langle (dx)^2\right\rangle = \left\langle (db)^2\right\rangle + \langle \hat{v}^2\rangle dt^2 + \langle \hat{a}.\hat{v}\rangle dt^3 + o(dt^3), \tag{6.5}$$

where $\left\langle (db)^2\right\rangle = 2\left\langle b^2\right\rangle$ in the case of a white noise [24, 42]. In Eq. (6.5) $\left\langle (dx)^2\right\rangle$ is a function of dt so that one can recover the value of the velocity variance $\langle \hat{v}^2\rangle$ by calculating time increments of $\left\langle (dx)^2\right\rangle (dt)$ over different values of dt followed by a simple polynomial fit in the form of Eq. (6.5). If the noise is coloured, $\left\langle (db)^2\right\rangle = 2\left\langle b^2\right\rangle - 2\langle b(t)b(t+dt)\rangle$. In this case, the method requires the noise to be correlated over short times when compared to the signal correlation time. As a result, only the lowest values of $\left\langle (dx)^2\right\rangle (dt)$ are biased by $\langle b(t)b(t+dt)\rangle$ and a fit still successfully allows for the evaluation of the root mean square (rms) velocity, $\hat{v}' = \sqrt{\langle \hat{v}^2\rangle}$. For an experimentally measured signal x, equally spaced at an acquisition rate f_s, the minimal value of dt is $1/f_s$; we can then obtain the values of dx for different values of $dt = n/f_s$. For this method, a value of the acquisition rate f_s higher than usual is required, in order to be able to access derivatives of the signal without aliasing error.

We can extend the previous calculation to higher order derivative statistics by considering higher order increments. The second-order increment $d^2x = x(t + dt) + x(t - dt) - 2x(t)$, which is related to the acceleration variance $\langle \hat{a}^2 \rangle$ here, yields, for instance:

$$\left\langle (d^2x)^2 \right\rangle = \left\langle (d^2b)^2 \right\rangle + \langle \hat{a}^2 \rangle dt^4 + \frac{1}{6}\left\langle \hat{a}\frac{d^2\hat{a}}{dt^2} \right\rangle dt^6 + o(dt^6), \tag{6.6}$$

where $\left\langle (d^2b)^2 \right\rangle = 6\left\langle b^2 \right\rangle$ in the case of a white noise [24, 42], but otherwise introduces additional noise correlation terms which are functions of dt.

6.4.1 *Lagrangian Auto-Correlation Functions*

The approach developed above is not restricted to one-time statistics of the signal derivatives but can be generalised to estimate the noiseless first- and second-order derivative auto-correlation functions of the signal $C_{\hat{v}\hat{v}} = \left\langle \hat{v}(t)\hat{v}(t+\tau) \right\rangle$ and $C_{\hat{a}\hat{a}} = \left\langle \hat{a}(t)\hat{a}(t+\tau) \right\rangle$. This is done by considering the correlations of first- and second-order increments $\left\langle dx(t)dx(t+\tau) \right\rangle$ and $\left\langle d^2x(t)d^2x(t+\tau) \right\rangle$ which are functions of dt and τ. Noiseless velocity and acceleration correlation functions are estimated, respectively, for each time lag τ using a polynomial fit of the signal time increment dt with the following expressions:

$$\begin{cases} C_{dxdx}(\tau, dt) = C_{\hat{v}\hat{v}}(\tau)dt^2 + \frac{1}{2}\left(C_{\hat{v}\hat{a}}(\tau) + C_{\hat{a}\hat{v}}(\tau)\right)dt^3 + \\ \qquad + C_{dbdb}(\tau, dt) + o(dt^3) \\ C_{d^2xd^2x}(\tau, dt) = C_{\hat{a}\hat{a}}(\tau)dt^4 + \frac{1}{12}\left(C_{\hat{a}(d^2\hat{a}/dt^2)}(\tau) + C_{(d^2\hat{a}/dt^2)\hat{a}}(\tau)\right)dt^6 + \\ \qquad + C_{d^2bd^2b}(\tau, dt) + o(dt^6), \end{cases} \tag{6.7}$$

where $C_{fg} = \left\langle f(t)g(t+\tau) \right\rangle$ is a cross-correlation function. It can be noted that the case of the rms values corresponds to $\tau = 0$ and it is noted that $\langle (dx)^2 \rangle$ and $\langle (d^2x)^2 \rangle$ are functions of dt. In the previous expressions and in the case of a white noise, we can write auto-correlation functions of the first- and second-order increments of the noise. With the signal sampled at a frequency f_s, one has $dt = n/f_s$ and $\tau = m/f_s$. The correlation functions of the digitised noise increments are written as:

$$\begin{cases} C_{dbdb}\left(\tau=\dfrac{m}{f_s},dt=\dfrac{n}{f_s}\right) = \left\langle b^2\right\rangle(2\delta_{m,0}-\delta_{m,n}), \\ C_{d^2bd^2b}\left(\tau=\dfrac{m}{f_s},dt=\dfrac{n}{f_s}\right) = \left\langle b^2\right\rangle(6\delta_{m,0}-4\delta_{m,n}+\delta_{m,2n}), \end{cases} \tag{6.8}$$

where $\delta_{m,n}$ is the Kronecker symbol. For both derivatives, the white noise magnitude in the first-order derivative auto-correlation functions is the highest for $\tau = 0$ and is an additive term. The noise then yields a negative term for $m = n$. In the case of second-order derivatives (for acceleration in the case of Lagrangian tracks), the noise magnitude has a larger weight and the noise also contributes to a third time point of the function ($m = 2n$) with a positive term of smaller amplitude. Considering white noise terms up to dt^6, all other values of τ will directly yield the function without noise.

Results

The method has been applied to the material particle trajectories from Ref. [43]. It has been tested successfully for different particle diameters (from 6 to 24 mm), Reynolds numbers ($350 < Re_\lambda < 520$), and two density ratios (0.9 and 1.14), as well as for neutrally buoyant particles from Ref. [44]. We will focus only on the case of particles 6 mm in diameter and of density ratio 1.14 at a Reynolds number $Re_\lambda = 520$ in this example. The position trajectories are obtained by stereo-matching of successive image pairs obtained, thanks to two cameras and ambient lighting. The particles appear as large, bright discs on a uniform dark background which yields sub-pixel noise for the trajectories (the apparent particle diameter is about 20 pixels) and is not correlated with the particle position as the background is uniform (nor with its velocity as the exposure time is short enough to fix the particles on the images). In practical situations, the presence of sub-pixel displacements can lead to a short-time correlation of the noise, typically over a few frames.

Figure 6.8 shows the evolution of $\left\langle (dx)^2\right\rangle$ and $\left\langle (d^2x)^2\right\rangle$ with dt. A simple linear function of dt^2 is enough for $\left\langle (dx)^2\right\rangle$ and a sixth-order one suits better $\left\langle (d^2x)^2\right\rangle$. The first points of $\left\langle (d^2x)^2\right\rangle$ do not follow Eq. (6.6), which may be due to the fact that we are not dealing with a purely white noise as will be shown in Fig. 6.9b. Using the estimated values of the rms acceleration, a', and $\langle (d^2b)^2\rangle$, we can define a noise-to-signal ratio $b' f_s^2/a' = 11.9$, where we have defined $b' = \sqrt{\langle (d^2b)^2\rangle/6}$ by analogy with the white noise case. When considering the noise weight on the velocity signals, we of course find a much smaller magnitude $b' f_s/v' = 0.14$ as it is only a first-order derivative (v' being the rms of the velocity estimated with this method).

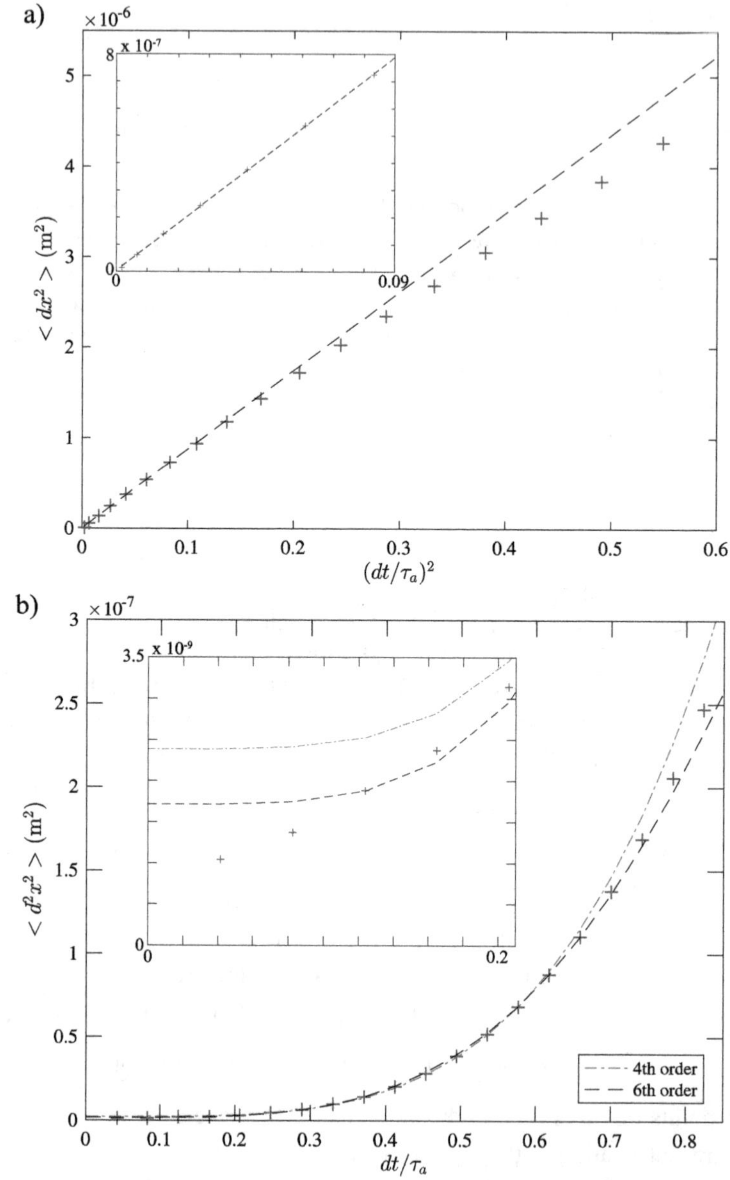

Fig. 6.8 (**a**) Evolution of $\left\langle (dx)^2 \right\rangle$ with $(dt/\tau_a)^2$, where $\tau_a = 8.1$ ms is the particle acceleration time scale (integral of the positive part of the particle acceleration auto-correlation function). The dashed line is a linear fit over the range $0 < dt/\tau_a \leq 0.25$. (**b**) Evolution of $\left\langle (d^2x)^2 \right\rangle$ with dt/τ_a. The dashed-dotted and dashed lines are fourth and sixth order fits ($\alpha + \beta(dt/\tau_a)^4$ resp. $\alpha + \beta(dt/\tau_a)^4 + \gamma(dt/\tau_a)^6$) over the range $0 < dt/\tau_a \leq 0.62$. The insets are zooms on low values of dt/τ_a

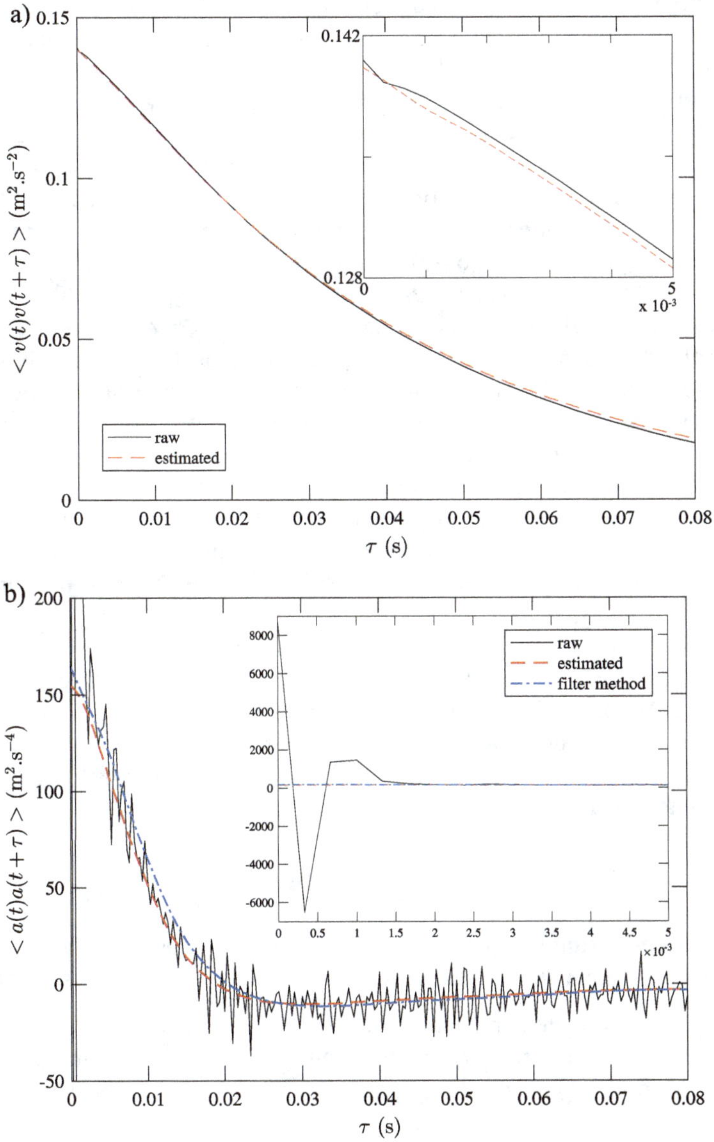

Fig. 6.9 (**a**) Auto-correlation functions of the velocity or acceleration (**b**) estimated from the proposed method (dashed line) and directly computed by differentiating the position signal obtained by PTV (continuous line). The insets are zooms on the low values of τ. The fit ranges used to obtain the functions are the same that used in Fig. 6.8. The dashed-dotted line in figure (**b**) is the correlation estimated from filtered trajectories using a Gaussian kernel $K = A_w \exp(-t^2/2w^2)$, where $w = 12$ points and A_w is a normalisation factor

Figure 6.9 shows the auto-correlation function of both the velocity and acceleration estimated with the proposed method, compared to the raw functions. With the low level of noise in this configuration, the velocity is almost unbiased and both functions are indistinguishable except for the first points of the raw function that are offset by the noise. On the second-order derivative, it can be observed in Fig. 6.9b that the raw acceleration auto-correlation function is biased for more than the three first points only (see inset). This is because the noise is not white but has a short correlation time compared to the signal. Combined with the finite duration of the trajectories, the raw correlation function is noisy over the whole range of time lags τ. This curve is plotted together with the one estimated with the method, fitting the coefficient up to $dt = 5$ ms which corresponds to 30% correlation loss in acceleration signals (same range as in Fig. 6.8b, but the precise choice is not critical). Although the signal-to-noise ratio is poor, the estimated correlation function seems to be following the median line between the peaks caused by noise and crosses zero at the location that seems to be indicated by the raw function. It is also close to the auto-correlation function from Ref. [43], estimated by filtering the data with a Gaussian kernel $K = A_w \exp(-t^2/2w^2)$ (with $w = 12$ points and a compact support of width $2w$, A_w is a normalisation factor). It should be stressed that the value $w = 12$ was chosen arbitrarily as a compromise between suppressing oscillations at small lags without altering too much the shape of the function at larger lags.

With the new method, we compute an acceleration time scale $\tau_a = 8.1$ ms and an acceleration magnitude $a' = 12.4\,\text{mmss}^{-2}$, which is close to the values $\tau_a = 8.8$ ms and $a' = 12.9\,\text{mmss}^{-2}$ found for the filtered data [43]. However, in the latter case, the value of a' depends strongly on the choice of the filter width w, so that one usually estimates a' by computing it for different filter widths which can then allow to extrapolate a best estimate value (as introduced in [5]).

Discussion

The present de-noising method estimates moments and auto-correlation functions of experimental signal derivatives. This method relies on two main assumptions:

1. The signal is correlated on a longer time scale than the noise.
2. The sampling frequency, f_s, is high enough so that the first and second derivatives of the signal can be computed by taking increments over several (N) points.

We have tested the method in the context of Lagrangian particle tracks in turbulence for which the noise is correlated on times much shorter than the signal, considering both first- and second-order derivatives of a time dependent signal. The results are in good agreement to what is obtained by classical filtering processes which require a long bias study specific to the data type [5, 45], and we believe them to be more accurate. The method avoids subjective tuning of the filter width and choice of filter type while yielding unbiased quantities by requiring data fits in an appropriate range.

While the fit range is still an adjustable parameter, we observed its impact on the results to be smaller than when filtering the data. Another advantage of the method is an easy access to the noise magnitude. While building a new experimental setup, one can gather just enough statistics to converge second-order moments to estimate the noise magnitude and try and improve the setup iteratively.

6.4.2 Eulerian Structure Functions

Method

The method presented above can be extended to compute Eulerian statistics, such as structure functions, from the collection of tracks (that can be two-frame displacement vectors in PIV). From particle positions $\mathbf{x}$, which are measured with some noise $\mathbf{b}$ ($\mathbf{x} = \hat{\mathbf{x}} + \mathbf{b}$, where $\hat{\mathbf{x}}$ are the actual positions), we define a 3D Lagrangian displacement field between two consecutive images taken at instants t and $t + dt$ is then $d\mathbf{x} = \mathbf{x}(t + dt) - \mathbf{x}(t) = d\hat{\mathbf{x}} + d\mathbf{b}$. This displacement field can be conditioned on a Cartesian grid so that its first moment

$$\langle d\mathbf{x} \rangle = \left\langle \hat{\mathbf{v}} \right\rangle dt + \left\langle \mathbf{b} \right\rangle + o(dt^2) \tag{6.9}$$

is computed in each bin of the grid to compute the mean flow $\left\langle \hat{\mathbf{v}} \right\rangle$. We then compute the centred second-order moment of the displacement field

$$\left\langle (d\mathbf{x} - \langle d\mathbf{x} \rangle)^2 \right\rangle = \left\langle \hat{\mathbf{v}}'^2 \right\rangle dt^2 + 2\left\langle \mathbf{b}'^2 \right\rangle + \left\langle \hat{\mathbf{a}}' \cdot \hat{\mathbf{v}}' \right\rangle dt^3 + o(dt^3), \tag{6.10}$$

where the prime stands for fluctuating quantities. Note that this formula is easily extended to centred cross-component second-order moments which are linked to the components of Reynolds stress tensor in each point of the grid.

The de-noising strategy is applied to data obtained from a pair of images taken with standard PIV cameras, one experimental set corresponds to a single value of dt. The moments $\langle d\mathbf{x} \rangle$ and $\left\langle (d\mathbf{x})'^2 \right\rangle = \left\langle (d\mathbf{x} - \langle d\mathbf{x} \rangle)^2 \right\rangle$ are then calculated for multiple experimental sets where images of the particles in the flow are collected at increasing values of dt. When the evolution of $\left\langle (d\mathbf{x})'^2 \right\rangle$ with dt is fitted by a polynomial of the form $c_1 dt^2 + c_2$ in each bin, the leading coefficient is the field $\left\langle \hat{\mathbf{v}}'^2 \right\rangle$. The third-order correction is negligible because dimensional analysis gives $\left\langle \hat{\mathbf{v}}'^2 \right\rangle / \left\langle \hat{\mathbf{a}}' \cdot \hat{\mathbf{v}}' \right\rangle \tau_\eta \sim Re_\lambda$, where $\tau_\eta = \sqrt{\nu/\varepsilon}$ is the dissipative time and Re_λ is the Reynolds number at the Taylor length scale. In turbulent flows, $\left\langle \hat{\mathbf{a}}' \cdot \hat{\mathbf{v}}' \right\rangle$ is well approximated by the dissipation rate ε. Taking dt smaller than the dissipative time ensures that the displacement field variance is well approximated. The advantage of this method is that it uses all the measurements taken at different values of

dt without having to choose any particular dt, as would be done in a classical PIV experiment. And unlike PIV, there is no filtering of the data in the form of windowing.

This method can be extended to higher order moments of the displacement field, as well as to recover increment statistics, for example, the longitudinal second-order structure function of the velocity ($\hat{S_2} = \left\langle [(\hat{\mathbf{v}}(\mathbf{x}+\mathbf{r}) - \hat{\mathbf{v}}(\mathbf{x})) \cdot \mathbf{e}_r]^2 \right\rangle$, with $\mathbf{e}_r = d\mathbf{x}/|d\mathbf{x}|$), by fitting the evolution of $\left\langle [(d\mathbf{x}(\mathbf{x}+\mathbf{r}) - d\mathbf{x}(\mathbf{x})) \cdot \mathbf{e}_r]^2 \right\rangle$ with a polynomial $\hat{S_2}(|\mathbf{r}|)dt^2 + c_2$:

$$\begin{aligned}\left\langle [(d\mathbf{x}(\mathbf{x}+\mathbf{r}) - d\mathbf{x}(\mathbf{x})) \cdot \mathbf{e}_r]^2 \right\rangle &= \left\langle [(d\mathbf{b}(\mathbf{x}+\mathbf{r}) - d\mathbf{b}(\mathbf{x})) \cdot \mathbf{e}_r]^2 \right\rangle + \\ &+ \left\langle [(\hat{\mathbf{v}}(\mathbf{x}+\mathbf{r}) - -\hat{\mathbf{v}}(\mathbf{x})) \cdot \mathbf{e}_r]^2 \right\rangle dt^2 + \cdots + \\ &+ \langle [(\hat{\mathbf{v}}(\mathbf{x}+\mathbf{r}) - \hat{\mathbf{v}}(\mathbf{x})) \cdot \mathbf{e}_r][(\hat{\mathbf{a}}(\mathbf{x}+\mathbf{r}) - \hat{\mathbf{a}}(\mathbf{x})) \cdot \mathbf{e}_r] \rangle dt^3 + o(dt^3).\end{aligned} \tag{6.11}$$

Note that the structure function computation does not require the conversion of displacements to Eulerian coordinates, but rather to bin the inter-particle distance $|\mathbf{r}|$. This means that measuring structure functions is possible at arbitrarily small separations $|\mathbf{r}|$, without any requirements on the Eulerian spatial binning. This method requires only a statistical convergence in the number of particles N at a certain range of inter-particle distance (a number that is proportional to N^2). This represents a significant advantage over methods for structure function computation that carry an associated increase in measurement noise at small separations $|\mathbf{r}|$.

The second-order moment of the velocity fluctuations and second-order structure function are presented here as examples of what the expansion of statistical moments, combined with data collected at different dt can achieve. Higher order moments for the velocity fluctuations and higher order structure functions can be easily computed by this method with reduced noise, although they will contain residual noise from the computation of lower order moments ($o(dt^3)$ terms above).

Results

Particle displacements measured in a homogeneous, isotropic turbulence experiment [33, 46] are used to demonstrate the validity and accuracy of the method. Two CMOS cameras with a resolution of 2048×1088 pixels were used in a stereoscopic arrangement. Images were collected in double-frame mode, separated by a time-step dt from $0.05\tau_\eta$ to $0.2\tau_\eta$. Alternatively, using a very fast acquisition/illumination rate using high-speed camera and kHz pulsed lasers allows us to collect a single image sequence and then take a variable dt in the analysis by skipping an increasing number of images in the sequence. Measurements were obtained in a volume of $10 \times 10 \times 1\ \mathrm{cm}^3$ using a Nd:YAG laser. For each experiment, approximately 10,000

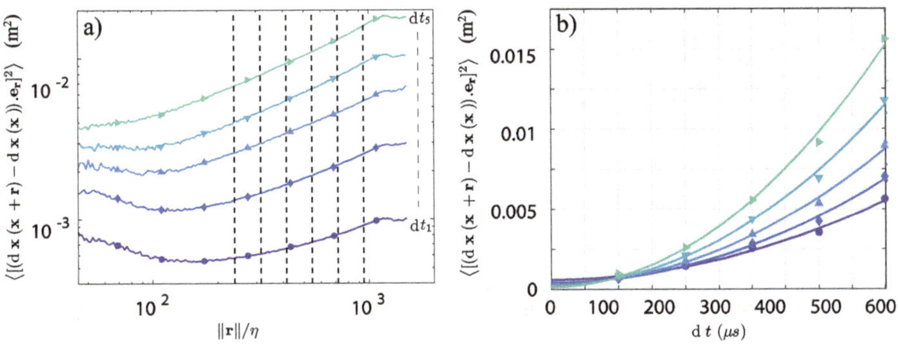

Fig. 6.10 (**a**) Longitudinal second-order structure functions of the raw displacement field against the separation $|\mathbf{r}|$ normalised by the Kolmogorov length scale η for different values of dt_i equally spaced from 0.05 to 0.2 τ_η at $Re_\lambda = 291$. (**b**) Same quantities but plotted at a given separation $|\mathbf{r}|$ (indicated by the vertical dashed lines on (**a**); ascending order is for different values of dt), as a function of the inter-frame time-step value dt. The lines are fits of the form $c_1 dt^2 + c_2$

pairs of image sets per time-step (each set providing the 3D position of several hundred particles in the flow) were collected to ensure statistical convergence.

The different results for the longitudinal second-order structure function of $d\mathbf{x}$ (Fig. 6.10a) at different time-steps, dt, show a strong dependency on how the noise affects the signal for different values of dt. The displacement correlation plotted at fixed separations (five different values) are all quadratic in dt (Fig. 6.10b), showing that this approximation is robust for different levels of measurement noise. The trend $c_1 dt^2 + c_2$ from Eq. (6.11) is followed at different values of the separation $|\mathbf{r}|$, with the positive values of c_2 being proportional to the variance of the noise (Eq. (6.10)). The quadratic coefficient c_1 is the second-order function of the velocity with the noise removed. The presence of the inertial range is highlighted by the 2/3 slope in Fig. 6.11a, over approximately one decade, in good agreement with the prediction of Kolmogorov for the second-order structure function in homogeneous isotropic turbulence ($\hat{S}_2 \sim \varepsilon^{2/3}|\mathbf{r}|^{2/3}$) [47]. Turbulence variables extracted from velocity measurements would be subject to a significant level of uncertainty and inaccuracy (seen in Fig. 6.10a) if the noise were not removed by the method proposed here.

Figure 6.11b shows the estimation of the dissipation rate of turbulent kinetic energy, $\varepsilon_r = \hat{S}_2^{3/2}/|\mathbf{r}|$ for three different Reynolds numbers studied in this experimental implementation of this de-noising method. The plateau values obtained confirm the presence of the inertial range and their values correspond to the ensemble average of the local dissipation rate. The estimations of ε, as well as u' (spatial average of the fluctuating velocity map), for different Reynolds numbers compare well with those in [33], obtained by 2D3C PIV, confirming the accuracy of the method. In fact, the values of u' and ε are slightly lower than those obtained by PIV. This discrepancy can be explained, qualitatively, based on the physics of the measurements and the effect of the noise on these metrics when it is not eliminated from the displacement measurements. Previous velocity measurements in the same

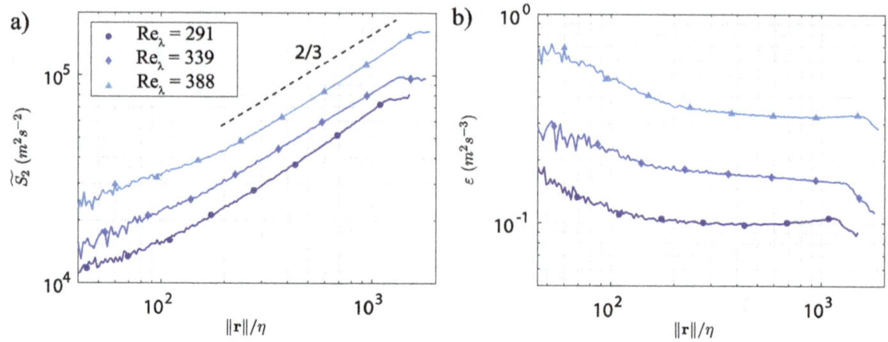

Fig. 6.11 (**a**) Second-order structure functions of the velocity extracted with the proposed method for different Re_λ. The black dashed line corresponds to a power law of exponent 2/3. (**b**) Energy dissipation rate estimated as $\varepsilon_r = \hat{S_2}^{3/2}/|\mathbf{r}|$

experiment, conducted by traditional PIV [33], corresponded well with the actual velocity measured with this technique, but with the noise variance retained. The structure function (and hence ε) measured with traditional techniques was also subject to this erroneous increase in the value due to the contribution of noise to the computation of this statistical value. Equation (6.11) shows that the term $\left\langle [(d\mathbf{b}(\mathbf{x}+\mathbf{r}) - d\mathbf{b}(\mathbf{x})) \cdot \mathbf{e_r}]^2 \right\rangle$ will increase the value ε due to noise. To determine the importance of this term, it is expanded into $4\left\langle \mathbf{b}^2 \right\rangle (1 - C_b(|\mathbf{r}|))$, where $C_b(|\mathbf{r}|)$ is the noise spatial correlation, bounded between $(-1, 1)$. Regardless of the value of C_b, it will erroneously increase the value of the structure function yielding a higher value of ε. As the value of C_b depends on spatial separation, it will not uniformly raise it for all values of $|\mathbf{r}|$ and the slope of the structure function may evolve with separation, making the value of ε noisier.

Discussion

The comparison of the flow statistics with a previous 2D3C PIV study [33] allows for the validation of the proposed method. In fact, the measurements show better results, with no need to tune arbitrary filtering parameters to remove noise (the interrogation window size, for instance). The only parameters that must be chosen for the method proposed here are the different values of dt that are accessible for a given flow and camera/illumination available, the form of the fit function, and finally the binning in space to compute the Eulerian average and fluctuating velocities (if so desired), and in separation distance to compute the structure function.

The values of dt are subject to two limitations. They must be high enough so that particles move more than the measurement error while keeping the large displacements associated with highest dt from interfering with the ability of the particle tracking algorithm to identify individual particles [48]. As mentioned, a maximum value of $dt \lesssim \tau_\eta$ ensures that the third-order correction remains small, $\left\langle \hat{\mathbf{v}}'^2 \right\rangle / \left\langle \hat{\mathbf{a}}' \cdot \hat{\mathbf{v}}' \right\rangle \tau_\eta \sim Re_\lambda$. This was verified in the present experimental setup and we found this correction to be negligible when compared to the second-order term. This was also the case for the structure function provided the separation lies in the inertial range $|\mathbf{r}| \gg \eta$. In such cases, the best agreement between fit functions and the data overall was found when using a quadratic function of dt. As for the number of time-step values needed, the value of ε when using only the three larger values of dt was only 5% lower than when using all five datasets. Using only the lowest value and largest values of dt allowed for a simple calculation of ε that was only 2% higher than with the full experimental set.

The displacement vector field obtained from particle tracking in this multiple time-step method is computed in a Lagrangian frame of reference. To compute the values of $\left\langle (d\mathbf{x})^2 \right\rangle$ against dt, the displacement field must be binned into a spatial grid, converting it to an Eulerian frame of reference. Although the number of particles per image, or Eulerian grid cell, is relatively small in these PTV images, the velocity is estimated independently for each particle pair. Thus, the statistical convergence in the method is reached relatively quickly (without the need for a very large number of image pairs). The computation of the structure functions highlights this advantage. As pointed out above, the structure function could in principle be computed to arbitrarily small separation between particles. However, great care should be taken in doing so because: (1) it is difficult to achieve statistical convergence in finding particles with small separations; (2) the second and third-order terms in Eq. (6.11) are of the same order of magnitude when the separation is in the dissipation range ($|\mathbf{r}| \sim 10\eta$). These reasons explain why an increase of the structure functions at small separations is observed in Fig. 6.10a.

6.5 Conclusions

We have presented recent developments in the characterisation of flows in laboratory experiments using particle tracking velocimetry, one of the most accurate techniques in experimental fluid mechanics. By tracking simultaneously hundreds of particles in 3D, it allows the experimentalist to address crucial questions related, for instance, to mixing and transport properties of flows.

The main aspects of particle tracking are addressed. A new optical calibration procedure based on a plane-by-plane transformation, without any camera model, is presented. It is at least as precise as Tsai model though more versatile as it naturally accounts for optical distortions and can be used in very complex configurations (such as Scheimpflug arrangement, for instance). Tracking algorithms are at the heart of PTV, and the practical implementation of two of their recent development is described: shadow particle tracking velocimetry using parallel light and trajectory reconstruction based on a four frames best estimate method (4BE) with improved initialisation. While the former was developed originally to access the size, orientation, or shape of the tracked particles, the latter is a natural extension of classical PTV setup and can be easily implemented as an add-on of any existing code.

A drawback of particle imaging techniques, as opposed to direct methods such as hot wire anemometry, is that they rely on measuring particle displacement. They are inevitably subject to noise amplification when computing spatial or temporal derivatives. We present recent developments addressing this important question which are based on computing statistics of the particle displacement with increasing time lag. They do not require any kind of filtering, and allow for the estimation of noiseless statistical quantities both in the Lagrangian framework (velocity and acceleration time correlation functions) and in the Eulerian framework (statistics of spatial velocity increments).

We conclude by mentioning that experimental techniques in fluid mechanics are continuously being improved, as new ideas combined with technological advances increase the resolution and the range of existing methods: for instance, cameras are becoming ever faster and sensors better resolved; an important breakthrough in high-resolution optical tracking is expected in the coming years, thanks to FPGA (field programmable gate array) technology which allows to process images on-board and hence to increase the effective data rate. Such a technique has been pioneered by Chan et al. [49] and further developed by Kreizer et al. [50] to achieve on-board particle detection, allowing to directly stream particle positions to the hard drive of the control computer, avoiding the usual memory limitation of high-speed cameras, and making the recording duration virtually unlimited.

References

1. R.J. Adrian, Particle-imaging techniques for experimental fluid mechanics. Annu. Rev. Fluid Mech. **23**(1), 261 (1991)
2. C. Tropea, A.L. Yarin, J.F. Foss, *Springer Handbook of Experimental Fluid Mechanics* (Springer, Berlin, 2007)
3. K. Yamamoto, Lagrangian measurement of fluid-particle motion in an isotropic turbulent field. J. Fluid Mech. **175**, 183 (1987)
4. M. Virant, T. Dracos, 3D PTV and its application on Lagrangian motion. Meas. Sci. Technol. **8**(12), 1539 (1997)
5. G.A. Voth, A. La Porta, A.M. Crawford, J. Alexander, E. Bodenschatz, Measurement of particle accelerations in fully developed turbulence. J. Fluid Mech. **469**, 121 (2002)
6. R.Y. Tsai, A versatile camera calibration technique for high-accuracy 3D machine vision metrology using off-the-shelf TV cameras and lenses. IEEE J. Robot. Autom. **3**(4), 323 (1987)
7. N. Ouellette, H. Xu, E. Bodenschatz, A quantitative study of three-dimensional Lagrangian particle tracking algorithms. Exp. Fluids **40**, 301 (2006)
8. H. Xu, Tracking Lagrangian trajectories in position-velocity space. Meas. Sci. Technol. **19**(7), 075105 (2008)
9. R.K. Lenz, R.Y. Tsai, Techniques for calibration of the scale factor and image center for high accuracy 3-D machine vision metrology. IEEE Trans. Pattern Anal. Mach. Intell. **10**, 713–720 (1988)
10. Z. Zhang, A flexible new technique for camera calibration. IEEE Trans. Pattern Anal. Mach. Intell. **22**, 1330 (2000)

11. J. Heikkila, O. Silven, in *Computer Vision and Pattern Recognition, 1997* (IEEE, Piscataway, 1997), pp. 1106–1112
12. D. Claus, A.W. Fitzgibbon, in *Computer Vision and Pattern Recognition, 2005*, vol. 1 (IEEE, Piscataway, 2005), pp. 213–219
13. F. Remondino, C. Fraser, Digital camera calibration methods: considerations and comparisons, in *ISPRS Commission V Symposium on Image Engineering and Vision Metrology*, vol. 36 (2006), pp. 266–272
14. H.G. Maas, D.P. A. Gruen, Particle tracking velocimetry in three-dimensional flows. part 1. photogrammetric determination of particle coordinates. Exp. Fluids **15**, 133 (1993)
15. N.A. Malik, D.P. Th. Dracos, Particle tracking velocimetry in three-dimensional flows. Exp. Fluids **15**, 279 (1993)
16. Y.G. Guezennec, R.S. Brodkey, N. Trigui, J.C. Kent, Algorithms for fully automated three dimensional tracking velocimetry. Exp. Fluids **17**, 209 (1994)
17. N.T. Ouellette, H. Xu, E. Bodenschatz, A quantitative study of three-dimensional Lagrangian particle tracking algorithms. Exp. Fluids **40**(2), 301 (2006)
18. P. Huck, N. Machicoane, R. Volk, in *Procedia IUTAM*, vol. 20 (Elsevier, Amsterdam, 2017), pp. 175–182
19. P.D. Huck, N. MacHicoane, R. Volk, Production and dissipation of turbulent fluctuations close to a stagnation point. Phys. Rev. Fluids **2**(8), 084601 (2017)
20. N.T. Ouellette, Probing the statistical structure of turbulence with measurements of tracer particle tracks. Ph.d. thesis, Cornell University, Ithaca (2006)
21. G.P. Romano, R.A. Antonia, T. Zhou, Evaluation of LDA temporal and spatial velocity structure functions in a low Reynolds number turbulent channel flow. Exp. Fluids **27**(4), 368 (1999)
22. O. Chanal, B. Chabaud, B. Castaing, B. Hébral, Intermittency in a turbulent low temperature gaseous helium jet. Eur. Phys. J. B **17**(2), 309 (2000)
23. N. Machicoane, M. López-Caballero, M. Bourgoin, A. Aliseda, R. Volk, A multi-time-step noise reduction method for measuring velocity statistics from particle tracking velocimetry. Meas. Sci. Technol. **28**(10), 107002 (2017)
24. N. Machicoane, P.D. Huck, R. Volk, Estimating two-point statistics from derivatives of a signal containing noise: application to auto-correlation functions of turbulent Lagrangian tracks. Rev. Sci. Instrum. **88**, 065113 (2017)
25. J.P. Salazar, J. De Jong, L. Cao, S.H. Woodward, H. Meng, L.R. Collins, Experimental and numerical investigation of inertial particle clustering in isotropic turbulence. J. Fluid Mech. **600**, 245 (2008)
26. J.S. J. Katz, Applications of holography in fluid mechanics and particle dynamics. Annu. Rev. Fluid Mech. **42**, 531 (2010)
27. D. Chareyron, J.L. Marié, C. Fournier, J. Gire, N. Grosjean, L. Denis, M. Lance, L. Méès, Testing an in-line digital holography 'inverse method' for the Lagrangian tracking of evaporating droplets in homogeneous nearly isotropic turbulence. New J. Phys. **14**(4), 043039 (2012)
28. D. Schanz, S. Gesemann, A. Schröder, Shake-The-Box: Lagrangian particle tracking at high particle image densities. Exp. Fluids **57**(5), 70 (2016)
29. R. Zimmermann, Y. Gasteuil, M. Bourgoin, R. Volk, A. Pumir, J.F. Pinton, Rotational intermittency and turbulence induced lift experienced by large particles in a turbulent flow. Phys. Rev. Lett. **15**, 154501 (2011)
30. R. Zimmermann, Y. Gasteuil, M. Bourgoin, R. Volk, A. Pumir, J.F. Pinton, Tracking the dynamics of translation and absolute orientation of a sphere in a turbulent flow. Rev. Sci. Instrum. **82**(3), 033906 (2011)
31. N. Machicoane, A. Aliseda, R. Volk, M. Bourgoin, A simplified and versatile calibration method for multi-camera optical systems in 3D particle imaging. Rev. Sci. Instrum. **90**, 035112 (2019)
32. A. Goshtasby, Image registration by local approximation methods. Image Vis. Comput. **6**(4), 255 (1988)

33. L. Fiabane, R. Zimmermann, R. Volk, J.F. Pinton, M. Bourgoin, Clustering of finite-size particles in turbulence. Phys. Rev. E **86**(3), 035301 (2012)
34. G.S. Settles, *Schlieren and Shadowgraph Techniques – Visualizing Phenomena in Transparent Media* (Springer, New York, 2001)
35. N. Machicoane, J. Bonaventure, R. Volk, Melting dynamics of large ice balls in a turbulent swirling flow. Phys. Fluids **25**(12), 125101 (2013)
36. P.K. Yeung, S.B. Pope, Lagrangian statistics from direct numerical simulations of isotropic turbulence. J. Fluid Mech. **207**, 531 (1989)
37. F. Ravelet, A. Chiffaudel, F. Daviaud, Supercritical transition to turbulence in an inertially driven von Kármán closed flow. J. Fluid Mech. **601**, 339 (2008)
38. R. Monchaux, F. Ravelet, B. Dubrulle, A. Chiffaudel, F. Daviaud, Properties of steady states in turbulent axisymmetric flows. Phys. Rev. Lett. **96**(12), 124502 (2006)
39. D. Faranda, F.M.E. Pons, B. Dubrulle, F. Daviaud, B. Saint-Michel, É. Herbert, P.P. Cortet, Modelling and analysis of turbulent datasets using auto regressive moving average processes. Phys. Fluids **26**(10), 105101 (2014)
40. A. Clark, N. Machicoane, A. Aliseda, A quantitative study of track initialization of the 4-frame best estimate algorithm for 3D Lagrangian particle tracking. Meas. Sci. Technol. **30**(4), 045302 (2019)
41. J. Graham, K. Kanov, X.I.A. Yang, M. Lee, N. Malaya, C.C. Lalescu, R. Burns, G. Eyink, A. Szalay, R.D. Moser, C. Meneveau, A web services accessible database of turbulent channel flow and its use for testing a new integral wall model for les. J. Turbulence **17**, 181 (2016)
42. N.G.V. Kampen, *Stochastic Processes in Physics and Chemistry* (Elsevier, Amsterdam, 1992)
43. N. Machicoane, R. Volk, Lagrangian velocity and acceleration correlations of large inertial particles in a closed turbulent flow, Phys. Fluids **28**(3), 035113 (2016)
44. N. Machicoane, R. Zimmermann, L. Fiabane, M. Bourgoin, J.F. Pinton, R. Volk, Large sphere motion in a nonhomogeneous turbulent flow. New J. Phys. **16**(1), 013053 (2014)
45. R. Volk, E. Calzavarini, E. Lévêque, J.F. Pinton, Dynamics of inertial particles in a turbulent von Kármán flow. J. Fluid Mech. **668**, 223 (2011)
46. R. Zimmermann, H. Xu, Y. Gasteuil, M. Bourgoin, R. Volk, J.F. Pinton, E. Bodenschatz, The Lagrangian exploration module: an apparatus for the study of statistically homogeneous and isotropic turbulence. Rev. Sci. Instrum. **81**(5), 055112 (2010)
47. A.N. Kolmogorov, The local structure of turbulence in incompressible viscous fluid for very large Reynolds numbers. Proc. R. Soc. A Math. Phys. Eng. Sci. **434**(1890), 9 (1991)
48. N.T. Ouellette, H. Xu, M. Bourgoin, E. Bodenschatz, An experimental study of turbulent relative dispersion models. New J. Phys. **8**(6), 109 (2006)
49. G.A.V. K.-Y. Chan, D. Stich, Real-time image compression for high-speed particle tracking. Rev. Sci. Instrum. **78**, 023704 (2007)
50. M. Kreizer, D. Ratner, A. Liberzon, Real-time image processing for particle tracking velocimetry. Exp. Fluids **48**(1), 105 (2010)

7

Simulating Motion of Active Particles through Brownian Dynamics and Finite-Difference Equations

Agnese Callegari and Giovanni Volpe

7.1 Introduction

Active particles differ from their passive counterparts for their ability to propel themselves. In Fig. 7.1 examples of active particles are given, classified with respect to their size and their propulsion speed. Living microorganisms propel themselves for different purposes such as finding food, escaping from predators or other dangers, and patrolling a territory [1]. Inspired by these microorganisms, researchers have recently developed several artificial particles capable of self-propelled motion activated by localised light, concentration, temperature gradients [2]. Despite the variety of possible self-propulsion mechanisms, we can identify some key features to describe the motion of a self-propelling micro- or nanosized particle: (1) *directionality* over a characteristic time interval, (2) *orientational noise*, and (3) *absence of inertia*. We note that in the case of living organisms the self-propulsion mechanism often implies a deformation in their shape; however, for simplicity, this aspect will not be taken into account in this chapter. Even if active particles are obviously three-dimensional and their motion also happens in 3D, here we will mainly consider motion in two dimensions, as in many real situations the motion of active particles is in a *quasi*-2D environment; for example, the case of motile bacteria moving above the lower horizontal surface of a sample slide. However, we will provide details on how to handle the motion of active particles in 3D as well. It is worth noting that in some cases 2D-confinement on active particles can give rise

A. Callegari (✉)
Bilkent University and UNAM, Ankara, Turkey
e-mail: callegari@fen.bilkent.edu.tr

G. Volpe
Gothenburg University, Gothenburg, Sweden

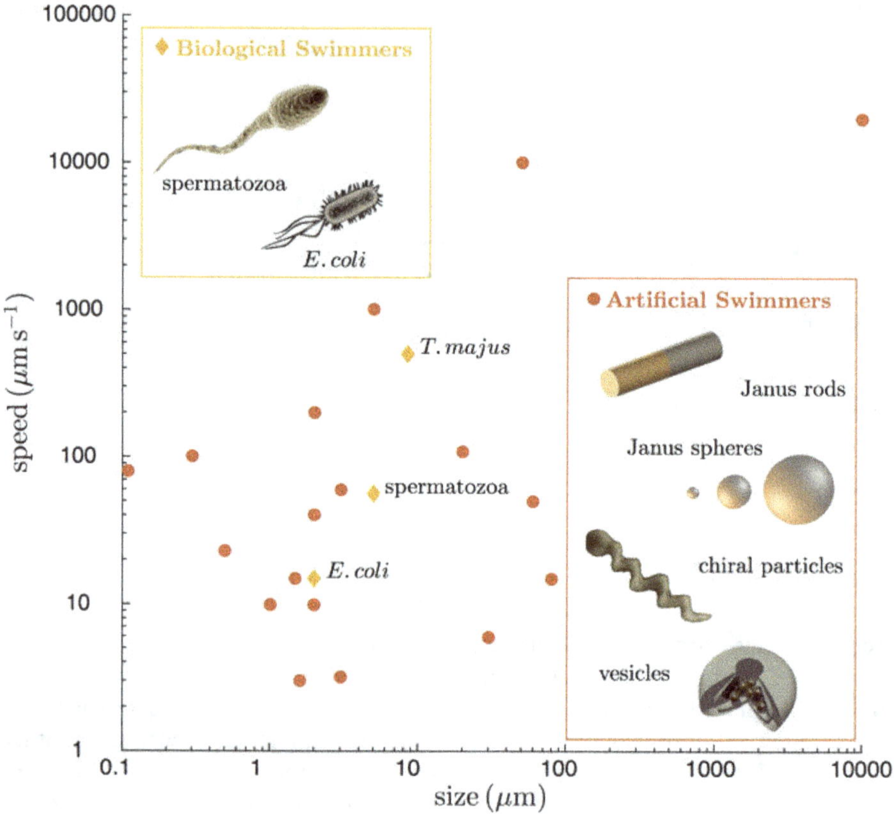

Fig. 7.1 Examples of active particles. Both living organisms and man-made particles are capable of propelling themselves. Here are represented biological and artificial particles of micro- and nanoscopic size. Their speed ranges from a few μm s^{-1} to mm s^{-1}. Adapted from Bechinger et al. [2]

to unexpected features that are not present in the case the active particles are moving in the bulk of a solution, i.e., full 3D motion with no confinement [3].

7.2 Passive Brownian Motion

Typical active particles are motile bacteria or artificial self-propelling microparticles, which perform their motion in a liquid environment. Since they are immersed in a fluid, active particles are subject to viscous force, always opposite to their velocity, and to thermal noise that is generated by the molecules constituting the fluid, which, because of the microscopic size of the particles, results in a non-negligible effect. Therefore, before entering into the details of how various models can describe the motion of active particles, we will discuss how the *passive* Brownian motion of a spherical particle can be described using Langevin equations and simulated using finite-difference equations [4].

Numerical simulations of Brownian dynamics date back to the 1970s and 1980s. Seminal works in this field are [5–7]. A comprehensive review is [8] and fundamental reference books are [9–11]. A reference for the numerical methods is [12], where the method with finite differences, better known as Euler–Maruyama scheme, is explained among other schemes. Here, we report the essential basics of passive Brownian dynamics and we invite the reader to refer to the milestone works cited above for a deeper insight and a comprehensive review.

Let us assume we have a spherical microscopic particle (for example, a transparent silica particle of 3 μm diameter) floating in a droplet of liquid solution (for example, a water solution) deposited on a microscopic glass slide. The mass of the particle is $m \approx 10^{-14}$ kg. If we observe it with a microscope, we will see that the particle moves erratically, hovering above the flat glass surface of a microscope slide. If we track the particle by recording its position at times regularly separated by a fixed time interval Δt, we will find that its translational motion is purely diffusive, with translation diffusion constant for each of the two main directions D_{t} given by

$$D_{\mathrm{t}} = \frac{k_{\mathrm{B}}T}{\gamma_{\mathrm{t}}}, \tag{7.1}$$

where k_{B} is the Boltzmann's constant, T is the absolute temperature, and γ_{t} is the friction coefficient of the particle for translational displacements (in the bulk of a liquid solution, $\gamma_{\mathrm{t}} = 6\pi\eta R$, where η is the viscosity of the fluid and R the radius of the particle). This equation is the simplest expression of Einstein's fluctuation–dissipation relation.

In the case of a homogeneous spherical particle with a perfectly smooth surface, it is not easy to experimentally detect the particle orientation. However, if we manage to measure the particle orientation, we will find that, in addition to an erratic translational motion, also the orientation of the particle changes randomly. If we can record the orientation, we will see that also the rotational motion is purely diffusive, this time with a different constant, the rotational diffusion constant D_{r} given by

$$D_{\mathrm{r}} = \frac{k_{\mathrm{B}}T}{\gamma_{\mathrm{r}}}, \tag{7.2}$$

where $\gamma_{\mathrm{r}} = 8\pi\eta R^3$ is the rotational friction coefficient of the particle.

The cause of these translational and rotational erratic motions lies in the interactions of the suspended colloidal particle with the molecules constituting the fluids, which are affected by the temperature and, at equilibrium, present a velocity partition function distributed according to Maxwell's distribution [13]. Because of the collisions with the fluid molecules, the particle experiences a *force* and a *torque* that perturb its motion (*thermal noise*).

The translational dynamics of a particle in a fluid environment is described by the Langevin equation:

$$m\mathbf{a} = -\gamma_t \mathbf{v} + \mathbf{F}_{th}, \tag{7.3}$$

where the term $-\gamma_t \mathbf{v}$ is the viscous friction force of the fluid, and $\mathbf{F}_{th}$ is the stochastic thermal force, which has zero average and variance $2\, k_B T \gamma_t$.

Because of the tiny mass of a microscopic particle, often inertia can be neglected. In fact, the characteristic time needed to forget the inertial effects is the relaxation time $\tau_{rel} = m/\gamma_t$, which increases with the mass of the particle m, and decreases with the friction coefficient γ_t. For microscopic particles like the prototype silica particle of 2 μm diameter, the relaxation time τ_{rel} is of the order of magnitude of 0.1 μs. Such relaxation time is several orders of magnitude below the time intervals typically sampled in experiments (for example, the time interval between two frames in an acquisition via a standard CMOS camera is of the order of some milliseconds). Therefore, Eq. (7.3) can be simplified to the overdamped Langevin equation:

$$\gamma_t \mathbf{v} = \mathbf{F}_{th}, \tag{7.4}$$

where the inertial term $m\mathbf{a}$ on the right side of Eq. (7.3) has been dropped. In all systems where $\Delta t \gg \tau_{rel}$, Eq. (7.4) is sufficient to capture the relevant measurable physical features. In fact, it is possible to demonstrate that the solution of Eq. (7.3) converges to Eq. (7.4) in the limit $m \to 0$ [14].

Often, Langevin equation (7.4) is rewritten as:

$$d\mathbf{r} = \sqrt{2D_t}\, d\boldsymbol{W}, \tag{7.5}$$

where $d\boldsymbol{W}$ is the derivative of a Wiener process, with zero average and variance 1 [4, 15]. The simplest, though effective, way to a numerical solution of Eq. (7.5) is to use a finite-difference approach. In 2D, the variables in Eq. (7.5) can be explicitly written as:

$$\begin{cases} dx = \sqrt{2D_t}\, dW_x \\ dy = \sqrt{2D_t}\, dW_y \end{cases}. \tag{7.6}$$

By writing the velocities as $v_x = \Delta x/\Delta t$ and $v_y = \Delta y/\Delta t$, Eq. (7.6) takes the form [4, 15]

$$\begin{cases} \Delta x = \sqrt{2D_t \Delta t}\; W_x \\ \Delta y = \sqrt{2D_t \Delta t}\; W_y \end{cases}, \tag{7.7}$$

where W_x and W_y are realisations of independent stochastic processes with average 0 and standard deviation 1 [4, 15]. By writing explicitly $\Delta x = x_{n+1} - x_n$ and $\Delta y = y_{n+1} - y_n$, we are led to the finite-difference equation

$$\begin{cases} x_{n+1} = x_n + \sqrt{2D_t\Delta t}\ W_{x,n} \\ y_{n+1} = y_n + \sqrt{2D_t\Delta t}\ W_{y,n} \end{cases}, \tag{7.8}$$

where, for a given Δt, we obtain the sequence $\{x_n, y_n\}$ representing the trajectory of the passive Brownian particle in the plane.

7.3 Active Particles

7.3.1 Active Brownian Motion

One of the simplest models of active motion is *active Brownian motion*. Let us consider a spherical particle that self-propels with a constant speed v along a given internal orientation direction in 2D. Just like a passive Brownian particle, this particle is also affected by thermal noise, which affects both its translation and rotation. The configuration of the active Brownian particle is described by three variables: two spatial coordinates x and y for the position according to the lab reference frame, and one rotational coordinate θ for the orientation of the particle with respect to the lab reference frame [16]. The equations determining the dynamics are

$$\begin{cases} \dot{x} = v\ \cos\theta\ + \xi_x \\ \dot{y} = v\ \sin\theta\ + \xi_y \\ \dot{\theta} = \xi_\theta \end{cases}. \tag{7.9}$$

The finite-difference equations relative to Eqs. (7.9) are therefore the following:

$$\begin{cases} x_{n+1} = x_n + v\ \cos\theta\ \Delta t + \sqrt{2D_t\Delta t}\ W_{x,n} \\ y_{n+1} = y_n + v\ \sin\theta\ \Delta t + \sqrt{2D_t\Delta t}\ W_{y,n} \\ \theta_{n+1} = \theta_n + \sqrt{2D_r\Delta t}\ W_{\theta,n} \end{cases}. \tag{7.10}$$

Figure 7.2 depicts typical trajectories for an active Brownian particle with 3 μm diameter with characteristic self-propulsion speed between $v = 0\,\mu\mathrm{m\,s^{-1}}$ (passive Brownian particle) and $12\,\mu\mathrm{m\,s^{-1}}$.

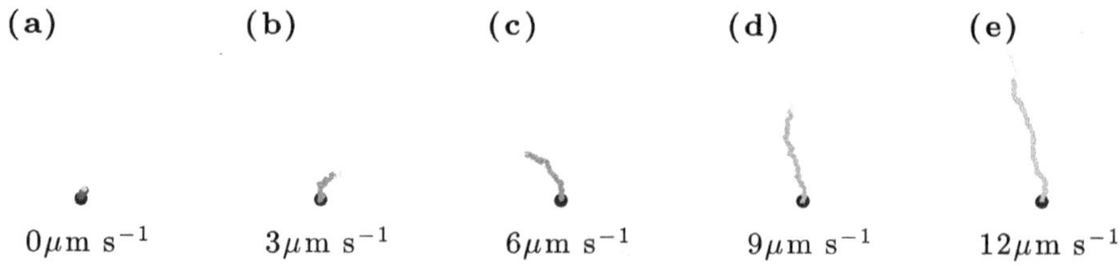

Fig. 7.2 From passive to active Brownian motion. (**a**) Passive Brownian motion ($v = 0\,\mu\text{m}\,\text{s}^{-1}$) and (**b–e**) active Brownian motion for increasing self-propulsion speeds (from $v = 3\,\mu\text{m}\,\text{s}^{-1}$ to $v = 12\,\mu\text{m}\,\text{s}^{-1}$). All trajectories last for the same amount of time

An important quantity for characterising the motion of microscopic systems is the mean square displacement (MSD). Considering a 2D motion, the MSD of a particle as a function of the time lapse t is

$$\Delta^2(t) = \left\langle (x(t_1 + t) - x(t_1))^2 + (y(t_1 + t) - y(t_1))^2 \right\rangle, \tag{7.11}$$

where the average is often performed over time.[1] The MSD over a given time t represents, therefore, the average quadratic displacement from the position the particle had a time t before. From the MSD, we can gain a lot of insights about the dynamics of a system.

The MSD of an active Brownian particle, i.e., for a trajectory governed by Eq. (7.9), is

$$\Delta^2(t) = \left(4D_\text{t} + v^2 t_\text{r}\right) t + \frac{v^2 t_\text{r}^2}{2}\left(e^{-\frac{2t}{t_\text{r}}} - 1\right), \tag{7.12}$$

where $t_\text{r} = D_\text{r}^{-1}$ is the characteristic time scale for the rotational diffusion, that, for a prototype particle with $3\,\mu\text{m}$ diameter, is of about 20 s. If we write explicitly the expression above for the limits $t \ll t_\text{r}$ and $t \gg t_\text{r}$, we have

$$\Delta^2(\tau) = \begin{cases} 4\,D_\text{t}\,t + v^2\,t^2 & t \ll t_\text{r} \\ \left(4\,D_\text{t} + v^2\,t_\text{r}\right) t & t \gg t_\text{r} \end{cases}. \tag{7.13}$$

For a prototype particle with $3\,\mu\text{m}$ diameter, the translational diffusion coefficient $D_\text{t} \approx 0.3\,\mu\text{m}^2\,\text{s}^{-1}$. If we consider an active particle with speed $v = 5\,\mu\text{m}\,\text{s}^{-1}$, then $v^2 = 25\,\mu\text{m}^2\,\text{s}^{-2}$. If we consider the case $t = 0.01 t_\text{r} = 0.2\,\text{s} \ll t_\text{r}$, then the diffusive contribution to the MSD ($4\,D_\text{t}\,t \approx 0.1\,\mu\text{m}^2$) is an order of magnitude

[1] In ergodic systems, the time average of a quantity coincides with the ensemble average, i.e., the average over all possible configurations, which, in this case, are all possible realisation of a trajectory.

smaller than the velocity contribution to the MSD ($v^2\ t^2 \approx 1\,\mu\text{m}^2$). The two contributions become comparable, when $t \approx 0.001 t_\text{r} = 0.02$ s. For smaller values of the time lapse t, the diffusive contribution prevails. However, if the reference active speed is faster than the $v = 5\,\mu\text{m}\,\text{s}^{-1}$, as it happens in many cases (for example, compare typical speeds given in Fig. 7.1), and the size is bigger than 3 µm diameter (again, compare typical swimmers sizes given in Fig. 7.1), then the range for the time lapse where the velocity contribution prevails starts from few hundredths of second.

Therefore, the experimentally observed dependence on the time interval t is essentially ballistic (i.e., quadratic, $\propto\ v^2 t^2$) for time scales smaller than the rotational diffusion time scale t_r, and diffusive (i.e., linear, $\propto\ t$) for time scales much longer than t_r. In the latter case, the rotational diffusion plays a role in the randomisation of the propulsion direction over long times, and acts as an effective enhancement of the diffusion proportional to $v^2\ t_\text{r}$, so that the long-term effective diffusion coefficient is

$$D_\text{eff} = D_\text{t} + \frac{v^2\ t_\text{r}}{4}\ . \tag{7.14}$$

If there is no self-propulsion (i.e., $v = 0$), the dynamics is that of a passive Brownian particle and the MSD is linear at all time scales. Figure 7.3 represents the MSD corresponding to trajectories obtained from the numerical integration of Eq. (7.9), for different values of the self-propulsion velocity v.

It is an interesting exercise to calculate the MSD also for the cases of run-and-tumble motion, chiral active Brownian motion, and active motion with Gaussian

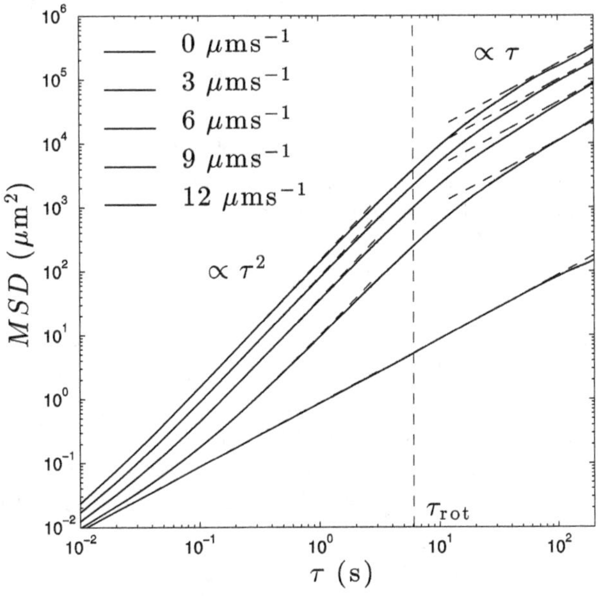

Fig. 7.3 Mean square displacement as a function of the propulsion velocity, from $v = 3\,\mu\text{m}\,\text{s}^{-1}$ to $v = 12\,\mu\text{m}\,\text{s}^{-1}$. The quadratic and diffusive regions are indicated. The vertical dashed line represents the rotational diffusion time scale t_r. The MSD for the passive Brownian particle ($v = 0\,\mu\text{m}\,\text{s}^{-1}$) is also represented, and is linear over all the range of time delays

noise reorientation mechanism, which will be described in the following subsections. Even though the microscopic locomotion mechanism differs at short time scales, the resulting MSDs present some universal features such as a ballistic behaviour at short time scales and an enhanced diffusive behaviour at long time scales.

7.3.2 Run-and-Tumble Motion

In the case of living organisms like motile bacteria, the observed motion can be described as a sequence of rectilinear forward steps and, occasionally, a sudden stop in the motion followed by a reorientation and by rectilinear motion along the new direction. Such kind of motion is known as *run-and-tumble motion* [1]. During the "runs", the bacterium moves forward because of the rotational motion of its flagella, which are tangled in a bunch. During the "tumbles", one of the flagella changes rotation direction breaking the bunch and the bacterium reorients itself. After this reorientation, the flagella form again a bunch and a new run starts. Run-and-tumble motion is a typical strategy of chemotactic organisms, i.e., organisms that calibrate their motion according to the presence or absence of determined chemical substances, usually an attractant or a repellent, often following the concentration gradient. The mechanism of run-and-tumble motion has been thoroughly studied in *E. coli* [17, 18] and various models have been developed to simulate it, from simple ones [19] to more complex ones [20].

In the absence of the chemical substance to which the chemotactic organism is responding, the reorientation events happen with a timing well-described by a Poisson process, characterised by the probability distribution:

$$P_\lambda(N = n) = e^{-\lambda}\frac{\lambda^n}{n!}, \tag{7.15}$$

where N is the number of events observed in the time interval Δt, λ is the average number of events expected in Δt, and n is a natural number. In this framework, the probability that tumbling happens is

$$P_{\text{tumble}} = 1 - P_\lambda(N = 0) = 1 - e^{-\lambda}\,. \tag{7.16}$$

If we want to describe the system dynamics in a finite-difference equations approach, we have to add another variable to the set (x, y, θ), because we have to know if the bacterium is running or tumbling. Therefore, we add then the discrete variable ϱ that keeps track of the status of the bacterium (1: run, 0: tumble), and each time step Δt has a probability P_{tumble} to be set to 0 and probability $P_{\text{run}} = 1 - P_{\text{tumble}}$ to be set to 1. The set of finite-difference equations is then:

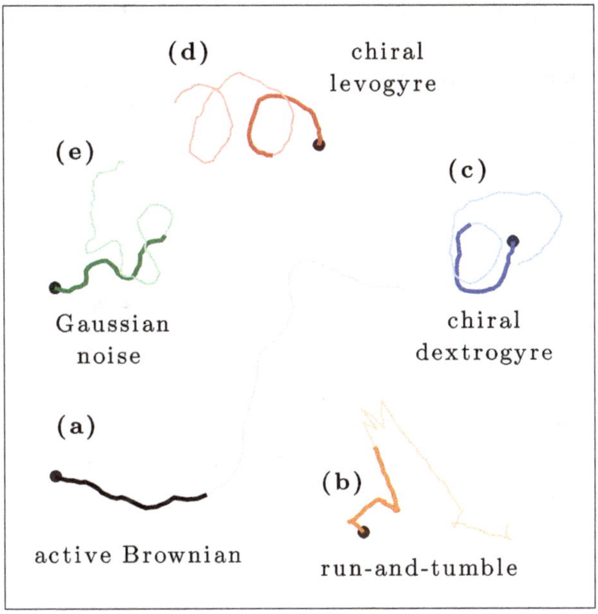

Fig. 7.4 Models for active motion: (**a**) active Brownian motion; (**b**) run-and-tumble motion; (**c**) dextrogyre and (**d**) levogyre chiral active Brownian motion; (**e**) Gaussian noise reorientation motion

$$\begin{cases} x_{n+1} = x_n + \varrho_n \; v \; \cos\theta \; \Delta t + \sqrt{2D_{\mathrm{t}}\Delta t} \; W_{x,n} \\ y_{n+1} = y_n + \varrho_n \; v \; \sin\theta \; \Delta t + \sqrt{2D_{\mathrm{t}}\Delta t} \; W_{y,n} \\ \theta_{n+1} = \theta_n + (1-\varrho_n) \; \Delta\Theta_{\mathrm{tumble,n}} \\ \varrho_{n+1} = 0 \text{ or } 1 \qquad \text{with probability } 1-e^{-\lambda} \text{ and } e^{-\lambda} \text{ each time step} \end{cases} . \tag{7.17}$$

An example of the appearance of a *run-and-tumble* trajectory is given in Fig. 7.4b.

7.3.3 *Chiral Active Brownian Motion*

It is not uncommon to observe that bacteria explore their environment by moving in circles. For example, *E. coli* bacteria have been shown to prefer to perform their quasi-circular motion in a clockwise fashion, when swimming close to a solid boundary, while they move counterclockwise when swimming near to an interface (for example, an air–liquid interface) [21–25]. The resulting motion is a chiral active Brownian motion. The term *chiral*, referring to the lack of symmetry of an object under mirror-reflection, is used to describe the tendency of swimmers to swim in circles either in a clockwise or a counterclockwise fashion. Therefore, the term *chiral* used in combination with the terms *active Brownian motion* indicates instead that the active particle tends to move in a well-defined way (clockwise or counterclockwise) with respect to a defined normal direction to the surface where the motion happens. Chiral motion is not only observed in living microorganisms, but also in artificial swimmers [26]. If we want to describe the dynamics of such a chiral active particle, we have to take into account that now the orientation varies with a defined *angular velocity* ω:

$$\begin{cases} \dot{x} = v\ \cos\theta\ + \xi_x \\ \dot{y} = v\ \sin\theta\ + \xi_y \\ \dot{\theta} = \omega\ + \xi_\theta \end{cases} . \tag{7.18}$$

According to the standard convention on the direction of angles, $\omega > 0$ will be associated with a *counterclockwise* motion, while $\omega < 0$ will characterise a *clockwise* motion. When translated to a finite-difference equation formalism, we have

$$\begin{cases} x_{n+1} = x_n + v\ \cos\theta\ \Delta t + \sqrt{2D_\mathrm{t}\Delta t}\ W_{x,n} \\ y_{n+1} = y_n + v\ \sin\theta\ \Delta t + \sqrt{2D_\mathrm{t}\Delta t}\ W_{y,n} \\ \theta_{n+1} = \theta_n + \omega\ \Delta t + \sqrt{2D_\mathrm{r}\Delta t}\ W_{\theta,n} \end{cases} . \tag{7.19}$$

An example of the appearance of a *chiral* trajectory is given in Fig. 7.4c, d, for dextrogyre and levogyre chiralities, respectively.

7.3.4 Gaussian Noise Reorientation Model

Active motion is not only related to particles that are able to propel themselves in a strict sense. In fact, it has been observed that passive colloids in an *active bath* (for example, in a solution containing motile bacteria) present an effective dynamics that is quite different from standard passive Brownian dynamics. The presence in the solution of motile living microorganisms changes the motion of the suspended colloidal particles in such a way that they behave as *effective active colloids* for which the reorientation mechanism can present an enhanced diffusion constant [27–30]. The model commonly used for describing such a situation is

$$\begin{cases} \dot{x} = v\ \cos\theta\ + \xi_x \\ \dot{y} = v\ \sin\theta\ + \xi_y \\ \dot{\theta} = \Xi_\theta \end{cases} . \tag{7.20}$$

This equation is practically Eq. (7.9), with the only difference that the noise term Ξ_θ is not characterised by the rotational diffusion constant $D_\mathrm{r} = k_\mathrm{B}T/(8\pi\eta R^3)$, but by a different $\tilde{D}_\mathrm{r}$, usually larger than the one dictated by the size of the particle. Because the dependence of the noise term is still a Gaussian, the model with enhanced diffusion constant for the orientation is referred as *Gaussian noise* model. An example of the appearance of a *Gaussian noise* trajectory is given in Fig. 7.4e.

7.4 More Complex Models

Until now, we have learned to model the active motion of a single spherically symmetric particle in a 2D homogeneous environment. In this section, we will extend the model to include (1) the 3D case of a single spherically and non-spherically symmetric particle, (2) the presence of external fields, (3) the presence of multiple interacting particles, and (4) the presence of a multiplicative noise (that is, noise that depends on the state of the system and not only on external variables like the temperature). All these extensions are used widely in simulations for the description of the behaviour of real active systems. For example, one can have a colloidal solution of Janus particles that are concentrated enough to come close to each other, so that their (steric, electrostatic, hydrodynamic) interaction should be considered for a correct description of their collective behaviour [31–33] or a colloidal particle in a bacterial bath in the presence of an external optical potential inducing on the particles a driving force in the direction of the light intensity gradient [30]. Another common case are active dynamics of non-spherical particles like elongated rods [34–37] or chiral particles [38–40] in 3D or the presence of a boundary [41, 42] that alters the value of the diffusion constant in its proximity via a diffusion gradient [43].

7.4.1 Non-Spherical Particles

If the particle is non-spherical, then the effect of the thermal noise is described by a diffusion matrix $\mathbb{D}$ of dimension 6×6, which takes into account all the translational and rotational modes of the non-spherical particle and possible correlations between them, including purely translational and purely rotational modes [44, 45]. The diffusion matrix $\mathbb{D}$ is always symmetrical ($\mathbb{D} = \mathbb{D}^T$) and is represented as:

$$\mathbb{D} = \begin{bmatrix} \mathbb{D}_\mathrm{tt} & \mathbb{D}_\mathrm{tr} \\ \mathbb{D}_\mathrm{rt} & \mathbb{D}_\mathrm{rr} \end{bmatrix}, \tag{7.21}$$

where $\mathbb{D}_\mathrm{tt}$ is the diffusion term for the purely translational modes, $\mathbb{D}_\mathrm{rr}$ is the diffusion term for the purely rotational modes, and $\mathbb{D}_\mathrm{tr}$ and $\mathbb{D}_\mathrm{rt}$ are the off-diagonal terms describing roto-translational effect of the thermal agitation that might arise for particular particles shapes breaking mirror symmetry. With the same diffusion matrix formalism, we can describe the Brownian motion of a spherical particle, in which case $\mathbb{D}$ is a diagonal matrix, with $\mathbb{D}_{\mathrm{tt}} = D_{\mathrm{t}}\mathbb{I}_{3\times3}$ and $\mathbb{D}_{\mathrm{rr}} = D_{\mathrm{r}}\mathbb{I}_{3\times3}$.

In case the passive motion of the spherical particle is in three dimensions, an equation analogous to Eq. (7.5) describes the dependence along all the three translation coordinates. However, because purely rotational and roto-translational terms may affect the orientation of the particle, it is a good practice to consider all the 6 degrees of freedom together. The analogue of Eq. (7.5) in 3D is

$$\begin{bmatrix} \dot{\mathbf{r}} \\ \dot{\boldsymbol{\theta}} \end{bmatrix} = \begin{bmatrix} \xi_{\mathrm{t}} \\ \xi_{\mathrm{r}} \end{bmatrix} \tag{7.22}$$

that, rewritten in the finite-difference formalism, becomes

$$\begin{bmatrix} \Delta\mathbf{r} \\ \Delta\boldsymbol{\theta}' \end{bmatrix} = \begin{bmatrix} \Xi_{\mathrm{t}} \\ \Xi_{\mathrm{r}} \end{bmatrix} \tag{7.23}$$

where the noise terms $\left[\ \Xi_{\mathrm{t}},\ \ \Xi_{\mathrm{r}}\ \right]$ are generated each step through a multi-varied Gaussian random number distribution satisfying the requirement of average equal to (0, 0, 0, 0, 0, 0) and variance matrix equal to $2\,\mathbb{D}\,\Delta t$.

The increments $\Delta\mathbf{r}$ represent the displacement of the centre of mass of the particle with respect to the previous position.

The angular increments $\Delta\boldsymbol{\theta}$ need to be handled with more care. The rotation of the particle is a free 3D rotation. Therefore, the increment $\Delta\boldsymbol{\theta}$ represents a set of three increment, one for each rotation axis (that is, one for each of the unit vectors defining the set of axis of the particle reference frame) [45]. As the composition of rotations among different axes is not commutative, one should choose a time step Δt such that the various increments $\Delta\boldsymbol{\theta}$ are small enough to ensure commutativity, within a certain error range. Each time step a rotation $\mathbb{R} = \mathbb{R}_x\,\mathbb{R}_y\,\mathbb{R}_z$ will be applied to the unit vectors determining the particle reference frame in order to find the rotated configuration. The rotation leaves the position of the centre of mass unaltered and, for the dynamics to make sense, should be such that the composition of the rotation matrices $\mathbb{R}_x$, $\mathbb{R}_y$, and $\mathbb{R}_z$ is the same, within the order of Δt.

A cleaner approach to the issue takes into account the algebra of the generators of the rotation matrix group. A rotation of an angle ϕ around the x, y, z axis is written, in 3D, as a matrix acting on the component of a vector along the base unit vectors, and in the specific case is

$$\mathbb{R}_x(\phi) = \begin{bmatrix} 1 & 0 & 0 \\ 0 & \cos\phi & -\sin\phi \\ 0 & \sin\phi & \cos\phi \end{bmatrix} \quad \mathbb{R}_y(\phi) = \begin{bmatrix} \cos\phi & 0 & \sin\phi \\ 0 & 1 & 0 \\ -\sin\phi & 0 & \cos\phi \end{bmatrix}$$
$$\mathbb{R}_z(\phi) = \begin{bmatrix} \cos\phi & -\sin\phi & 0 \\ \sin\phi & \cos\phi & 0 \\ 0 & 0 & 1 \end{bmatrix} \tag{7.24}$$

Each of this matrices can be written as an exponential of the generator matrices $\mathbb{G}_x, \mathbb{G}_y, \mathbb{G}_z$:

$$\mathbb{G}_x = \begin{bmatrix} 0 & 0 & 0 \\ 0 & 0 & -1 \\ 0 & 1 & 0 \end{bmatrix} \quad \mathbb{G}_y = \begin{bmatrix} 0 & 0 & 1 \\ 0 & 0 & 0 \\ -1 & 0 & 0 \end{bmatrix} \quad \mathbb{G}_z = \begin{bmatrix} 0 & -1 & 0 \\ 1 & 0 & 0 \\ 0 & 0 & 0 \end{bmatrix} \tag{7.25}$$

in the following way:

$$\mathbb{R}_x(\phi) = e^{\phi\ \mathbb{G}_x} = \sum_{n=0}^{+\infty} \frac{\phi^n}{n!}\ \mathbb{G}_x^n; \quad \mathbb{R}_y(\phi) = e^{\phi\ \mathbb{G}_y} = \sum_{n=0}^{+\infty} \frac{\phi^n}{n!}\ \mathbb{G}_y^n;$$
$$\mathbb{R}_z(\phi) = e^{\phi\ \mathbb{G}_z} = \sum_{n=0}^{+\infty} \frac{\phi^n}{n!}\ \mathbb{G}_z^n. \tag{7.26}$$

Instead of performing a small finite rotation about one axis at a time, where we might have issues because of non-commutativity, it is wiser to perform directly a rotation around the axis individuated by the vector $\boldsymbol{\omega}$

$$\boldsymbol{\omega} = (\omega_x, \omega_y, \omega_z) = \left(\frac{\Delta\theta_x}{\Delta t}, \frac{\Delta\theta_y}{\Delta t}, \frac{\Delta\theta_z}{\Delta t}\right) \tag{7.27}$$

of the angle $\theta = \sqrt{(\Delta\theta_x)^2 + (\Delta\theta_y)^2 + (\Delta\theta_z)^2} = \Delta t\ |\boldsymbol{\omega}|$ that is exactly the rotation acting on the particle. Such rotation matrix can be written as the exponential of the skew-symmetric matrix $\boldsymbol{\theta}_\times$:

$$\mathbb{R}_{\hat{\boldsymbol{\omega}}}(\theta) = e^{\boldsymbol{\theta}_\times} = \mathbb{I} + \sum_{n=1}^{+\infty} \frac{1}{n!}\ \boldsymbol{\theta}_\times^n, \tag{7.28}$$

where

$$\boldsymbol{\theta}_\times = \Delta t \begin{bmatrix} 0 & -\omega_z & \omega_y \\ \omega_z & 0 & -\omega_x \\ -\omega_y & \omega_x & 0 \end{bmatrix} = \begin{bmatrix} 0 & -\Delta\theta_z & \Delta\theta_y \\ \Delta\theta_z & 0 & -\Delta\theta_x \\ -\Delta\theta_y & \Delta\theta_x & 0 \end{bmatrix} \tag{7.29}$$

Because of the properties of $\boldsymbol{\theta}_\times$, namely $\boldsymbol{\theta}_\times^3 = -\theta^2\,\boldsymbol{\theta}_\times$, can be written as:

$$\mathbb{R}_{\hat{\boldsymbol{\omega}}}(\theta) = e^{\boldsymbol{\theta}_\times} = \mathbb{I} + \frac{\sin\theta}{\theta}\,\boldsymbol{\theta}_\times + \frac{1-\cos\theta}{\theta^2}\,\boldsymbol{\theta}_\times^2. \tag{7.30}$$

Equation (7.30) is the Rodrigues formula [46] for the rotation of an angle θ around a direction $\hat{\boldsymbol{\omega}}$ that is exactly the rotation of the axes of the particle reference frame due to a rotational noise term (stochastic rotational torque) of $\boldsymbol{\xi}_\mathrm{r}$.

7.4.2 External Fields

In many situations the particles feel the effects of external force or torque fields. In the case of colloidal particles suspended in a solution, these external fields can be due to the optical force generated by the presence of an optical potential [47], the presence of a hydrodynamic flux [48], the combined effect of weight and buoyancy that, for particles with an asymmetric mass distribution, can give rise to a torque leading to gravitaxis [49, 50], or the presence on an external magnetic field for paramagnetic particles [51–53]. Moreover, in many current realisations of artificial microswimmers electric [54], magnetic [55, 56], and acoustic fields [57, 58], or a combination of them [59, 60], play an important role to activate the self-propelling mechanism, to control the swimming behaviour, or to confine the active particles [61]. A variety of models have been developed in order to understand the mechanism of the self-propulsion and its proper simulation [62–65]. When writing the equations of motion for the total force $\mathbf{F}_\mathrm{tot}$ and torque $\mathbf{T}_\mathrm{tot}$ acting on a particle, we need to include the contributions of the external fields $\mathbf{F}_\mathrm{ext}$ and $\mathbf{T}_\mathrm{ext}$ as:

$$\begin{cases} \mathbf{F}_\mathrm{tot} = -\gamma_\mathrm{t}\,\mathbf{v} + \mathbf{F}_\mathrm{ext} + \mathbf{F}_\mathrm{thermal} \\ \mathbf{T}_\mathrm{tot} = -\gamma_\mathrm{r}\,\boldsymbol{\omega} + \mathbf{T}_\mathrm{ext} + \mathbf{T}_\mathrm{thermal} \end{cases} \tag{7.31}$$

that, in the overdamped limit, become

$$\begin{cases} \mathbf{v} = \dfrac{\mathbf{F}_\mathrm{ext}}{\gamma_\mathrm{t}} + \boldsymbol{\xi}_\mathrm{t} = \dfrac{D_\mathrm{t}}{k_\mathrm{B}T}\mathbf{F}_\mathrm{ext} + \boldsymbol{\xi}_\mathrm{t} \\ \boldsymbol{\omega} = \dfrac{\mathbf{T}_\mathrm{ext}}{\gamma_\mathrm{r}} + \boldsymbol{\xi}_\mathrm{r} = \dfrac{D_\mathrm{r}}{k_\mathrm{B}T}\mathbf{T}_\mathrm{ext} + \boldsymbol{\xi}_\mathrm{r} \end{cases}. \tag{7.32}$$

In the case of an active Brownian particle in 2D, the presence of an external force $\mathbf{F}_{\text{ext}}$ and/or torque $T_{\text{ext},z}$ is included in the equations as follows:

$$\begin{cases} \dot{x} = v\,\cos\theta + \dfrac{D_{\text{t}}}{k_{\text{B}}T}F_{\text{ext},x} + \xi_x \\ \dot{y} = v\,\sin\theta + \dfrac{D_{\text{t}}}{k_{\text{B}}T}F_{\text{ext},y} + \xi_y \\ \dot{\theta} = \dfrac{D_{\text{r}}}{k_{\text{B}}T}T_{\text{ext},z} + \xi_\theta \end{cases} . \tag{7.33}$$

In the general case of an active particle in a 3D environment, the effect of an external force $\mathbf{F}_{\text{ext}}$ and torque T_{ext} is included as follows:

$$\begin{bmatrix} \dot{\mathbf{r}} \\ \dot{\boldsymbol{\theta}} \end{bmatrix} = \frac{\mathbb{D}}{k_{\text{B}}T}\begin{bmatrix} \mathbf{F}_{\text{ext}} \\ \mathbf{T}_{\text{ext}} \end{bmatrix} + \begin{bmatrix} \boldsymbol{\xi}_{\text{t}} \\ \boldsymbol{\xi}_{\text{r}} \end{bmatrix} . \tag{7.34}$$

This formalism takes into account the effect of possible roto-translational effects, and reduces to the standard set of separable equations in the case translational and rotational motion are independent from each other. Besides the mathematical formalism, the presence of an external field might induce important features in the behaviour of a system of active Brownian particles. We will see an example of this in more details after discussing the modalities of simulating systems with more than one particle.

7.4.3 Interacting Particles

The study of the behaviour of a single active Brownian particle is an important starting point for understanding the behaviour of a system of multiple particles. When multiple particles are present, interactions among the particles may significantly change the dynamics. Such interactions, moreover, can affect the collective behaviour of the system, by determining the emergence of cooperative phenomena, like, for example, phase separation or the formation of dynamic clusters. Moreover, the presence of activity itself might change dramatically the behaviour of a system: the same interactions present in a system of active or passive particles may give rise to totally different outcomes, just because in one case we have the additional feature of self-propulsion.

Usually interactions among active particles are divided into two main categories: *non-aligning* and *aligning* interactions. *Non-aligning* interactions might be attractive or repulsive, might depend on the relative position of the particles, but are *independent* on their *direction of motion*. *Aligning* interactions instead depend on the particles' direction of motion as well. Often such interactions tend to align the particles in their direction of motion, favouring phenomena such as swarming, like, for example, in the Vicsek model [66]. Theoretically, it has been predicted that active particles responding to chemical signalling or to hydrodynamic interactions may interact mutually with an effective aligning interaction [67–70].

We start our brief overview of interacting active particles with steric interactions, which prevent the particles from occupying the same volume, and we continue with two kind of aligning interaction, one characterising the Vicsek model, and another one giving a torque on particles at very short distance from one another.

Steric Interactions

Usually colloidal particles have a well-defined, rigid shape, and it is not possible for them to overlap. This steric interaction, that is present between passive and active colloids, together with the presence of active particles might give rise to interesting phenomena, like the formation of metastable clusters, even though the interaction itself is not attractive. For example, a set of passive colloids does not form spontaneously any cluster except that in the presence of a strong attractive interaction or of a driving force that pushes the passive particles all in the same space [30]. A set of active particles, instead, because of their activity can form metastable cluster in dilute suspensions even in the presence of repulsive mutual interactions [71–74]. In fact, depending on the propulsion velocity v and on the rotational diffusion time τ_r, two active particles might collide together and stay locked because of the persistency of their active motion. Such clusters then can break apart over a time scale dictated by τ_r because the reorientation process makes one of the particle to point away [75, 76].

For what concerns the simulations of the cases where the particles are rigid and have a finite dimension, steric interactions can be implemented via the so-called *hard-sphere correction*, shown in Fig. 7.5, in order to avoid non-physical situations where the particles overlap and share in part the same volume. This is

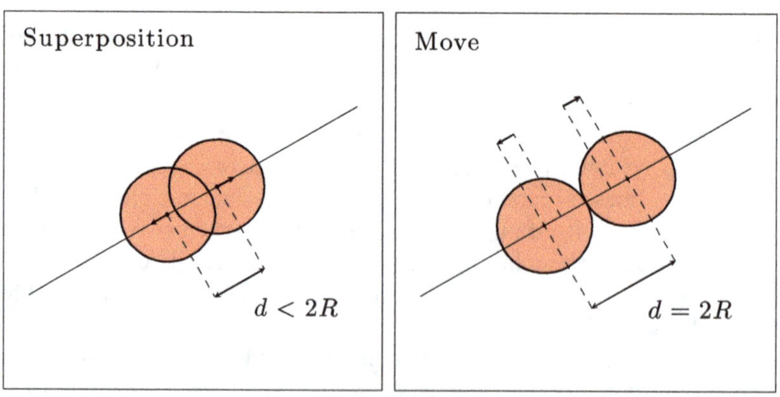

Fig. 7.5 Steric interactions. Schematic for hard-spheres interaction between two particles, where the particles are displaced away from each other whenever they get superposed

done by checking the mutual distance between couples of particles after each time step: if an overlap is found (for example, when the distance between the particles' centres is smaller than the diameter of the particles), then both particles are moved away from each other along the direction connecting the centres in such a way that their distance becomes exactly one diameter, so they touch each other, but are not superposed. This procedure has been used in [77] and, although it does not implement an elastic collision among particles like the other approaches [9] where conservation of energy and momentum is enforced, it is computationally lighter. However, the allowed superposition among the superposing particles is usually a small fraction of the particle volume, so, the time step of the simulation should be accordingly small, not to lead to unphysical effects, like a state of excessive superposition or to the unphysical conditions of two nearby particles that, instead of colliding, miss each other because of an excessive value of the calculated displacement. In case numerous particles are in close contact to each other, like, for example, when they are part of a cluster, one should check iteratively each couple of neighbouring particles, so to ensure that, after each readjusting move, the condition of non-superposition remains valid. Such implementation has been shown to be equivalent to introduce a short-range repulsive potential that prevents superposition. However, computationally the hard-spheres correction is preferred, because for the correct dynamics with the short-range, repulsive potential one would have to use a much smaller time step, which is unpractical.

Vicsek Model

The Vicsek model [66] is one of the simplest models to feature alignment and swarming in a system of active particles. In the Vicsek model, the particles move with constant speed and interact via the following aligning interaction: each particle can sense the orientation of the particles within a determined flocking radius R_{flocking}, and at each step reorients itself according to the average value of the neighbouring particles' orientation (Fig. 7.6). If, for each particle i in the system, we define $\mathcal{S}_i$ as the set of neighbouring particles at the considered instant in time, then the equations describing the system are

$$\begin{cases} \dot{x}_i = v \cos\theta_i \\ \dot{y}_i = v \sin\theta_i \\ \theta_i = \langle\theta_j\rangle_{j\in\mathcal{S}_i} + \xi_{\theta,i} \end{cases} . \tag{7.35}$$

Varying the range of the parameters describing the system, within this model one can obtain a phase transition from undirected motion to unidirectional motion, as a function of the particles' density: beyond a given density, whatever the initial conditions on position and orientation, the particles will come to move in the same direction, thanks to the aligning interaction. If we include in Vicsek model the steric interaction, we can obtain crystallisation at high densities and low-noise intensity.

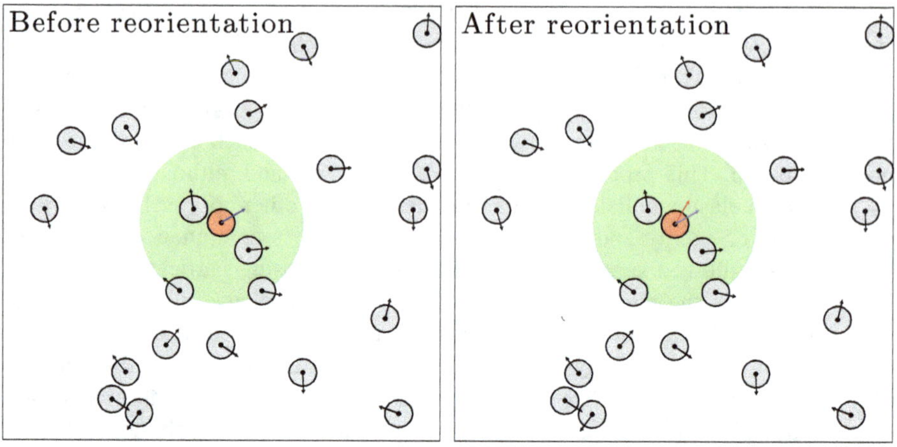

Fig. 7.6 Vicsek model: reorientation mechanism. The status at a given instant is represented. For establishing the orientation at the following instant of the particle represented in orange, one has to take the average of the orientations of the particles lying within the flocking radius (green area). In the second panel the new orientation is shown in red. The previous orientation is shown in light blue, for a comparison

Short-Range Aligning Interactions

Here we present another mechanism of aligning interaction [78], which can describe the effective behaviour of particles interacting with some aligning hydrodynamic interactions, or of bacteria moving in a background of colloidal particles, or of people moving in a crowd. We have a set of finite-size active particles in 2D which, when closer than a given distance R_{align}, interact by means of a torque. The behaviour featured by a particle is a tendency to reorient the direction of its active displacement towards the particles in the forward direction of motion, and away from the particles that are lying backwards. If $\hat{\mathbf{k}}$ is the unit vector direction perpendicular to the plane of motion, the torque acting on a given particle will be along $\hat{\mathbf{k}}$. The torque acting on a given particle n because of the presence of another particle i within the interaction distance is proportional to the cosine of the angle formed by the direction connecting the centre of the particle to the particle i ($\hat{\mathbf{r}}_{ni}$) with the direction of the velocity of the particle n ($\hat{\mathbf{v}}_n$), and inversely proportional to the square of the distance among the particles.

In order to implement the behaviour described above, the torque should be along the vector $\hat{\mathbf{v}}_n \times \hat{\mathbf{r}}_{ni}$, and proportional to $(\hat{\mathbf{v}}_n \cdot \hat{\mathbf{r}}_{ni})/r_{ni}^2$. Such coefficient can be positive, negative, or zero, depending on $\hat{\mathbf{v}}_n \cdot \hat{\mathbf{r}}_{ni}$, and this is the ingredient that gives the wanted behaviour of reorienting towards particles set in the front, and turning away from the particles lying at the back (Fig. 7.7).

Expressing the total torque on a particle using the coefficient along the given $\hat{\mathbf{k}}$ direction, we have that the total torque felt by the particle n can be modelled according to:

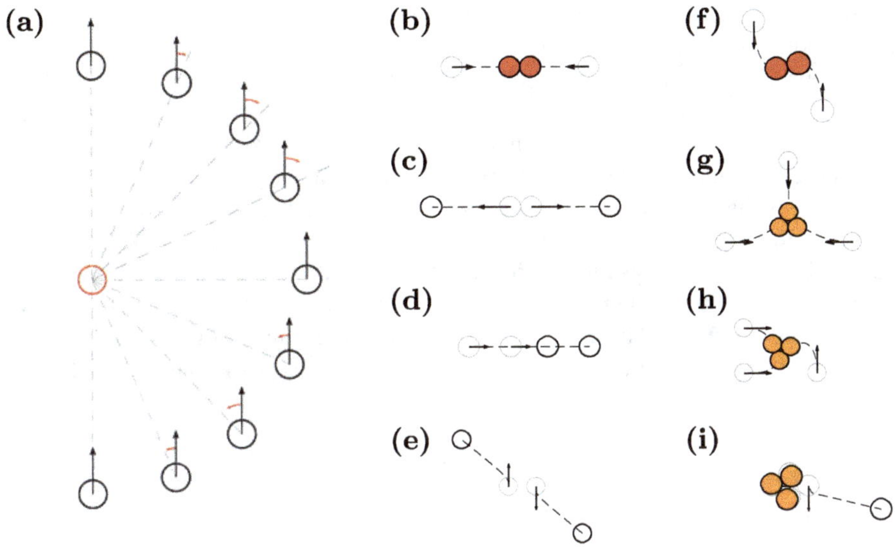

Fig. 7.7 Short-range aligning interaction: reorientation mechanism. In (**a**) a finite set of relative positions and relative motions are shown. The particle represented in red (centre of the scheme) exerts a torque on the particle in black. Such torque obeys the mechanism in Eq. (7.36) and the reorientation angle it induces is represented, for the various case considered, by a red oriented arc. In (**b–e**) a few cases characteristic of the induced dynamics with two and three particles are represented. In particular, it is shown the situations leading to stable clusters of 2 particles (**b, f**) and 3 particles (**g–i**). In the cases represented in (**c–e**), the relative arrangement of the particles does not allow to form any cluster, but part away from each other (**c, e**) or proceed at constant speed and constant distance, with one of the particles following the other one. Reproduced from Nilsson and Volpe [78].

$$T_n = T_0 \sum_{i \in \mathcal{S}} \frac{\hat{\mathbf{v}}_n \cdot \hat{\mathbf{r}}_i}{r_i^2} \hat{\mathbf{v}} \times \hat{\mathbf{r}}_i \cdot \hat{\mathbf{k}}, \tag{7.36}$$

where $\mathcal{S}$ is the set of particles within the radius of interaction R_{align} from particle n, and T_0 is a parameter setting the strength of the interaction. So, in general, the model is described by the following equations:

$$\begin{cases} \dot{x}_n = v \ \cos\theta_n \ + \xi_{x,n} \\ \dot{y}_n = v \ \sin\theta_n \ + \xi_{y,n} \ . \\ \dot{\theta}_n = T_n + \xi_{\theta,n} \end{cases} \tag{7.37}$$

where T_n is the torque obtained from Eq. (7.36) for each particle n. We note that in this specific model the rotational friction coefficient that usually is present in front of the torque is set to 1.

Such a model, though simple, displays a rich set of behaviours. In fact, depending on the rotational noise conditions and on the particle concentration, we can find a gaseous phase, where the particles are moving independently, a phase where metastable clusters composed by few particles are present, and a more pronounced clustering phase where clusters of more particles are formed. It has been shown [78] that the clustering transition occurs for the critical noise level of $T_0/4R^2_{\text{align}}$. We expect also that the density should have an effect on the clustering transition; however, this aspect has not been investigated in the original paper. It has been also shown that in a mixed system of a few active and many passive particles, where the active particles interact with an alignment term as in Eq. (7.36) with the other active particles and with a term of the opposite sign where interacting with the passive particles, the interaction can lead to the formation of metastable channels [78].

7.4.4 Multiplicative Noise

In all the previous examples we have dealt with a uniform diffusion constant and therefore with noise conditions uniform in time, independent from the status of the system. However, in many real systems the noise may depend on the configuration of the system: for example, it is known that the presence of a rigid planar wall induces a gradient in the thermal diffusion coefficient, in such a way that the closer the particle to the planar boundary, the smaller the diffusion coefficient [79–82]. In particular, for the direction perpendicular to the planar surface, the diffusion coefficient $D_\perp$ becomes zero when the particle finds itself in contact with the wall [79, 83]. An analogous effect happens also when two particles in the bulk of a solution come close to each other: the presence of one of the particle in the proximity of the other alters the diffusion coefficient of the second particle, and vice versa [84].

In general, a multiplicative noise is described as:

$$\xi = \sqrt{(2\,D(x))}W, \tag{7.38}$$

where the diffusion constant depends on the value of the variable describing the status of the system. To fix our ideas, let us suppose it indicates the distance of a particle from the planar wall. In such cases, the full Langevin equation for the particle would be Eq. (7.3), with the only difference that the noise term has a dependence on the variable x. The corresponding equation is [15][2]

$$\dot{x} = \frac{D(x)}{k_{\text{B}}T}F_{\text{ext}} + \frac{\text{d}D(x)}{\text{d}x} + \xi(x), \tag{7.39}$$

[2]The equation here is written in the Îto form. Alternative forms are possible, see details in Ref. [15].

where the additional term $\frac{dD}{dx}$ is the *spurious drift* and is necessary for the correct convergence of the solution to the original Langevin equation with multiplicative noise [14]. Finally, the corresponding finite-difference equation describing a particle in a gradient of diffusion is

$$x_{n+1} = x_n + \frac{dD(x_n)}{dx}\Delta t + \sqrt{2D(x_n)\Delta t}\, W_n. \quad (7.40)$$

7.5 Numerical Examples

Here we show two examples of collective behaviours emerging in a system of self-propelling particles.

7.5.1 *Living Crystals*

We have performed a simulation of a system of $N = 100$ active particles considering steric interaction and a short-range phoretic attraction on the line of the model employed in [85], which was proposed to reproduce the experimentally observed living crystals emerging in a solution with light-activated colloids. In this model, the phoretic attractive force, due to the advective flow generated by the decomposition of hydrogen peroxide on the exposed hematite surface, induced an attractive velocity $v_P(r)$ for the active particles that scales according to the inverse of the square distance among the particles:

$$v_P(r) = v_{P0}\,\frac{r_0^2}{r^2}. \quad (7.41)$$

Such dependence is the expected behaviour for a phoretic attraction to a reaction source, and fit well the observed experimental behaviour [85].

Varying the strength of this attractive interaction, which is the magnitude v_{P0} of the attractive velocity at a given reference distance r_0, and the self-propulsion velocity of the particle v, in the case particles are clustering, we can observe the formation of clusters of different average sizes. Such clusters are actively changing and rearranging themselves, and their stability depends also on the relative ratio of the interaction velocity at short distances (comparable with the diameter of the particles) with the self-propulsion velocity of the single active particle. For a given active velocity v, if the attractive reference velocity v_{P0} for a given r_0 equal to the diameter of the particles is comparable to (or bigger than) the self-propulsion velocity v, then the clusters will tend to form. The bigger the reference value v_{P0} of the attractive phoretic interaction, the more stable and bigger the cluster formed.

In Fig. 7.8 we provide two examples of simulations of active Brownian particles interacting via an attractive phoretic force that induces an attractive phoretic velocity given by Eq. (7.41). The radius of the particles is set to $R = 1.0\,\mu\text{m}$, and their speed is set to $v = 2\,\mu\text{m}\,\text{s}^{-1}$. In both the simulation presented, the initial configuration of positions and orientations is randomly generated. We consider a square arena with a side of $L = 100\,\mu\text{m}$ (concentration of $\phi_V = 0.03$) with periodic boundary conditions; the concentration of the particles can be adjusted by varying the size of the arena. The time step of the simulation is set to $\Delta t = 0.1$ s. With these parameters we observe the formation of metastable clusters as in Ref. [85]. By decreasing the strength of the interaction, we can see that the clusters become less stable. In Fig. 7.8 a few screenshots of the time evolution are presented.

In the dynamics represented in Fig. 7.8a–c, the reference phoretic velocity v_{P0} is set to $2\,\mu\text{m}\,\text{s}^{-1}$ at a distance between colloids of $r_0 = 2\,\mu\text{m}$, and the radius of the

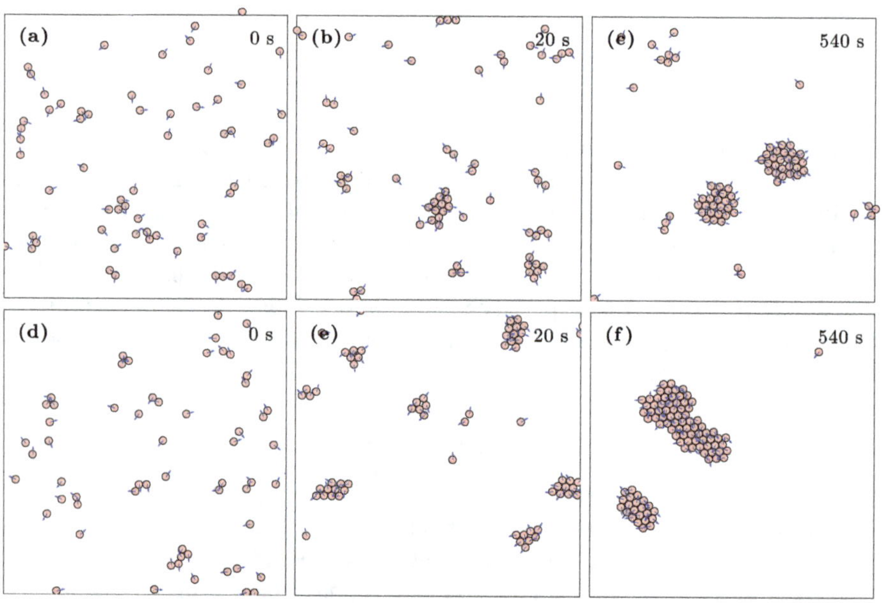

Fig. 7.8 Living crystals. Simulation of 100 active particles of radius $R = 1.0\,\mu\text{m}$ with phoretic, short-range attractive interaction. The system forms metastable clusters. Depending on the particle concentration, on the noise conditions, and on the strength of the attractive interaction, the clusters may become more or less stable. In the simulation, the noise affecting the position and the orientation is set to the thermal noise felt by a spherical particle at equilibrium in a thermal bath. In (**a–c**), the phoretic interaction is set in such a way to give a phoretic attractive velocity equal to the speed of the active particle at a reference distance of $2.0\,\mu\text{m}$ between the particles; the clusters formed are of the size of 10–20 particles, and are not too tightly bond and visibly evolve in the time scale of the simulation. In (**d–f**), the phoretic interaction was set to give a phoretic attractive velocity equal to the speed of the active particle at a reference distance of $2.5\,\mu\text{m}$ between the particles; in this case the clusters formed are more compact, and, as the time pass, tend to form bigger clusters

interaction is set to $R_{\mathrm{int,phor}} = 10\,\mu\mathrm{m}$ [85]. We can observe that there is a tendency to form clusters, but they are not bigger than 20–30 particles each, and a consistent fraction of the particles are still freely moving. In the dynamics represented in Fig. 7.8d–f, instead, the reference phoretic velocity v_{P0} is set to $2\,\mu\mathrm{m\,s^{-1}}$ at a reference distance between colloids of $r_0 = 2.5\,\mu\mathrm{m}$, i.e., the attractive interaction is stronger. Also in this case the radius of the interaction is set to $R_{\mathrm{int,phor}} = 10\,\mu\mathrm{m}$. As the attractive interaction is stronger, now the particles tend to form bigger clusters and less particles are freely moving independently.

7.5.2 *Colloids with Short-Range Aligning Interaction*

We consider a system of active particles interacting via the aligning interaction described in Eq. (7.36) and with a dynamics described by Eq. (7.37), following [78], where we explicitly take into account the proper value of the friction coefficient of a spherical particle in water for calculating the effect of the torque. We perform a simulation with $N = 400$ particles in a low-noise condition, with colloids of dimension $R = 1.0\,\mu\mathrm{m}$, setting the interaction radius $R_{\mathrm{align}} = 2.5\,\mu\mathrm{m}$, a speed of $v = 0.1\,\mu\mathrm{m\,s^{-1}}$, and torque $T_0 = 0.5\,\mathrm{rad}\,\mu\mathrm{m}^2$ (Eq. (7.36)). With a simulation time step $\Delta t = 0.05\,\mathrm{s}$, we observe that clusters start forming and, because of the low-noise level, persist for a long time. In Fig. 7.9 a few screenshots of the time evolution are presented. In the chosen conditions of noise, range of the interaction, and strength of the interaction, we observe that mainly small clusters of two, three particles are formed. We observe also that some simple configurations like the ones shown in the circled regions of Fig. 7.9b–c (cluster of two particles facing each other with opposite orientation, and cluster of three particles organised in an equilateral triangle configuration with their orientation facing the centre of the triangle) are quite stable in time, and they do not move, and this happens because of

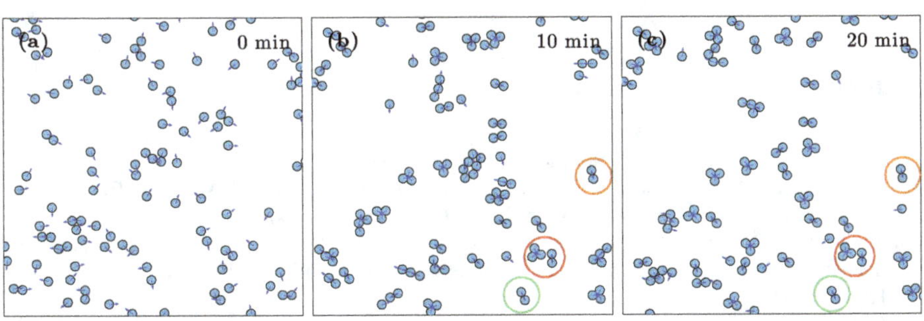

Fig. 7.9 Short-range aligning interactions. Screenshot of a simulation with initial random distribution of position and velocities. Stable clusters up to five particles form in the time of the simulation. As expected, small clusters of 2 and 3 particles are quite stable in time, given the low-noise level condition chosen, as can be seen comparing the configurations of the clusters evidenced in the circles

the interaction mechanisms that tends to preserve the configuration and the relative low level of noise. Such small stable clusters break apart or transform only when a travelling particle or cluster comes in their proximity and, driven by the aligning interaction, targets them, as it is shown in the Supplementary Movies of Ref. [78]. Varying the conditions of noise, the strength of the interaction, the range of the interaction, one can reproduce various regimes, from a gaseous phase, with mainly small clusters and single isolated particles, to a cluster phase, with mainly clusters of various dimensions, the distribution of the size of the cluster being determined by the chosen condition of noise, strength of the interaction, and range of the interaction.

7.6 Conclusions

We have provided a concise, self-contained introduction to the simulation of active Brownian systems using Brownian dynamics and finite-difference equations. We have illustrated how to describe the dynamics of a micron-sized active particle starting from a model spherical particle, how to account for white thermal noise for translational and rotational motion, and active displacement, starting from a motion in two dimensions. We have briefly explained the way to proceed to simulate the full 3D motion, with the proper formalism also for non-spherically symmetric objects, and reviewed a few type of interactions that have been used in the description of the dynamics of active systems, and, depending on the choice of the parameters, present a variety of complex behaviours. We have tried to provide an overview that gives hints of the up-to-date research on active system, where simulation is often an important tool to test the understanding of the key mechanism determining the behaviour of the particular system under investigation. At the same time, we have kept the discussion as simple as possible and given a few numerical examples to provide an effective starting point for the students approaching numerical simulation of Brownian dynamics for the first time.

References

1. H.C. Berg, *E. coli in Motion* (Springer, New York, 2004)
2. C. Bechinger, R. Di Leonardo, H. Löwen, C. Reichhardt, G. Volpe, G. Volpe, Active particles in complex and crowded environments. Rev. Mod. Phys. **88**, 45006 (2016)
3. K. Dietrich, D. Renggli, M. Zanini, G. Volpe, I. Buttinoni, L. Isa, Two-dimensional nature of the active Brownian motion of catalytic microswimmers at solid and liquid interfaces. New J. Phys. **19**, 65008 (2017)
4. G. Volpe, G. Volpe, Simulation of a Brownian particle in an optical trap. Am. J. Phys. **81**, 224 (2013)
5. D.L. Ermak, J.A. McCammon, Brownian dynamics with hydrodynamic interactions. J. Chem. Phys. **69**(4), 1352 (1978)
6. G. Ciccotti, J.P. Ryckaert, On the derivation of the generalized Langevin equations for interacting Brownian particles. J. Stat. Phys. **26**, 73 (1981)

7. W.F. van Gunsteren, H.J.C. Berendsen, Algorithms for Brownian dynamics. Mol. Phys. **45**, 637 (1982)
8. J.F. Brady, G. Bossis, Stokesian dynamics. Annu. Rev. Fluid Mech. **20**, 111 (1988)
9. M.P. Allen, D.J. Tildesley, *Computer Simulation of Liquids*, 1st edn. (Oxford Science Publications; Clarendon Press, Oxford, 1987)
10. P.E. Kloeden, E. Platen, *Numerical Solution of Stochastic Differential Equations* (Springer, New York, 1992)
11. H.J.C. Berendsen, *Simulating the Physical World* (Cambridge University Press, New York, 1987)
12. E. Platen, An introduction to numerical methods for stochastic differential equations. Acta Numer. **8**, 197–246 (1999)
13. E. Nelson, *Dynamical Theories of Brownian Motion* (Princeton University Press, Princeton, 1967)
14. S. Hottovy, A. McDaniel, G. Volpe, J. Wehr, The Smoluchowski–Kramers limit of stochastic differential equations with arbitrary state-dependent friction. Commun. Math. Phys. **336**, 1259 (2015)
15. G. Volpe, J. Wehr, Effective drifts in dynamical systems with multiplicative noise: a review of recent progress. Rep. Prog. Phys. **79**, 53901 (2016)
16. G. Volpe, S. Gigan, G. Volpe, Simulation of the active Brownian motion of a microswimmer. Am. J. Phys. **82**, 659 (2014)
17. B.E. Scharf, K.A. Fahrner, L. Turner, H.C. Berg, Control of direction of flagellar rotation in bacterial chemotaxis. Proc. Natl. Acad. Sci. **95**, 201 (1998)
18. U. Alon, L. Laura Camarena, M.G. Surette, B. y Arcas, Y. Liu, S. Leibler, J.B. Stock, Response regulator output in bacterial chemotaxis. EMBO J. **17**, 4238 (1998)
19. C.C.N. Wang, K.L. Ng, Y.C. Chen, P.C.Y. Sheu, J.J.P. Tsai, in *2011 IEEE 11th International Conference on Bioinformatics and Bioengineering*, vol. 1 (IEEE Computer Society, Los Alamitos, 2011), pp. 228–233
20. J.B. Kirkegaard, R.E. Goldstein, The role of tumbling frequency and persistence in optimal run-and-tumble chemotaxis. IMA J. Appl. Math. **83**, 700 (2018)
21. J. Yuan, K.A. Fahrner, L. Turner, H.C. Berg, Asymmetry in the clockwise and counterclockwise rotation of the bacterial flagellar motor, Proc. Natl. Acad. U.S.A. **107**, 12846 (2010)
22. L. Lemelle, J.F. Palierne, E. Chatre, C. Place, Counterclockwise circular motion of bacteria swimming at the air-liquid interface. J. Bacteriol. **192**, 6307 (2010)
23. S. van Teeffelen, H. Löwen, Dynamics of a Brownian circle swimmer. Phys. Rev. E **78**, 20101 (2008)
24. B. ten Hagen, S. van Teeffelen, H. Löwen, Brownian motion of a self-propelled particle. J. Phys. Condens. Matter **23**, 194119 (2011)
25. R. Di Leonardo, D. Dell'Arciprete, L. Angelani, V. Iebba, Swimming with an image. Phys. Rev. Lett. **106**, 38101 (2011)
26. F. Kümmel, B.T. Hagen, R. Wittkowski, I. Buttinoni, R. Eichhorn, G. Volpe, H. Löwen, C. Bechinger, Circular motion of asymmetric self-propelling particles. Phys. Rev. Lett. **110**, 198302 (2013)
27. X.L. Wu, A. Libchaber, Particle diffusion in a quasi-two-dimensional bacterial bath. Phys. Rev. Lett. **84**, 3017 (2000)
28. G. Grégoire, H. Chaté, Y. Tu, Active and passive particles: modeling beads in a bacterial bath. Phys. Rev. E **64**, 11902 (2001)
29. A. Argun, A.R. Moradi, E. Pinçe, G.B. Bagci, A. Imparato, G. Volpe, Non-Boltzmann stationary distributions and nonequilibrium relations in active baths. Phys. Rev. E **94**, 62150 (2016)
30. E. Pince, S.K.P. Velu, A. Callegari, P. Elahi, S. Gigan, G. Volpe, G. Volpe, Disorder-mediated crowd control in an active matter system. Nat. Commun. **7**, 10907 (2016)
31. M.J. Huang, J. Schofield, R. Kapral, Chemotactic and hydrodynamic effects on collective dynamics of self-diffusiophoretic Janus motors. New J. Phys. **19**, 125003 (2017)

32. P. Bayati, A. Najafi, Dynamics of two interacting active Janus particles. J. Chem. Phys. **144**, 134901 (2016)
33. E.V. Novak, E.S. Pyanzina, S.S. Kantorovich, Behaviour of magnetic Janus-like colloids. J. Phys. Condens. Matter **27**, 234102 (2015)
34. J. Chena, N. Claya, H. Kong, Non-spherical particles for targeted drug delivery. Chem. Eng. Sci. **125**, 20 (2015)
35. K.L. Lee, L.C. Hubbard, S. Hern, I. Yildiz, M. Gratzl, N.F. Steinmetz, Shape matters: the diffusion rates of TMV rods and CPMV icosahedrons in a spheroid model of extracellular matrix are distinct. Biomater. Sci. **1** (2013)
36. W. Uspal, M. Popescu, M. Tasinkevych, S. Dietrich, Shape-dependent guidance of active Janus particles by chemically patterned surfaces. New J. Phys. **20** (2018)
37. S. Michelin, E. Lauga, Geometric tuning of self-propulsion for Janus catalytic particles. Sci. Rep. **7**, 42264 (2017)
38. B.Q. Ai, Ratchet transport powered by chiral active particles. Sci. Rep. **6**, 18740 (2016)
39. J. Chen, Y. Hua, Y. Jiang, X. Zhou, L. Zhang, Rotational diffusion of soft vesicles filled by chiral active particles. Sci. Rep. **7**, 15006 (2017)
40. E. Tjhung, M.E. Cates, D. Marenduzzo, Contractile and chiral activities codetermine the helicity of swimming droplet trajectories. Proc. Natl. Acad. U.S.A. **114**, 4631 (2017)
41. S. Das, A. Garg, A.I. Campbell, J. Howse, A. Sen, D. Velegol, R. Golestanian, S.J. Ebbens, Boundaries can steer active Janus spheres. Nat. Commun. **6**, 8999 (2015)
42. Z. Shen, A. Würger, J.S. Lintuvuoria, Hydrodynamic interaction of a self-propelling particle with a wall—comparison between an active Janus particle and a squirmer model. Eur. Phys. J. E **41**, 39 (2018)
43. A. Mozaffari, N. Sharifi-Mood, J. Koplik, C. Maldarelli, Self-propelled colloidal particle near a planar wall: a Brownian dynamics study. Phys. Rev. Fluids **3**, 14104 (2018)
44. H. Brenner, Coupling between the translational and rotational Brownian motion of rigid particles of arbitrary shape. J. Colloid Interface Sci. **23**, 407 (1967)
45. M.X. Fernandes, J.G. de la Torre, Brownian dynamics simulation of rigid particles of arbitrary shape in external fields. Biophys. J. **83**, 3039 (2002)
46. J.S. Dai, Euler-Rodrigues formula variations, quaternion conjugation and intrinsic connections. Mech. Mach. Theory **92**, 144 (2015)
47. P.H. Jones, O.M. Maragò, G. Volpe, *Optical Tweezers: Principles and Applications* (Cambridge University Press, London, 2015)
48. M. Yang, M. Ripoll, Thermophoretically induced flow field around a colloidal particle. Soft Matter **9**, 4661 (2013)
49. A.I. Campbell, S.J. Ebbens, Gravitaxis in spherical Janus swimming devices. Langmuir **29**, 14066 (2013)
50. B. ten Hagen, F. Kümmel, R. Wittkowski, D. Takagi, H. Löwen, C. Bechinger, Gravitaxis of asymmetric self-propelled colloidal particles. Nat. Commun. **5**, 4829 (2014)
51. N. Casic, N. Quintero, R. Alvarez-Nodarse, F.G. Mertens, L. Jibuti, W. Zimmermann, T.M. Fischer, Propulsion efficiency of a dynamic self-assembled helical ribbon. Phys. Rev. Lett. **110**, 168302 (2013)
52. .E. Koser, N.C. Keim, P.E. Arratia, Structure and dynamics of self-assembling colloidal monolayers in oscillating magnetic fields. Phys. Rev. E **88**, 62304 (2013)
53. J.E. Martin, A. Snezhko, Driving self-assembly and emergent dynamics in colloidal suspensions by time-dependent magnetic fields. Rep. Prog. Phys. **76** (2013)
54. G. Loget, A. Kuhn, Electric field-induced chemical locomotion of conducting objects. Nat. Commun. **2**, 535 (2011)
55. L. Baraban, D. Makarov, O.G. Schmidt, G. Cuniberti, P. Leiderer, A. Erbef, Control over Janus micromotors by the strength of a magnetic field. Nanoscale **5**, 1332 (2013)
56. G. Grosjean, G. Lagubeau, A. Darras, M. Hubert, G. Lumay, N. Vandewalle, Remote control of self-assembled microswimmers. Sci. Rep. **5**, 16035 (2015)
57. M. Kaynak, A. Ozcelik, A. Nourhani, P.E. Lammert, V.H. Crespi, T.J. Huang, Acoustic actuation of bioinspired microswimmers. Lab Chip **17**, 395 (2017)

58. D. Ahmed, M. Lu, A. Nourhani, P.A. Lammert, Z. Stratton, H.S. Muddana, V.H. Crespi, T.J. Huang, Selectively manipulable acoustic-powered microswimmers. Sci. Rep. **5**, 1 (2015)
59. A.F. Demirörs, M.T. Akan, E. Poloni, A.R. Studart, Active cargo transport with Janus colloidal shuttles using electric and magnetic fields. Soft Matter **14**, 4741 (2018)
60. C. Wyatt Shields, O.D. Velev, The evolution of active particles: toward externally powered self-propelling and self-reconfiguring particle systems. Chem **3**, 539 (2017)
61. S.C. Takatori, R. De Dier, J. Vermant, J.F. Brady, Acoustic trapping of active matter. Nat. Commun. **7**, 10694 (2016)
62. J. de Graaf, H. Menke, A.J.T.M. Mathijssen, M. Fabritius, C. Holm, T.N. Shendruk, Lattice-Boltzmann hydrodynamics of anisotropic active matter. J. Chem. Phys. **144**, 134106 (2016)
63. P. Kreissl, C. Holm, J. de Graaf, The efficiency of self-phoretic propulsion mechanisms with surface reaction heterogeneity. J. Chem. Phys. **144**, 204902 (2016)
64. J. de Graaf, A.J.T.M. Mathijssen, M. Fabritius, H. Menke, C. Holm, T.N. Shendruk, Understanding the onset of oscillatory swimming in microchannels. Soft Matter **12**, 4704 (2016)
65. A.T. Brown, W.C.K. Poon, C. Holm, J. de Graaf, Ionic screening and dissociation are crucial for understanding chemical self-propulsion in polar solvents. Soft Matter **13**, 1200 (2017)
66. T. Vicsek, A. Czirók, E. Ben-Jacob, I. Cohen, O. Shochet, Novel type of phase transition in a system of self-driven particles. Phys. Rev. Lett. **75**, 1226 (1995)
67. W. Marth, A. Voigt, Collective migration under hydrodynamic interactions: a computational approach. Interface Focus. **6**, 20160037 (2016)
68. K. Yeo, E. Lushi, P.M. Vlahovska, Collective dynamics in a binary mixture of hydrodynamically coupled microrotors. Phys. Rev. Lett. **114**, 188301 (2015)
69. B. Liebchen, H. Löwen, Synthetic chemotaxis and collective behavior in active matter. Acc. Chem. Res. **51**(122), 982–2990 (2018)
70. F. Schmidt, B. Liebchen, H. Löwen, G. Volpe, Light-controlled assembly of active colloidal molecules. J. Chem. Phys. **150**, 094905 (2019)
71. I. Buttinoni, J. Bialké, F. Kümmel, H. Löwen, C. Bechinger, T. Speck, Dynamical clustering and phase separation in suspensions of self-propelled colloidal particles. Phys. Rev. Lett. **110**, 238301 (2013)
72. M.E. Cates, J. Tailleur, Motility-induced phase separation. Annu. Rev. Condens. Matter Phys. **6**, 219 (2015)
73. J. Stenhammar, R. Wittkowski, D. Marenduzzo, M.E. Cates, Activity-induced phase separation and self-assembly in mixtures of active and passive particles. Phys. Rev. Lett. **114**, 18301 (2015)
74. F. Kümmel, P. Shabestari, C. Lozano, G. Volpe, C. Bechinger, Formation, compression and surface melting of colloidal clusters by active particles. Soft Matter **11**, 6187 (2015)
75. N. de Macedo Biniossek, H. Löwen, T. Voigtmann, F. Smallenburg, Static structure of active Brownian hard disks. J. Phys. Condens. Matter, Spec. Issue Liq. Matter **30**, 074001 (2018)
76. F. Ginot, I. Theurkauff, I. Detcheverry, C. Ybert, C. Cottin-Bizonne, Aggregation-fragmentation and individual dynamics of active clusters. Nat. Commun. **9**, 696 (2018)
77. W. Schaertl, H. Sillescu, Brownian dynamics simulation of colloidal hard spheres. effect of sample dimensionality on self-diffusion. J. Stat. Phys. **74**, 687 (1994)
78. S. Nilsson, G. Volpe, Metastable clusters and channels formed by active particles with aligning interactions. New J. Phys. **19**, 115008 (2017)
79. H. Brenner, The slow motion of a sphere through a viscous fluid towards a plane surface. Chem. Eng. Sci. **16**, 242 (1961)
80. A. Banerjee, K.D. Kihm, Experimental verification of near-wall hindered diffusion for the Brownian motion of nanoparticles using evanescent wave microscopy. Phys. Rev. E **72**, 42101 (2005)
81. C.K. Choi, C.H. Margraves, K.D. Kihm, Examination of near-wall hindered Brownian diffusion of nanoparticles: experimental comparison to theories by Brenner (1961) and Goldman et al. (1967). Phys. Fluids **19**, 103305 (2007)
82. D.S. Sholl, M.K. Fenwick, E. Atman, D.C. Prieve, Brownian dynamics simulation of the motion of a rigid sphere in a viscous fluid very near a wall. J. Chem. Phys. **113**, 9268 (2000)

83. T. Brettschneider, G. Volpe, L. Helden, J. Wehr, C. Bechinger, Force measurement in the presence of Brownian noise: equilibrium-distribution method versus drift method. Phys. Rev. E **83**, 41113 (2011)
84. G.K. Batchelor, Brownian diffusion of particles with hydrodynamic interaction. J. Fluid Mech. **74**, 1 (1976)
85. J. Palacci, S. Sacanna, A.P. Steinberg, D.J. Pine, P.M. Chaikin, Living crystals of light-activated colloidal surfers. Science **339**, 936 (2013)

8

Active Particles and their Statistical Description: Application of UCNA

Umberto Marini Bettolo Marconi, Claudio Maggi, and Alessandro Sarracino

8.1 Introduction

The aim of this chapter is to provide an overview of some recent advances and open problems in the statistical description of active particles. In particular, we shall illustrate a theoretical approach based on the so-called unified coloured noise approximation (UCNA).

Active matter is composed of systems which are able to convert energy from the environment into directed motion. Every element of an active matter system can be considered out of equilibrium, in contrast to boundary driven systems, like those subject to a concentration gradient which are locally equilibrated [1–3].

Active systems abound in nature, ranging from flock of birds, structure-forming cytoskeletons of cells to bacterial colonies, but can also be man-made in a laboratory using biological building blocks or synthetic components. Being at the crossroads between biology, chemistry and physics, the subject has drawn the attention of scientists of different areas. In this article, we shall discuss active systems whose behaviour is assimilable to that of some bacteria or self-propelled particles and whose constituents are driven by an external random force and constantly spend energy to move through a viscous medium.

U. M. B. Marconi (✉)
Scuola di Scienze e Tecnologie, Università di Camerino, Camerino, Italy
e-mail: umberto.marinibettolo@unicam.it

C. Maggi
NANOTEC-CNR, Institute of Nanotechnology, Soft and Living Matter Laboratory, Lecce, Italy

A. Sarracino
CNR-ISC, Dipartimento di Fisica, Sapienza Università di Roma, Roma, Italy
Dipartimento di Ingegneria, Universitá degli Studi della Campania "L. Vanvitelli", Caserta, Italy

Run and tumble [4, 5] and active Brownian particle (ABP) models [6] have been initially proposed to interpret experiments conducted on bacterial suspensions. More recently, the Gaussian coloured noise (GCN) model has gained a lot of attention. It was introduced with the idea of capturing the peculiar aspect of run and tumble and ABP models (i.e. the persistence of the trajectories of the active particles) and of reducing their mathematical complexity. In the GCN the components of the active force have a Gaussian distribution and are exponentially correlated in time with a characteristic time, τ. By applying to the GCN model an adiabatic elimination of the fast degrees of freedom one obtains the UCNA [7, 8].

The UCNA [9] has the special property that its configurational steady state distribution is known, and that many stationary properties can be estimated. Employing this approximation, we present a description of a model of N mutually interacting active particles in the presence of external fields and characterise its steady state behaviour. Within the UCNA, we show that it is possible to develop a statistical mechanical approach similar to the one employed in the study of equilibrium liquids and to obtain the explicit form of the many-particle distribution function by means of the multidimensional unified coloured noise approximation. Such a distribution plays a role analogous to the Gibbs distribution in equilibrium statistical mechanics and provides a complete information about the microscopic steady state of the system. From here we develop a method to determine the one- and two-particle distribution functions in the spirit of the Born–Green–Yvon (BGY) equations of equilibrium statistical mechanics [10]. The resulting equations which contain extra-correlations induced by the activity allow determining the stationary density profiles in the presence of external fields, the pair correlations and the pressure of active fluids. In the low-density regime we obtain the effective pair potential ϕ acting between two isolated particles separated by a distance, r, showing the existence of an effective attraction. We apply the equations to different problems ranging from the study of the swim pressure, its relation to the mobility, to the investigation of the stationary state induced by a moving object in a "bath" of active particles.

Before closing this short introduction, we mention the fact that the UCNA method has been applied to the study of the effect of self-propulsion on a mean-field order–disorder transition [11]. Starting from a ϕ^4 scalar field theory subject to an exponentially correlated noise, the UCNA allows us to map the non-equilibrium active dynamics onto an effective equilibrium one. One can study the evolution of the second-order critical point as a function of the noise parameters: the correlation time, τ, and the noise strength, D. Our results suggest that the universality class of the ϕ^4 equilibrium model remains unchanged.

8.1.1 *The Genesis of the UCNA Model of Active Particles*

In order to understand the physical motivations of the model we shall discuss, it is necessary to give a brief historical account. In modern times, H.C. Berg was the first to introduce a model to describe the motion of bacteria in a viscous medium at small Reynolds number, the so-called run and tumble model, where the bacteria swim with constant velocity until a random tumble event suddenly decorrelates the orientation [4]. The active Brownian particles (ABP) model introduced to make analytical progress describes particles swimming at fixed speed u that rotates by slow angular diffusion. The two models have been shown to possess the same coarse-grained fluctuating hydrodynamics by Cates and Tailleur [12]. An advantage of the ABP is the possibility of taking into account external fields acting on the bacteria such as obstacles or gravity and interactions among them. For the n-th particle one has

$$\dot{\mathbf{r}}_n(t) = v_0 \mathbf{e}_n(t) - \frac{\nabla_n U}{\gamma}, \tag{8.1}$$

where U represents the total potential energy of an N particle systems, whereas $\gamma v_0 \mathbf{e}_i$ is the so-called active force, whose modulus is fixed, but whose direction $\mathbf{e}_n(t)$ changes in time by rotational diffusion according to the law

$$\dot{\mathbf{e}}_n(t) = \sqrt{D_r}\, \boldsymbol{\eta}_n \times \mathbf{e}_n(t), \tag{8.2}$$

where $\boldsymbol{\eta}_n(t)$ are Gaussian distributed with zero mean and have time correlations $\langle \boldsymbol{\eta}_n(t)\boldsymbol{\eta}_m(t')\rangle = 2\mathbf{I}\delta_{mn}\delta(t-t')$, where D_r is a rotational diffusion coefficient.

In spite of the great progress achieved using the ABP, the so-called active Ornstein–Uhlenbeck (AOU) or Gaussian coloured model has gained a great popularity because it has a simpler mathematical structure and lends itself to some analytical treatments due to the Gaussian character of the fluctuations of the active force [13]. The governing equations of such model are very similar to Eq. (8.1)

$$\dot{\mathbf{r}}_n(t) = \mathbf{u}_n(t) - \frac{\nabla_n U}{\gamma} \tag{8.3}$$

and

$$\dot{\mathbf{u}}_n(t) = -\frac{1}{\tau}\mathbf{u}_n(t) + \frac{\sqrt{D}}{\tau}\boldsymbol{\eta}_n(t) \tag{8.4}$$

with the difference that the active force $\gamma v_0 \mathbf{e}_n(t)$ is replaced by $\gamma \mathbf{u}_n(t)$, where the components of $\mathbf{u}_n(t)$ vary between $-\infty$ and ∞.

$$\langle \mathbf{u}_n(t)\rangle = 0, \quad \langle \mathbf{u}_n(t)\mathbf{u}_m(t')\rangle = \delta_{mn}\mathbf{I}\frac{D}{\tau}e^{-|t-t'|/\tau}. \tag{8.5}$$

Within the AOUP we can obtain a series of useful results and in some cases we can solve exactly the equations, As, for instance, in the case of harmonic potentials where the equilibrium distribution is known. In the free-particle case $U = 0$ the free mean squared displacement is $\langle (r(t)-r(0))^2\rangle = 2D\tau[t+\tau(1-e^{-t/\tau})]$. Thus, a free particle moves ballistically with typical speed $v = \sqrt{D/\tau}$ at short times ($t \ll \tau$) and diffusively with diffusion constant D at long times ($t \gg \tau$).

The typical distance travelled by a particle during a ballistic flight is the persistence length $\mathcal{L} = \sqrt{D\tau}$. If one observes the system on scales larger than $\mathcal{L}$ its properties will be almost indistinguishable from those of a system subject to standard thermal noise with an effective temperature $T = D\gamma$.

8.2 The Unified Coloured Noise Approximation (UCNA)

In the following, we consider the evolution equation relative to the GCN and from this we shall derive the UCNA equation. For the sake of simplicity, we introduce the vector $\mathbf{x}$ of components x_i of index $i \equiv (\alpha, n)$, where α is the Cartesian component associated with the coordinate of the n-th particle. We first differentiate w.r.t. time Eq. (8.3) and eliminate the active force $\gamma \mathbf{u}_n(t)$ using Eq. (8.4). The resulting equation has the form of an underdamped Langevin equation

$$\tau \frac{d^2 x_i}{dt^2} + \sum_j \left(\delta_{ij} + \frac{\tau}{\gamma} \frac{\partial^2 U}{\partial x_i \partial x_j} \right) \frac{dx_j}{dt} = -\frac{1}{\gamma} \frac{\partial U}{\partial x_i} + \sqrt{D} \eta_i(t), \tag{8.6}$$

with space dependent friction matrix:

$$\Gamma_{ij} = \delta_{ij} + \frac{\tau}{\gamma} \frac{\partial^2 U}{\partial x_i \partial x_j}. \tag{8.7}$$

Neglecting the acceleration term in Eq. (8.6) we shall obtain the so-called unified coloured noise approximation, which is analogous to the Kramers to Smoluchowski reduction and is exact in the limits $\tau \to 0$ and $\tau \to \infty$.

One can derive the UCNA equation by the original method of Hänggi and Jung [7]: on a new time scale $s = t\tau^{-1/2}$ one can recast the Langevin equation into the form

$$\frac{d^2 x_i}{ds^2} + \sum_j \Gamma_{ij} \frac{dx_j}{ds} = -\frac{1}{\gamma} \frac{\partial U}{\partial x_i} + \frac{\sqrt{D}}{\tau^{1/4}} \eta_i(s), \tag{8.8}$$

with $\langle \eta_i(s) \eta_j(s') \rangle = 2\delta_{ij}\delta(s - s')$ and $\Gamma_{ij} = (\frac{1}{\tau^{1/2}}\delta_{ij} + \frac{\tau^{1/2}}{\gamma} \frac{\partial^2 U}{\partial x_i \partial x_j})$. If $det \Gamma_{ij}$ is positive definite the damping is large for both small and large correlation times τ and in both cases one can set $\frac{d^2 x_i}{ds^2} = 0$ and obtain a Markovian approximation of the coloured noise process of the form

$$\frac{dx_i}{ds} = -\sum_j \Gamma_{ij}^{-1} \left(\frac{1}{\gamma} \frac{\partial U}{\partial x_j} - \frac{\sqrt{D}}{\tau^{1/4}} \eta_j(s) \right), \tag{8.9}$$

which is to be interpreted in the Stratonovich sense.

8.2.1 *Kinetic Approach*

It is straightforward to write the equation of evolution for the N-particle probability distribution density of positions $\mathbf{x}$ associated with the overdamped limit $\frac{d^2x_i}{dt^2} = 0$. It reads

$$\frac{\partial P_N(\mathbf{x},t)}{\partial t} = \sum_{ij} \frac{\partial}{\partial x_i} \Gamma_{ij}^{-1}(\mathbf{x},t) \left(D \sum_k \frac{\partial}{\partial x_k} \Gamma_{jk}^{-1}(\mathbf{x},t) + \frac{1}{\gamma} \frac{\partial U}{\partial x_j} \right) P_N(\mathbf{x},t). \tag{8.10}$$

It is, however, instructive to derive such an equation from a kinetic argument. We consider Eq. (8.6), define the velocity variable $v_i = \dot{x}_i$ and write the following (stochastically equivalent to Eq. (8.6)) Kramers equation describing evolution of phase-space distribution of N particles, $\Phi_N(x_i, v_i, t)$:

$$\frac{\partial \Phi_N}{\partial t} + \sum_i v_i \frac{\partial \Phi_N}{\partial x_i} - \sum_i \frac{1}{\gamma\tau} \frac{\partial U}{\partial x_i} \frac{\partial \Phi_N}{\partial v_i} = \frac{1}{\tau} \sum_i \frac{\partial}{\partial v_i} \left(\frac{D}{\tau} \frac{\partial}{\partial v_i} + \sum_k \Gamma_{ik} v_k \right) \Phi_N. \tag{8.11}$$

This kind of equation occurs in the study of colloidal solutions and is treated by multiple time scale methods. In general, we cannot solve Eq. (8.11) which involves both the velocity and the position variables. However, we can attack the problem by assuming that the velocity degrees of freedom evolve much faster than the positional degrees of freedom. This type of assumption is done when one reduces the Kramers phase-space equation to the Smoluchowski configurational equation. In fact, we make the ansatz that the phase-space distribution factorises in a spatial part and a velocity part. We construct a time-independent trial phase-space distribution having a factorised form:

$$\Phi_N(\mathbf{x}, \mathbf{v}, t) = \Pi(\mathbf{v}|\mathbf{x}, t) P_N(\mathbf{x}, t), \tag{8.12}$$

where Π is the conditional velocity distribution when the particles positions are fixed at $\mathbf{x}$. $P_N(\mathbf{x})$ corresponds to the distribution of the particles and is the marginalised distribution giving the distribution of positions of the particles regardless their velocities:

$$P_N(\mathbf{x}, t) = \int d\mathbf{v}\, \Phi_N(\mathbf{x}, \mathbf{v}, t). \tag{8.13}$$

In order to determine P_N we integrate Eq. (8.11) with respect to all velocities and obtain the continuity equation relating the probability density P_N and the probability current J_i

$$\frac{\partial P_N(\mathbf{x}, t)}{\partial t} + \sum_i \frac{\partial J_i(\mathbf{x}, t)}{\partial x_i} = 0\,, \tag{8.14}$$

where the current $J_i(\mathbf{x}, t)$ is the dN-dimensional vector:

$$J_i(\mathbf{x}, t) = \int d\mathbf{v}\, v_i \Phi_N(\mathbf{x}, \mathbf{v}, t). \tag{8.15}$$

After multiplying Eq. (8.11) by v_i and integrating over the dN velocities, we obtain the momentum balance equations

$$\frac{\partial J_i(\mathbf{x}, t)}{\partial t} + \sum_k \frac{\partial p_{ik}(\mathbf{x}, t)}{\partial x_k} + \frac{1}{\gamma\tau}\frac{\partial U}{\partial x_i} P_N(\mathbf{x}, t) = -\frac{1}{\tau}\sum_k \Gamma_{ik}(\mathbf{x}, t) J_k(\mathbf{x}, t), \tag{8.16}$$

where $p_{ik}(\mathbf{x}, t) \equiv \int d\mathbf{v} v_i v_j \Phi_N$ Eqs. (8.14) and (8.16) are an $Nd + 1$ system which is not closed because one does not know the explicit form of the tensor $p_{ik}(\mathbf{x}, t)$ in terms of P_N and J_i. A simplifying ansatz is to assume that the velocities have a local distribution similar to the one they would have in an equilibrium system; this is the following multivariate Gaussian distribution:

$$\Pi(\mathbf{v}|\mathbf{x}) \approx \left(\frac{\tau}{2\pi D}\right)^{N/2} \sqrt{\det \Gamma}\, \exp\left(-\frac{\tau}{2D}\sum_{ij} v_i \Gamma_{ij}(\mathbf{x}) v_j\right). \tag{8.17}$$

It is important to notice that the variance depends on the positions of the particles in contrast with equilibrium systems and is consistent with the fact that the friction is position dependent. Within the Gaussian ansatz we can rewrite the balance Eq. (8.16) as

$$\frac{\partial J_i}{\partial t} + \frac{D}{\tau}\sum_k \frac{\partial}{\partial x_k}\Gamma^{-1}_{ik} P_N + \frac{1}{\gamma\tau}\frac{\partial U}{\partial x_i} P_N = -\frac{1}{\tau}\sum_k \Gamma_{ik} J_k. \tag{8.18}$$

Finally, we assume that the time derivative of the current vanishes on a faster time scale than the time derivative of the density so that dropping the time derivative in Eq. (8.18) and expressing J_i in terms of P_N in Eq. (8.14) we obtain Eq. (8.10)

$$\frac{\partial P_N(\mathbf{x},t)}{\partial t} = D\sum_{ij}\frac{\partial}{\partial x_i}\Gamma^{-1}_{ij}(\mathbf{x},t)\left(\sum_k \frac{\partial}{\partial x_k}\Gamma^{-1}_{jk}(\mathbf{x},t) + \frac{1}{D\gamma}\frac{\partial}{\partial x_j}U(\mathbf{x})\right)P_N(\mathbf{x},t)$$
$$= -\sum_i \frac{\partial}{\partial x_i}J_i(\mathbf{x},t). \tag{8.19}$$

The advantage of such a derivation is that we have obtained not only the distribution function of positions, but also the approximate form of the distribution of the velocities of the particles. The latter is peculiar because it depends on the positions of all the particles at variance with the equilibrium case.

8.2.2 Stationary Solution in the Absence of Current

Let us consider the configurational distribution function $P_N(\mathbf{x})$ in the steady state associated with Eq. (8.10). In order to realise the steady state there are two possibilities, namely when the divergence of the probability flux vanishes, $\sum_i \partial_i J_i = 0$, or when the flow, $\mathbf{J}$, itself vanishes. Since only the configurational space is considered in such a reduced description and the positional variables, x_i, are even under time-reversal transformation, the condition $J_i = 0$ for arbitrary i is equivalent to the detailed balance condition [14]. In detail, if the matrix Γ^{-1} is non-singular, $J_i = 0$ implies

$$D\sum_k \frac{\partial}{\partial x_k}\Gamma^{-1}_{ik}(\mathbf{x})P_N(\mathbf{x}) + \frac{1}{\gamma}\frac{\partial U}{\partial x_i}P_N(\mathbf{x}) = 0, \tag{8.20}$$

which can be rewritten as:

$$\frac{\partial}{\partial x_i}P_N(\mathbf{x}) = -\frac{1}{D\gamma}\sum_k \Gamma_{ik}(\mathbf{x})\frac{\partial U(\mathbf{x})}{\partial x_k}P_N(\mathbf{x}) + \frac{\partial}{\partial x_i}\ln\det\Gamma(\mathbf{x})P_N(\mathbf{x}). \tag{8.21}$$

The detailed balance implies a stronger condition than the one represented by having a stationary distribution, since it implies that there is no net flow of probability around any closed cycle of states. Such a situation is no longer true when we consider the phase-space probability density of the original GCN problem, as discussed in Appendix 2.

From the above equation, one can find an exact expression for the probability density, which reads

$$P_N(\mathbf{x}) = \frac{1}{Z_N} \exp\left\{ -\frac{1}{D\gamma} \left[U(\mathbf{x}) + \frac{\tau}{2\gamma} \sum_k^N \left(\frac{\partial U(\mathbf{x})}{\partial x_k} \right)^2 \right] + \ln \left| \det \left(\delta_{ik} + \frac{\tau}{\gamma} \frac{\partial^2 U(\mathbf{x})}{\partial x_i \partial x_k} \right) \right| \right\}, \tag{8.22}$$

where Z_N is a normalisation constant.

Such a formula, in principle, fully describes within the unified colour approximation the steady state distribution of a system of interacting particles subject to coloured noise. In the white-noise limit $\tau \to 0$ the formula reduces to the Boltzmann distribution corresponding to the potential U. For finite values of τ, instead, the distribution maintains a Boltzmann-like distribution but with the effective potential given by Eq. (8.22). The presence of the additional terms $\left(\frac{\partial U(\mathbf{x})}{\partial x_k}\right)^2$ and $\ln \det \Gamma$ has repercussions in the form of the steady state configuration. Such a form of distribution is at a first glance surprising since in equilibrium systems, energy is exchanged reversibly with the environment and the form of $P_N(x)$ is determined by the potential and the temperature of the environment. On the contrary, in non-equilibrium systems, energy is exchanged irreversibly with the environment and in general there is no one-to-one correspondence between potential and $P_N(x)$. The vanishing of all components of the probability current (see Eq. (8.20)) is tantamount of the existence of the detailed balance condition, i.e. of the microscopic reversibility in the dynamics of the active system. This is reflected in the Boltzmann-like form of the distribution function. One may ask whether this is an artefact of the UCNA treatment of the dynamics or is a genuine property of the system. As we shall discuss below, by considering an elementary case, the detailed balance condition is violated by the original GCN dynamics by terms proportional to the persistence time, τ.

For a total potential, U, consisting of the sum of purely repulsive pair potentials the overall result is to create a sort of effective attractive potential among the particles. The origin of such an attraction can be understood as follows: the drag force on each particle is determined by the bare friction with the solvent medium plus an additional contribution stemming from the interactions. The non-equilibrium force is an attraction between self-propelled particles causing them to cluster. In the case of $J_i \neq 0$ and $\sum_i \partial_i J_i = 0$, it is not possible in general to obtain explicit solutions apart from some special cases which will be discussed later.

8.2.3 Fox Approximation

The approximate treatment obtained by applying the UCNA method is not unique. An alternative method has been put forward by Fox [15] who employed functional calculus in order to derive the effective equation for the distribution function $P(\mathbf{x}, t)$ corresponding to the GCN model. The resulting equation of evolution is valid in the small τ regime and has been applied to active fluids by Farage et al. [16]. It reads

$$\frac{\partial P_N^{fox}(\mathbf{x}, t)}{\partial t} = D \sum_i \frac{\partial}{\partial x_i} \left(\sum_k \frac{\partial}{\partial x_k} \Gamma_{ik}^{-1}(\mathbf{x}, t) P_N^{fox}(\mathbf{x}, t) \right. \\ \left. + \frac{1}{D\gamma} \frac{\partial}{\partial x_i} U(\mathbf{x}) P_N^{fox}(\mathbf{x}, t) \right). \tag{8.23}$$

Interestingly, the Fox and the UCNA approaches in the case of a single coloured noise yield the same steady state distribution function, whereas the approach to such a solution is different in the two cases. In the case where the particles are subject to different types of noises, each characterised by its own relaxation time, the UCNA approximation does not give the correct equation of motion even in the small τ limit, whereas the Fox method correctly reproduces such a limit. Therefore, in order to describe mixtures of active particles or of passive and active particles it is convenient to apply Fox's approach in spite of the fact that it only describes the small τ regime [17, 18].

8.2.4 Entropy Production in UCNA

The detailed balance requires that the probability of making a transition forward in time equals the probability of making the reverse transition, backward in time, when the system is in the steady state. It is easy to verify that within the UCNA approximation the condition of detailed balance holds if the probability current vanishes, $\mathbf{J} = 0$ in the steady state. The vanishing of $\mathbf{J}$ implies the existence of an effective potential U_{eff} which fully determines the distribution. We shall see in Sect. 8.7 that this is not the case when $\mathbf{J} \neq 0$. This is the reason why the UCNA steady state distribution has a form similar to a Boltzmann distribution, although with an effective potential which depends on the persistence time.

A measure of the distance from thermodynamic equilibrium is provided by the entropy production, so that it is interesting to study such a quantity in the steady state of the UCNA evolution equation. To this purpose, let us consider the rate of change of the Shannon entropy (for the sake of simplicity we study the case with $N = 1$ of Eq. (8.10)).

$$\dot{S}(t) = -\int dx \frac{\partial}{\partial t} P(x, t) \ln P(x, t) \\ = \int dx \ln P(x, t) \, \nabla J(x, t) = -\int dx \frac{\nabla P}{P} J, \tag{8.24}$$

where we obtained the last equality by partial integration. We decompose $\dot{S}$ into two contributions:

$$\dot{S} = \dot{S}_s + \dot{S}_m,$$

where $\dot{S}_s$ is the entropy production due to irreversible processes occurring inside the system and $\dot{S}_m$ is the entropy flux from the environment to the system. We shall show that $\dot{S}_s$ is positive definite, whereas S_m can have either sign. In the steady state the rate of change of the entropy vanishes so that $\dot{S}_m = -\dot{S}_s$. From Eq. (8.19) for $N = 1$ we have

$$\frac{\partial P(x,t)}{\partial t} = \nabla\left[\frac{D}{\Gamma(x,t)}\left(\nabla\frac{P(x,t)}{\Gamma(x,t)} + \frac{1}{D\gamma}\nabla U(x)P(x,t)\right)\right] = -\nabla J(x,t), \tag{8.25}$$

with the following probability current:

$$J(x,t) = -\frac{D}{\Gamma(x,t)}\left(\nabla\frac{P(x,t)}{\Gamma(x,t)} + \frac{1}{D\gamma}\nabla U P(x,t)\right),$$

where

$$\Gamma(x) = 1 + \frac{\tau}{\gamma}\nabla^2 U(x). \tag{8.26}$$

We can eliminate ∇P from Eq. (8.24) and obtain the following expression in terms of the current:

$$\dot{S}(t) = \int dx \frac{(\Gamma(x,t)J(x,t))^2}{DP(x,t)} + \int dx\, J(x,t)\,\frac{\nabla U_{\text{eff}}(x,t)}{D\gamma}, \tag{8.27}$$

with

$$U_{\text{eff}}(x) = U + \frac{\tau}{2\gamma}(\nabla U)^2 - D\gamma \ln\Gamma.$$

We now identify the first term in Eq. (8.27)

$$\dot{S}_s(t) = \frac{1}{D}\int dx \frac{\Gamma^2(x,t)J^2(x,t)}{P(x,t)} \tag{8.28}$$

as an entropy production rate always non-negative, and the second term

$$\dot{S}_m(t) = \frac{1}{T}\int dx\, J(x,t)\,\nabla U_{\text{eff}}(x,t) \tag{8.29}$$

with the entropy flux due to heat exchanges between the system and the surroundings and the temperature $T = D\gamma$.

We identify the heat flux with the average change of effective potential energy, U_{eff}, of the system in the unit time evaluated as follows:

$$\begin{aligned}\langle \dot{Q}(t)\rangle &= \frac{d}{dt}\int dx\, U_{\text{eff}}(x)P(x,t)\\ &= \int dx\, U_{\text{eff}}(x)\dot{P}(x,t) = -\int dx\, U_{\text{eff}}(x)\nabla J(x,t).\end{aligned} \tag{8.30}$$

After an integration by parts we obtain

$$\langle \dot{Q}(t)\rangle = \int dx\, J(x,t)\, \nabla U_{\text{eff}}(x), \tag{8.31}$$

and by comparing Eqs. (8.31) and (8.29), we find the following relation:

$$\dot{S}_m(t) = \frac{1}{T}\langle \dot{Q}(t)\rangle. \tag{8.32}$$

Finally, we have

$$\dot{S}(t) = \frac{1}{T}\langle \dot{Q}(t)\rangle + \frac{1}{D}\int dx \frac{\Gamma^2(x,t)J^2(x,t)}{P(x,t)}. \tag{8.33}$$

Notice that now the temperature entering the formula connecting $\dot{S}_m$ and $\langle \dot{Q}\rangle$ is uniform and given by T.

Let us remark that due to the detailed balance condition both $\dot{S}_s$ and $\dot{S}_m$ vanish in the steady state UCNA, showing that the UCNA method maps the underlying GCN non-equilibrium description into an equilibrium one. At variance with the UCNA, in the GCN both $\dot{S}_s$ and $\dot{S}_m$ are non-vanishing in the steady state.

8.2.5 H-Theorem

The following calculation proves the approach to the stationary distribution in terms of the entropy functional. One sees immediately that the entropy flux

$$\dot{S}_m(t) = -\int dx\, J(x,t)\, \nabla \ln P_{\text{steady}}(x), \tag{8.34}$$

so that using Eq. (8.24) we can rewrite

$$\begin{aligned}\dot{S}_s(t) = \dot{S}(t) - \dot{S}_m(t) &= -\int dx\, J(x,t)\, \nabla \ln \frac{P(x,t)}{P_{\text{steady}}(x)} \\ &= -\int dx \ln \frac{P(x,t)}{P_{\text{steady}}(x)} \frac{\partial P(x,t)}{\partial t}.\end{aligned} \tag{8.35}$$

The quantity $\dot{S}_s$ is nothing else but the rate of change of the Kullback–Leibler entropy, $S_{KL} \equiv -\int dx\, P(x,t) \ln(P(x,t)/P_{\text{steady}}(x))$, which is positive due to the sign of $\dot{S}_s$ and vanishes at equilibrium

$$\dot{S}_{KL}(t) = \dot{S}_s(t) \geq 0. \tag{8.36}$$

Thus the Kullback–Leibler entropy of the UCNA process is an ever increasing function and satisfies an H-theorem. The relative entropy $S_{KL}(t)$ is a functional of the non-equilibrium probability distribution and generalises the ordinary thermodynamic entropy which is defined for equilibrium states.

8.3 Born–Green–Yvon Hierarchy in the Steady State

We go back now to the multidimensional case and adopt indices to specify components and particles. We focus attention on the steady state properties of the system as described by the UCNA. We must remark that formula, Eq. (8.22) refers to N particles and therefore is not of practical use when the particles are mutually interacting. We need to derive from it expressions for the one-body and two-body distribution functions. The procedure is similar to the one employed in equilibrium statistical mechanics. We shall use the steady condition, Eq. (8.20), to derive a set of equations similar to the BGY hierarchy for distribution functions in equilibrium systems. The hierarchy becomes of practical utility in conjunction with a suitable truncation scheme in order to eliminate the dependence on the higher order correlations.

In the following, the Cartesian components (from 1 to d) are identified by the indices α and β, and the particles are identified by Latin indices. The total potential is assumed to be the sum of the mutual pairwise interactions $w(\mathbf{r} - \mathbf{r}')$ between the particles and of the potential exerted by the external field $u(\mathbf{r})$: $U(\mathbf{r}_1, \ldots, \mathbf{r}_N) = \sum_{i>j}^{N} w(\mathbf{r}_i, \mathbf{r}_j) + \sum_i^N u(\mathbf{r}_i)$.

The hierarchy follows from Eq. (8.10) and considering the reduced probability distribution functions of order n:

$$P_N^{(n)}(\mathbf{r}_1, \mathbf{r}_2, \ldots, \mathbf{r}_n) \equiv \int d\mathbf{r}_{n+1} \ldots d\mathbf{r}_N\, P_N(\mathbf{r}_1, \mathbf{r}_2, \ldots, \mathbf{r}_N). \tag{8.37}$$

When we integrate Eq. (8.10) over $(N-2)$ coordinates we obtain an equation which relates the two-body marginal distribution $P_N^{(2)}(\mathbf{r}_1, \mathbf{r}_2)$ to marginal distributions of different orders

$$
\begin{aligned}
&-T\int\int d\mathbf{r}_3\ldots d\mathbf{r}_N\sum_\beta\sum_n\frac{\partial}{\partial r_{\beta n}}[\Gamma^{-1}_{\alpha 1,\beta n}(\mathbf{r}_1,\ldots,\mathbf{r}_N)P_N(\mathbf{r}_1,\ldots,\mathbf{r}_N)]\\
&=P_N^{(2)}(\mathbf{r}_1,\mathbf{r}_2)\Big(\frac{\partial u(\mathbf{r}_{\alpha 1})}{\partial r_{\alpha 1}}+\frac{\partial w(\mathbf{r}_1-\mathbf{r}_2)}{\partial r_{\alpha 1}}\Big)\\
&+\sum_{k>2}\int d\mathbf{r}_k P_N^{(3)}(\mathbf{r}_1,\mathbf{r}_2,\mathbf{r}_k)\frac{\partial w(\mathbf{r}_1-\mathbf{r}_k)}{\partial r_{\alpha 1}}.
\end{aligned}
\tag{8.38}
$$

The equation for $P_N^{(2)}$ has a structure similar to that of a standard equilibrium gas but for the term containing $\Gamma^{-1}_{\alpha 1,\beta n}$ and unless we introduce approximations it is of little practical use.

We write

$$
\Gamma_{\alpha i,\beta k}=\Big(\delta_{\alpha\beta}+\frac{\tau}{\gamma}u_{\alpha\beta}(\mathbf{r}_i)+\frac{\tau}{\gamma}\sum_j w_{\alpha\beta}(\mathbf{r}_i-\mathbf{r}_j)\Big)\delta_{ik}-\frac{\tau}{\gamma}w_{\alpha\beta}(\mathbf{r}_i-\mathbf{r}_k)(1-\delta_{ik}),
\tag{8.39}
$$

and we remark that in the limit of small (τ/γ) the matrix $\Gamma^{-1}_{\alpha 1,\beta n}$ can be approximated as [19]:

$$
\Gamma^{-1}_{\alpha i,\beta k}\approx\Big(\delta_{\alpha\beta}-\frac{\tau}{\gamma}u_{\alpha\beta}(\mathbf{r}_i)-\frac{\tau}{\gamma}\sum_{j\neq i}w_{\alpha\beta}(\mathbf{r}_i-\mathbf{r}_j))\Big)\delta_{ik}=\tilde{\Gamma}^{-1}_{\alpha\beta}(\mathbf{r}_i)\delta_{ik},
\tag{8.40}
$$

where $u_{\alpha\beta}\equiv\frac{\partial^2 u(\mathbf{r})}{\partial r_\alpha\partial r_\beta}$ and $w_{\alpha\beta}\equiv\frac{\partial^2 w(\mathbf{r})}{\partial r_\alpha\partial r_\beta}$. We substitute this approximation and recast Eq. (8.38) in terms of the n-th order density distributions $\rho^{(n)}(\mathbf{r}_1,\mathbf{r}_2,\ldots,\mathbf{r}_n)=\frac{N!}{(N-n)!}P_N^{(n)}(\mathbf{r}_1,\mathbf{r}_2,\ldots,\mathbf{r}_n)$ and find

$$
\begin{aligned}
&T\sum_\beta\frac{\partial}{\partial r_{\beta 1}}\Big[\rho^{(2)}(\mathbf{r}_1,\mathbf{r}_2)\delta_{\alpha\beta}-\frac{\tau}{\gamma}\Big(\rho^{(2)}(\mathbf{r}_1,\mathbf{r}_2)u_{\alpha\beta}(\mathbf{r}_1)+\rho^{(2)}(\mathbf{r}_1,\mathbf{r}_2)w_{\alpha\beta}(\mathbf{r}_1-\mathbf{r}_2)\\
&\quad+\int d\mathbf{r}_k\rho^{(3)}(\mathbf{r}_1,\mathbf{r}_2,\mathbf{r}_k)w_{\alpha\beta}(\mathbf{r}_1-\mathbf{r}_k)\Big)\Big]\\
&=-\rho^{(2)}(\mathbf{r}_1,\mathbf{r}_2)\Big(\frac{\partial u(\mathbf{r}_1)}{\partial r_{\alpha 1}}+\frac{\partial w(\mathbf{r}_1-\mathbf{r}_2)}{\partial r_{\alpha 1}}\Big)-\int d\mathbf{r}_k\rho^{(3)}(\mathbf{r}_1,\mathbf{r}_2,\mathbf{r}_k)\frac{\partial w(\mathbf{r}_1-\mathbf{r}_k)}{\partial r_{\alpha 1}},
\end{aligned}
\tag{8.41}
$$

which represents the BGY equation for the pair density distribution $\rho^{(2)}$. By integrating also over the coordinate 2 we find the BGY equation for the one-body density:

$$T_s \sum_\beta \frac{\partial}{\partial r_{\beta 1}} \Big[\delta_{\alpha\beta} \rho^{(1)}(\mathbf{r}_1) - \frac{\tau}{\gamma} \rho^{(1)}(\mathbf{r}_1) u_{\alpha\beta}(\mathbf{r}_1) \\ - \frac{\tau}{\gamma} \int d\mathbf{r}_2 \rho^{(2)}(\mathbf{r}_1, \mathbf{r}_2)\, w_{\alpha\beta}(\mathbf{r}_1 - \mathbf{r}_2) \Big] \\ = -\rho^{(1)}(\mathbf{r}_1) \frac{\partial u(\mathbf{r}_1)}{\partial r_{\alpha 1}} - \int d\mathbf{r}_2 \rho^{(2)}(\mathbf{r}_1, \mathbf{r}_2) \frac{\partial w(\mathbf{r}_1 - \mathbf{r}_2)}{\partial r_{\alpha 1}}, \tag{8.42}$$

that in the limit of $\tau \to 0$ is just the BGY equation for the single-particle distribution function.

The r.h.s. of Eq. (8.42) contains the coupling to the external field and the so-called direct interaction among the particles, whereas the l.h.s. besides the ideal gas term contains a term proportional to the activity parameter.

8.4 Active Pressure

A natural way to define the pressure in a system of active spherical particles driven by coloured noise is by using the virial theorem which relates the virial of the external forces confining the particles in a given volume to the pressure exerted on the walls by the particles themselves. The forces exerted by the bounding walls of the container are macroscopically described as external pressure [20, 21]. Each oriented area element $d\mathbf{S}$ exerts a force $-p(\mathbf{r})d\mathbf{S}$ so that

$$\sum_i^N \langle \mathbf{F}_i^{\text{ext}} \cdot \mathbf{r}_i \rangle = -\oint p(\mathbf{r})\mathbf{r} \cdot d\mathbf{S} = -\bar{p} V d, \tag{8.43}$$

where $\bar{p}$ is the average pressure over the boundary surface, $\mathbf{r}$ is the position vector of the surface element and the last equality follows from the divergence theorem ($\nabla \cdot \mathbf{r} = d$).

Now, in order to evaluate the external force virial, Eq. (8.43), we multiply Eq. (8.38) by $r_{\alpha 1}$, integrate over $\mathbf{r}_1$ and $\mathbf{r}_2$ and sum over indices. After an integration by parts we obtain the following equation:

$$\sum_i^N \langle (\mathbf{F}_i^{\text{ext}} + \mathbf{F}_i^{\text{int}}) \cdot \mathbf{r}_i \rangle + T \sum_{\alpha i} \langle \Gamma^{-1}_{\alpha i, \alpha i} \rangle = 0, \tag{8.44}$$

where the forces are separated in two parts: wall and interparticle forces, $F_{\alpha i}^{\text{ext}} = -\frac{\partial u(\mathbf{r}_1)}{\partial r_{\alpha 1}}$ and $F_{\alpha i}^{\text{int}} = -\sum_k \frac{\partial w(\mathbf{r}_1 - \mathbf{r}_k)}{\partial r_{\alpha 1}}$, respectively, and the symbol $\langle \cdot \rangle$ stands for an average over the stationary distribution P_N.

In the case where the confining vessel has constant curvature one finds: $\bar{p} = p(\mathbf{r})$. In general, when the linear size of the vessel is much larger than the persistence length, the standard virial definition of pressure based on the assumption of the constancy of the pressure on the boundary of the system is correct [20, 21]. In order to obtain a closed expression for the pressure, we write the term stemming from the internal forces as:

$$\sum_i^N \langle \mathbf{F}_i^{\rm int} \cdot \mathbf{r}_i \rangle = \frac{1}{2} \sum_i \sum_j^{\prime} \langle \mathbf{F}_{ij} \cdot (\mathbf{r}_i - \mathbf{r}_j) \rangle, \tag{8.45}$$

and approximate the average of the trace of Γ^{-1} as:

$$T \sum_i^N \sum_\alpha^d \langle \Gamma^{-1}_{\alpha i, \alpha i} \rangle \approx T_s \sum_\alpha^d \int d\mathbf{r}\, \tilde{\Gamma}^{-1}_{\alpha\alpha}(\mathbf{r}) \rho^{(1)}(\mathbf{r}), \tag{8.46}$$

where in the second equality we have used Eq. (8.40). We now write

$$\begin{aligned} p_v = & \frac{T}{dV} \sum_\alpha^d \int d\mathbf{r}\, \tilde{\Gamma}^{-1}_{\alpha\alpha}(\mathbf{r}) \rho^{(1)}((\mathbf{r}) \\ & - \frac{1}{2dV} \sum_\alpha^d \int d\mathbf{r} \int d\mathbf{r}' (r_\alpha - r'_\alpha) \rho^{(2)}(\mathbf{r}, \mathbf{r}') \frac{\partial w(\mathbf{r} - \mathbf{r}')}{\partial r_\alpha}, \end{aligned} \tag{8.47}$$

where the second term in Eq. (8.47) is analogous to the direct contribution to the pressure in passive fluids stemming from interactions. Finally, we obtain the explicit representation

$$\begin{aligned} \frac{T}{dV} \sum_\alpha^d \int d\mathbf{r}\, \tilde{\Gamma}^{-1}_{\alpha\alpha}(\mathbf{r}) \rho^{(1)}(\mathbf{r}) \approx & \frac{T}{V} \Big[N - \frac{1}{d} \frac{\tau}{\gamma} \int d\mathbf{r} \sum_\alpha u_{\alpha\alpha}(\mathbf{r}) \rho^{(1)}(\mathbf{r}) \\ & - \frac{1}{d} \frac{\tau}{\gamma} \int d\mathbf{r} \int d\mathbf{r}' \rho^{(2)}(\mathbf{r}, \mathbf{r}') \sum_\alpha w_{\alpha\alpha}(\mathbf{r} - \mathbf{r}') \Big]. \end{aligned} \tag{8.48}$$

The first term in the r.h.s. of Eq. (8.48) represents an ideal gas-like contribution to the pressure, TN/V, also referred to as the swim pressure, due to the rotational degrees of freedom. The second and third term in Eq. (8.48) represent indirect interaction contributions, and take into account the slowing down of active fluids near a boundary and in regions of high density, respectively. The indirect interaction pressure involves the interplay between the rotational degrees of freedom and the interparticle forces and is a non-equilibrium effect. In fact, in the limit of $\tau \to 0$ the quantity Eq. (8.48) reduces to $T_t N/V$, the ideal gas contribution to the pressure.

Besides the virial method, for the UCNA there exist two other approaches to evaluate not only the pressure but also the surface tension. In the first of approach, these can be identified with the volume and area derivatives, respectively, of the partition function associated with the stationary non-equilibrium distribution. The second alternative method is a mechanical approach and is related to the work necessary to deform the system. The pressure is obtained by comparing the expression of the work in terms of local stress and strain with the corresponding expression in terms of microscopic distribution. This work is determined from the force balance encoded in the Born–Green–Yvon equation and can be used to obtain a formula for the local pressure tensor and the surface tension even in inhomogeneous situations. Nicely, the three procedures lead to the same values of the pressure, and give support to the idea that the UCNA partition function is more than a formal property of the system, but determines the stationary non-equilibrium thermodynamics of the model. For further details the reader may consult ref. [9].

8.5 Velocity Correlations

The kinetic derivation of the UCNA has shown that a system of active particles displays velocity correlations. Within the present treatment these correlations have been approximated by means of a Gaussian multivariate distribution whose variance depends on the potential.

We consider N interacting particles in $1d$. We perform numerical simulations of systems with $N = 1000$ composed by GCN-driven particles interacting via the potential $\phi(\mathbf{x}) = \sum_{i>j}(x_i - x_j)^{-12}$ for several values of the density $\rho = N/L$, of D and τ.

The velocity variance depends on the configuration of the particles, so that by averaging it over the positions we obtain the overall velocity variance of a GCN-driven system:

$$\langle v^2 \rangle = \frac{1}{dN} \int d\mathbf{x}\, P_N(\mathbf{x}) \int d\mathbf{v}\, \mathbf{v} \cdot \mathbf{v}\, \Pi(\mathbf{v}|\mathbf{x}). \tag{8.49}$$

The $\langle v^2 \rangle$ computed numerically via Eq. (8.49) is plotted in Fig. 8.1b (full lines) as a function of the $1d$ density $\rho = N/L$ of the system and for several values of D (at fixed τ). In all these simulations we compute the variance $\langle \dot{x}^2 \rangle$ and report the results in Fig. 8.1 as connected symbols.

To test the validity of the Gaussian ansatz for the velocity distribution given by Eq. (8.17), we compute the average over positions in Eq. (8.49) directly from the coordinates obtained numerically, instead of using the theoretical P_N of Eq. (8.22). This is plotted in Fig. 8.1 as dashed lines and follows well the numerical curves, although some expected deviation [7] is observed upon increasing D to very high values. If we assume a uniform density and long-ranged interactions (mean-field approximation) the velocity distribution Eq. (8.17) simplifies substantially since all the out-of-diagonal term of $\partial_\alpha \partial_\beta w(\mathbf{r}_i - \mathbf{r}_j)$ are of order one and can be neglected with respect to the terms on the main diagonal that are of order N [19]. This yields

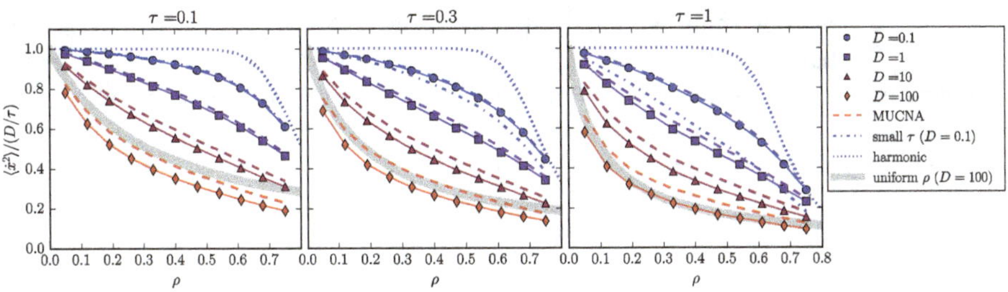

Fig. 8.1 Normalised velocity variance for a $1d$ system of many interacting active particles. Symbols are the results of numerical simulations for several values of τ and D (see legend). Dashed lines are the theoretical velocity variances obtained by averaging Eq. (8.49) over the coordinates obtained numerically. Thick lines are the result of a homogeneous density approximation. Dashed-dotted lines represent the small-τ approximation connecting the variance to the pair distribution function. Dotted lines are the velocity variances obtained by mapping the system onto a harmonic model (see ref. [22])

the density-dependent variance:

$$\frac{\langle v^2(\rho)\rangle}{(D/\tau)} = \frac{1}{1+\frac{\tau}{\gamma}w_2\rho} = \frac{1}{1+\rho\mathcal{L}}, \tag{8.50}$$

where $w_2 = \int_\sigma^\infty dx\, w''(x)$ is the mean potential curvature integrated from the diameter σ and $\mathcal{L} = \sqrt{D\tau}$ is the characteristic length of the active motion. In the last equality of Eq. (8.50) we have used the fact that, for a generic repulsive potential, σ corresponds roughly to the distance where the interaction force balances the self-propulsion force (i.e. $|w'(\sigma)| \approx \gamma v = \gamma\sqrt{D/\tau}$) and $w_2 = w'(\infty) - w'(\sigma) \approx \gamma\sqrt{D/\tau}$. This is plotted in Fig. 8.1 as a thick line for the largest D and follows well the data when $\mathcal{L}$ is large. To first order in τ we obtain the results plotted as dashed-dotted lines in Fig. 8.1 and the theory compares well with the numerical simulations. However, by fixing τ and increasing D this approximation deviates strongly from the simulations. We can now derive an expression for pressure by a kinetic argument by identifying it with

$$p = \rho\langle v^2(\rho)\rangle = \frac{D}{\tau}\frac{\rho}{1+\rho\mathcal{L}}. \tag{8.51}$$

8.6 Simple Applications

8.6.1 Active Elastic Dumbbells

Let us consider N mutually noninteracting elastic dumbbells, i.e. two point particles bound together by an elastic spring of constant α^2, moving in a vessel represented by a harmonic weak confining potential, of spring constant ω^2 [23, 24]. Such a model, similar to the harmonic trap model [25–27], was proposed long ago by Riddell and Uhlenbeck. It contains the minimal ingredients to observe the competition between internal forces and confining potential and can be solved without introducing further approximations. The potential energy reads [28]:

$$U(\mathbf{r}_1, \mathbf{r}_2) = w(\mathbf{r}_1 - \mathbf{r}_2) + u(\mathbf{r}_1) + u(\mathbf{r}_2)$$

with $w(\mathbf{r}) = \frac{1}{2}\alpha^2\mathbf{r}^2$. By setting $u(\mathbf{r}) = \frac{k}{2}\frac{\mathbf{r}^2}{L^2}$, one introduces a volume dependence in the spring constant associated with the confining potential and for simplicity of notation we shall use $\omega^2 = \frac{k}{L^2}$.

The virial pressure is obtained by applying the general formula Eq. (8.47)

$$pdL^d = -\langle \boldsymbol{F}_1^{\mathrm{ext}} \cdot \mathbf{r}_1 + \boldsymbol{F}_2^{\mathrm{ext}} \cdot \mathbf{r}_2 \rangle = dT\langle \Gamma_{11}^{-1} + \Gamma_{22}^{-1} \rangle + \langle \boldsymbol{F}_{12} \cdot (\mathbf{r}_1 - \mathbf{r}_2) \rangle. \tag{8.52}$$

By simple algebraic manipulations we find

$$p = \frac{T}{L^d}\left[\frac{1}{1 + \frac{\tau}{\gamma}\omega^2} + \frac{\omega^2}{\omega^2 + 2\alpha^2}\frac{1}{(1 + \frac{\tau}{\gamma}(\omega^2 + 2\alpha^2))}\right]. \tag{8.53}$$

In the limit of $\tau \to 0$ and $\alpha \to 0$ the pressure reduces to the expected ideal gas pressure of a system of $2N$ noninteracting particles in a vessel of volume L^d. On the other hand, one can see that the pressure decreases with increasing values of τ, i.e. if the persistence length $\mathcal{L} = \sqrt{D\tau}$ exceeds the typical size of the vessel the particles do not explore the whole space of the vessel, but remain localised at the bottom.

8.6.2 Pressure of N Noninteracting Active Particles Surrounded by Harshly Repulsive Walls

As a second example we consider an assembly of N noninteracting active particles constrained in a region of space near the origin by a spherically symmetric external potential in three dimensions. Using Eq. (8.42) one can derive the following exact formula expressing the mechanical balance condition:

$$\frac{d}{dr}p_N(r) + \frac{2}{r}(p_N(r) - p_T(r)) = -\rho^{(1)}(r)u'(r), \tag{8.54}$$

where the components of the pressure tensor normal (N) and tangential (T) to the walls are

$$p_N(r) = T_s \frac{\rho^{(1)}(r)}{1 + \frac{\tau}{\gamma} u''(r)}, \tag{8.55}$$

and

$$p_T(r) = T_s \frac{\rho^{(1)}(r)}{1 + \frac{\tau}{\gamma} \frac{u'(r)}{r}}. \tag{8.56}$$

The density profile according to Eq. (8.22) can be written explicitly as:

$$\rho^{(1)}(r) = \rho_0 \exp\left[-\frac{u(r)}{T_s} - \frac{\tau}{2\gamma T_s}(u'(r))^2 \right] \left(1 + \frac{\tau}{\gamma} u''(r)\right) \left(1 + \frac{\tau}{\gamma} \frac{u'(r)}{r}\right)^2, \tag{8.57}$$

so that we can fully determine the components of the pressure tensor.

8.7 Active Particles in a Time-Dependent Potential

The results presented in the previous sections concern static cases, where the external potential is constant in time. In this section we address the interesting issue of a time-dependent external potential [29]. In particular, we shall consider a shifting potential $U(x, t) = U(x - ct)$ in one dimension, moving at constant speed c and inducing a stationary current in the system. The potential barrier interacts with a fluid of active particle, see the sketch depicted in Fig. 8.2. The effect of a moving potential on a particle fluid is a general problem in modelling the motion of a driven obstacle in a medium, in several different fields, such as in the active microrheology of colloidal systems or in the translocation dynamics of polymer chains through nanopores.

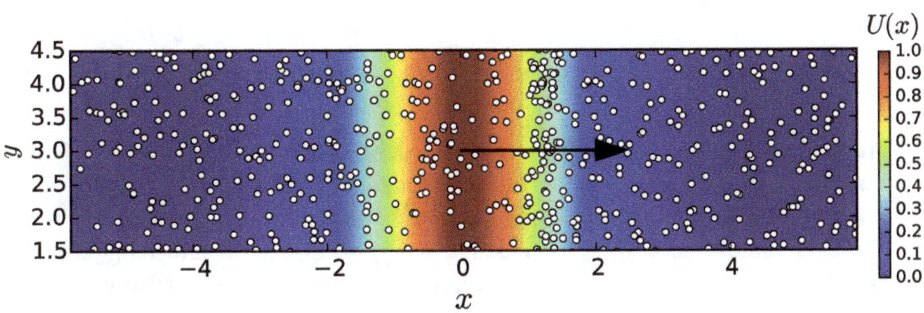

Fig. 8.2 Sketch of the system: a potential barrier moves at constant velocity c in a channel with a noninteracting active particle fluid, producing a density profile which is non-uniform along the x direction [29]

In the case of the GCN model, we will show that the coupling of self-propulsion (namely a finite persistence time τ) with the stationary current gives place to an effective dynamical potential, which vanishes in both the limits of $c \to 0$, and $\tau \to 0$ (passive particles). The main physical effect we observe in this model, which is accounted for by a generalised UCNA scheme, consists in a much enhanced accumulation of active particles at the interface fluid/obstacle, with respect to the static case or with respect to the behaviour of a passive particle fluid.

8.7.1 *Effective Potential*

As shown in Sect. 8.3, the first step in deriving UCNA equations is to take the time derivative in Eq. (8.3). For the sake of simplicity, we consider a one-dimensional system and a shifting time-dependent potential of the form $U(x,t) = U(x - ct)$. We have

$$\dot{x}(t) = v(t), \tag{8.58}$$

$$\dot{v}(t) = -\frac{1}{\tau}\Gamma(x - ct)v(t) + \frac{1}{\tau\gamma}F^*(x - ct) + \frac{\sqrt{D}}{\tau}\eta(t), \tag{8.59}$$

$$F^*(x) = F(x) - \tau c\frac{dF(x)}{dx} = -\frac{dU(x)}{dx} + \tau c\frac{d^2U(x)}{dx^2}, \tag{8.60}$$

$$\Gamma(x) = 1 + \frac{\tau}{\gamma}\frac{d^2U(x)}{dx^2}. \tag{8.61}$$

By comparing with Eq. (8.6) we note that a new term appears in Eqs. (8.61): an effective force $F^*(x)$, which reduces to $-dU/dx$ when $c = 0$. As we shall show, this additional contribution in the force term due to the finite velocity of the obstacle $c > 0$ is responsible for new dynamical effects.

8.7.2 *Dynamical UCNA and Particle Density Profile*

In order to show how these effects can be described within a generalised UCNA scheme, it is useful to consider the associated Fokker–Planck equation. It is time saving to adopt non-dimensional variables for positions, velocities and time, and rescale forces accordingly. We define $v_T = \sqrt{D/\tau}$, measure lengths using the characteristic length, ℓ, of the potential, and introduce the following non-dimensional variables:

$$\bar{t} \equiv t\frac{v_T}{\ell}, \quad \bar{v} \equiv \frac{v}{v_T}, \quad \bar{x} \equiv \frac{x}{\ell}, \quad \bar{F}(\bar{x},\bar{t}) \equiv \frac{\ell F(x,t)}{D\gamma},$$

$$\zeta = \frac{\ell}{\tau v_T}, \quad \bar{\Phi} = v_T \ell \, \Phi, \quad \bar{c} = \frac{c}{v_T}, \tag{8.62}$$

where ζ plays the role of a non-dimensional friction. To lighten the notation we shall drop the bar over the non-dimensional variables without incurring in ambiguities. For the probability distribution of position and velocity $\Phi(y, v)$ we thus obtain

$$\frac{\partial}{\partial t}\Phi(y, v) + v\frac{\partial}{\partial y}\Phi(y, v) + F^*(y)\frac{\partial}{\partial v}\Phi(y, v) = \zeta\frac{\partial}{\partial v}\left[\frac{\partial}{\partial v} + \Gamma(y)v\right]\Phi(y, v), \tag{8.63}$$

where we have introduced the shifted variable $y = x - ct$. We, now, look for an approximate solution to this equation. We start by eliminating the v dependence of the phase-space distribution $\Phi(y, v)$, by multiplying by powers of v and integrating w.r.t. v. Thus, one obtains a set of coupled first order ordinary differential equations, the so-called Brinkman hierarchy, whose first two members are the continuity equation and the momentum balance equation, respectively:

$$-c\frac{d\rho(y)}{dy} + \frac{dJ(y)}{dy} = 0, \tag{8.64}$$

$$-c\frac{dJ(y)}{dy} + \frac{d\Pi(y)}{dy} - F^*(y)\rho(y) + \zeta\Gamma(y)J(y) = 0. \tag{8.65}$$

Here we have introduced the density $\rho(y)$, the current $J(y)$ and the momentum current $\Pi(y)$, defined as:

$$\rho(y) = \int dv\Phi(y, v), \tag{8.66}$$

$$J(y) = \int dv v\Phi(y, v), \tag{8.67}$$

$$\Pi(y) = \int dv v^2\Phi(y, v). \tag{8.68}$$

According to the continuity Eq. (8.64) the current must be a linear function of the density, yielding

$$J(y) = c[\rho(y) - \bar{\rho}], \tag{8.69}$$

where $\bar{\rho}$ is a constant such that the solution is periodic at $\rho(L) = \rho(-L)$, where $2L$ is the system size. As we shall see later, for large systems $L \gg l$, $\bar{\rho} \approx \rho(\pm L)$ and the current is almost vanishing at the boundaries.

From the analysis of the case of static potentials, discussed in the previous sections, we know that the solution of Eq. (8.63) in regions where $F^*(y) = 0$ and $\Gamma(y) = 1$ can be written as (see Eq. (8.17)):

$$\Phi(y, v) = \left[\rho(y) - \bar{\rho}\right] H_0(v - c) + \bar{\rho}\, H_0(v), \tag{8.70}$$

where

$$H_0(v) = \sqrt{\frac{1}{2\pi}} \exp\left(-\frac{1}{2}v^2\right) \tag{8.71}$$

is a Hermite function of zero order. Indeed, by substituting the form Eq. (8.70) in Eq. (8.63) (with $F^* = 0$), we obtain a solution provided $\rho(y)$ satisfies the following condition:

$$\frac{d\rho(y)}{dy} = -\zeta c[\rho(y) - \bar{\rho}]. \tag{8.72}$$

Next, we insist in looking for a solution of Eq. (8.63) even in the region where $F^*(y) \neq 0$ of the form:

$$\Phi(y, v) = \left[\rho(y) - \bar{\rho}\right] H_0(y, v - c) + \bar{\rho}\, H_0(y, v), \tag{8.73}$$

where we have introduced the following (non-uniform) Hermite function, which is position dependent through the trial function $\beta(y)$:

$$H_0(y, v) = \sqrt{\frac{\beta(y)}{2\pi}} \exp\left(-\frac{\beta(y)}{2}v^2\right). \tag{8.74}$$

Substituting now the trial distribution Eq. (8.73) into Eq. (8.63), we get

$$\begin{aligned}
&H_1(y, v - c)\frac{1}{\sqrt{\beta(y)}}\Big\{\rho'(y) - \beta(y)[F^*(y) - \zeta\Gamma(y)c][\rho(y) - \bar{\rho}] \\
&\quad - \frac{\beta'(y)}{\beta(y)}[\rho(y) - \bar{\rho}]\Big\} - H_1(y, v)\frac{1}{\sqrt{\beta(y)}}\left[\beta(y)F^*(y)\bar{\rho} + \frac{\beta'(y)}{\beta(y)}\bar{\rho}\right] \\
&\quad + \zeta[\Gamma(y) - \beta(y)][(\rho(y) - \bar{\rho})H_2(y, v - c) + \bar{\rho}H_2(y, v)] \\
&\quad - \frac{\beta'(y)}{2\sqrt{\beta^3(y)}}\left[(\rho(y) - \bar{\rho})H_3(y, v - c) + \bar{\rho}H_3(y, v) - c\beta^{1/2}H_2(y, v)\right] = 0,
\end{aligned} \tag{8.75}$$

where prime denotes derivative w.r.t. y, and $H_1(y, v)$, $H_2(y, v)$ and $H_3(y, v)$ are the Hermite functions of order 1, 2 and 3, respectively, defined by the recursion relation:

$$H_{\nu+1}(y, v) = -\frac{1}{\sqrt{\beta(y)}}\frac{\partial H_\nu(y, v)}{\partial v}.$$

The trial solution fails to solve Eq. (8.63). However, if we limit ourselves to consider only the two lowest moments of the probability distribution, i.e. if after multiplying by $(v - c)$, we integrate Eq. (8.75) over v, we obtain the following condition which gives the equation for the density profile:

$$\frac{1}{\beta(y)}\frac{d\rho(y)}{dy} - [F(y) - \zeta c]\rho - \frac{\beta'(y)}{\beta^2(y)}\rho - \zeta c\Gamma(y)\bar{\rho} = 0. \tag{8.76}$$

If we continue the projection procedure beyond the first order in $(v - c)$ there will be an error in the equation for the second moment, which becomes inconsistent with the value of the second moment imposed by the trial distribution (which, in fact, is already fixed by the trial form and therefore does not contain enough parameters to satisfy the extra conditions.)

The ansatz for the phase-space distribution gives the following expression for the momentum flux:

$$\Pi(y) = \frac{\rho(y)}{\beta(y)} + c^2[\rho(y) - \bar{\rho}]. \tag{8.77}$$

Note that Eq. (8.76) is perfectly equivalent to Eq. (8.65) when the latter is endowed with a closure, indeed represented by Eq. (8.77). The static UCNA approximation is recovered by setting the arbitrary function $\beta(y) = \Gamma(y)$ and $c = 0$, (i.e. $J = 0$).

In order to deal with possible zeroes of the function $\beta(y)$ let us solve the non-linear differential equation for the profile using the auxiliary function:

$$n(y) = \frac{\rho(y)}{\beta(y)} \tag{8.78}$$

that satisfies the equation

$$\frac{dn(y)}{dy} = \left[F(y) - \zeta c\right]\beta(y)n(y) + \zeta c\Gamma(y)\bar{\rho}. \tag{8.79}$$

Then, defining the effective potential

$$w(y) = \int_{-L}^{y} ds\beta(s)\frac{dU(s)}{ds} + \zeta c\int_{-L}^{y} ds[\beta(s) - 1]$$

allows us to rearrange Eq. (8.79) as follows:

$$\frac{dn(y)}{dy} = \left[-\frac{d}{dy}w(y) - \zeta c\right]n(y) + \zeta c\Gamma(y)\bar{\rho}. \tag{8.80}$$

The solution of the inhomogeneous equation is then

$$n(y) = Ae^{-w(y)-\zeta cy} + \zeta c\bar{\rho}e^{-w(y)-\zeta cy}\int_{-L}^{y} dse^{w(s)+\zeta cs}\Gamma(s), \tag{8.81}$$

$$A = n(-L)e^{w(-L)-\zeta cL}. \tag{8.82}$$

By construction $w(-L) = 0$ and one may verify that $n(L) = n(-L)$, but $w(L) \neq w(-L)$. Eventually, one has

$$\begin{aligned} n(y) &= n(L)e^{-[w(y)-w(-L)]-c\zeta(y+L)} \\ &\times \left\{1 + [e^{2\zeta cL}e^{w(L)-w(-L)} - 1]\frac{\int_{-L}^{y} dse^{w(s)+c\zeta s}\Gamma(s)}{\int_{-L}^{L} dse^{w(s)+c\zeta s}\Gamma(s)}\right\}, \end{aligned} \tag{8.83}$$

and, from Eq. (8.78), the density profile reads

$$\begin{aligned} \rho(y) &= \frac{\rho(L)}{\beta(L)}\beta(y)e^{-[w(y)-w(-L)]-c\zeta(y+L)} \\ &\times \left\{1 + [e^{2\zeta cL}e^{w(L)-w(-L)} - 1]\frac{\int_{-L}^{y} dse^{w(s)+c\zeta s}\Gamma(s)}{\int_{-L}^{L} dse^{w(s)+c\zeta s}\Gamma(s)}\right\}, \end{aligned} \tag{8.84}$$

where $\rho(L)$ is fixed by the normalisation. The explicit expression for the density $\bar{\rho}$ is

$$\bar{\rho} = \frac{1}{\zeta c}\frac{\rho(L)}{\beta(L)}\frac{e^{w(L)+c\zeta L} - e^{w(-L)-c\zeta L}}{\int_{-L}^{L} dye^{w(y)+c\zeta y}\Gamma(y)}.$$

We empirically set $\beta(y) = \Gamma(y)$ in the regions where $\Gamma(y) \geq 0$, and $\beta(y) = 0$ otherwise. Then, the expression (8.84) can be evaluated numerically and in Fig. 8.3 we compare the analytical prediction with numerical simulations, in the case of the following external potential:

$$U(y) = U_0\left[\tanh\left(\frac{y+1}{\xi}\right) - \tanh\left(\frac{y-1}{\xi}\right)\right], \tag{8.85}$$

characterised by the steepness $1/\xi$.

8.7.3 Average Drag Force

Our analytical approach allows us to obtain an estimate for the average drag force exerted by the active fluid on the moving wall, defined as:

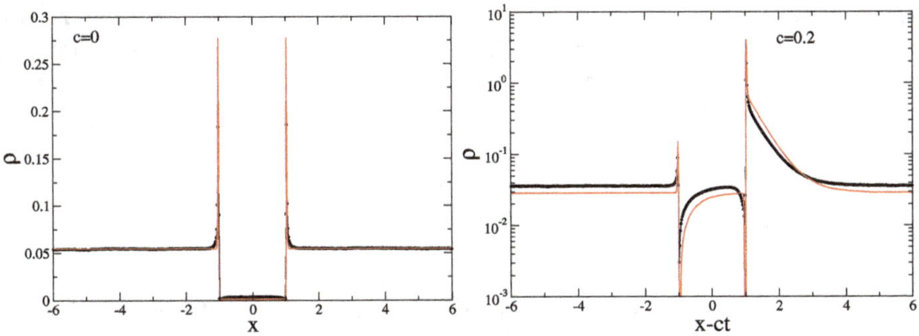

Fig. 8.3 Density profiles for the static case $c = 0$ (left) and for the moving potential with $c = 0.2$ (right). Red lines represent analytical predictions, while black dots are numerical simulations. Other parameters are $U_0 = 0.5$, $\xi = 0.1$, $\zeta = 2$ [29]

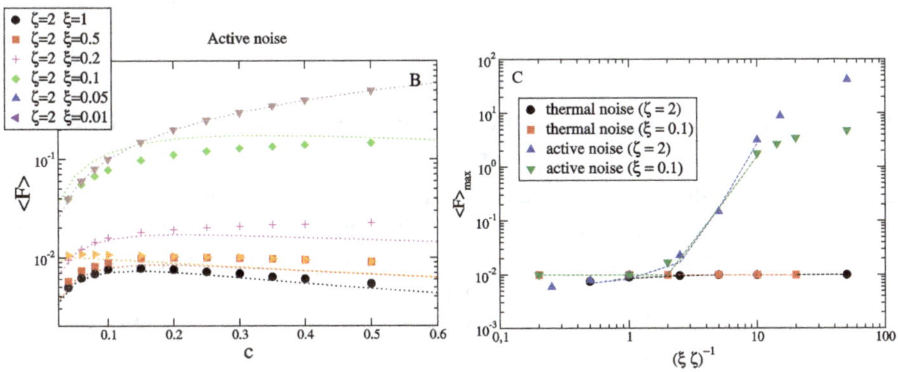

Fig. 8.4 Left: Comparison between the analytical predictions (dotted lines) and the numerical simulations (symbols) for the average drag force exerted by the active fluid on the moving wall. Right: Comparison between the maximal drag force computed in the active and passive case [29]

$$\langle F \rangle = \int_{-L}^{L} dy F(y)\rho(y). \tag{8.86}$$

The comparison of the analytical prediction with numerical simulations for the drag force is shown in Fig. 8.4. Note the non-monotonic behaviour of the force–velocity relation is characterised by a maximum value of the force $\langle F \rangle_{\max}$ for a particular velocity c^*.

In order to highlight the new physical effects arising due to the coupling of self-propulsion and a stationary current, it is useful to compare the behaviour observed in the active particle model with the one obtained in the case of a moving potential in a (passive) thermal bath. In the latter situation, the noise term acting in the stochastic equation for the particle velocity is a delta-correlated noise of amplitude $2/\zeta$. In the right panel of Fig. 8.4 we show the maximum value of the drag force $\langle F \rangle_{\max}$

as a function of $1/(\xi\zeta)$ in both models. The qualitative difference between the two behaviours relies on the observation that in the active case the average drag force can increase indefinitely by reducing the parameter ξ, which characterises the steepness of the travelling potential.

8.8 Conclusions

In this chapter, we have reviewed the recent developments of the theory of active particles driven by coloured noise within an approximate scheme, the UCNA. Such a method has the great advantage of providing predictions and equations much simpler with respect to other methods. The reason is that the adiabatic approximation at the basis of the method eliminates the faster degrees of freedom, the velocity in the case of the GCN and of the ABP.

Appendix 1: Entropy Production and Heat Flux in the GCN

Let us consider the elementary case of a single active particle in one dimension driven by Gaussian coloured noise, with phase-space distribution function $p(x, v, t)$. We derive the equations for the entropy production and entropy flux in phase-space (x, v) and we shall use small letters to distinguish probabilities and thermodynamic variables from the configurational variables of UCNA. We start from the Fokker–Planck equation

$$\frac{\partial p}{\partial t} + v\frac{\partial p}{\partial x} - \frac{1}{\gamma\tau}\frac{\partial U}{\partial x}\frac{\partial p}{\partial v} = \frac{1}{\tau}\frac{\partial}{\partial v}\left(\frac{D}{\tau}\frac{\partial}{\partial v} + \Gamma v\right)p. \tag{8.87}$$

Within the GCN, we consider the time derivative of the total Shannon entropy production defined as

$$\begin{aligned} \dot{s}(t) &= -\iint dxdv\frac{\partial}{\partial t}p(x, v, t)\ln p(x, v, t) \\ &= \iint dxdv \ln p(x, v, t)\, \mathrm{div}\mathbf{I} = \iint dxdv \mathrm{div}\left(\frac{\mathbf{I}}{p}\right)p, \end{aligned} \tag{8.88}$$

where the (x, v) components of the current vector $\mathbf{I}$ are:

$$I_x = vp(x, v, t) \tag{8.89}$$

$$I_v = -\frac{1}{\gamma\tau}\frac{\partial U}{\partial x}p - \frac{\Gamma}{\tau}vp - \frac{D}{\tau^2}\frac{\partial p}{\partial v} \tag{8.90}$$

Let us define the temperature $T = D/\tau$ and the local temperature $\theta(x) = T/\Gamma(x)$. Now, the total time derivative of the entropy Eq. (8.88) can be written as the sum, $\dot{s}(t) = \dot{s}_s(t) + \dot{s}_m(t)$ (see Sect. 2.4). Explicitly, we find after integrating by parts Eq. (8.88) the following expressions:

$$\dot{s}_s(t) = \frac{1}{\tau}\frac{1}{T}\iint dxdv\frac{1}{p}\left(\Gamma(x)vp + T\frac{\partial p}{\partial v}\right)^2 \tag{8.91}$$

and the entropy flux

$$\dot{s}_m(t) = -\iint dxdv\frac{1}{\theta(x)}\frac{\Gamma(x)}{\tau}\left[v^2 p(x,v,t) + \theta(x)v\frac{\partial}{\partial v}p(x,v,t)\right]. \tag{8.92}$$

The dimensional form of the total energy $\epsilon(x, v)$ and of the heat flux are, respectively:

$$\epsilon(t) = \frac{1}{2}v^2 + \frac{U(x)}{\tau\gamma}, \tag{8.93}$$

and

$$\langle\dot{q}(t)\rangle = \iint dxdv\epsilon(t)\frac{\partial}{\partial t}p(x,v,t) \tag{8.94}$$

$$\langle\dot{q}\rangle = -\iint dxdv\frac{\Gamma(x)}{\tau}\left[v^2 p(x,v,t) + \theta(x)v\frac{\partial}{\partial v}p(x,v,t)\right]. \tag{8.95}$$

It is suggestive to rewrite

$$\dot{q}(t) = \int dx\dot{\tilde{q}}(x,t), \tag{8.96}$$

$$\dot{s}_m(t) = \int dx\frac{1}{\theta(x)}\dot{\tilde{q}}(x,t), \tag{8.97}$$

with a local density of heat flux defined as

$$\begin{aligned}\dot{\tilde{q}}(x,t) &= -\frac{\Gamma(x)}{\tau}\int dv\left[v^2 p(x,v,t) + \theta(x)v\frac{\partial}{\partial v}p(x,v,t)\right]\\ &= -\frac{1}{\tau}\frac{T}{\theta(x)}n(x,t)\left[\langle v^2\rangle_x - \theta(x)\right],\end{aligned} \tag{8.98}$$

where $n(x,t) = \int dvp(x,v,t)$ and $n(x,t)\langle v^2\rangle_x = \int dvv^2 p(x,v,t)$, with $\langle v^2\rangle_x$ the mean squared velocity at given position. Expression Eq. (8.97) represents an interesting connection between the local entropy production of the medium (or entropy flux) and the local heat flux divided by the same local temperature $\theta(x) = T/\Gamma(x)$ featuring in the approximate detailed balance solution, Eq. (8.103).

Define the dissipative components of the current as:

$$\tilde{I}_x(x, v) = 0,$$

$$\tilde{I}_v(x, v) = -\frac{\Gamma}{\tau} v p - \frac{T}{\tau}\frac{\partial p}{\partial v}.$$

It is clear that in the GCN in the steady state while the time derivative of the entropy vanishes $\dot{s} = 0$, its two contributions are not necessarily zero.

$$\dot{s}_s(t) = \frac{\tau}{T}\iint dx dv \frac{\tilde{I}_v^2(x, t)}{p(x, v, t)} \tag{8.99}$$

$$\dot{s}_m(t) = \iint dx dv \, \frac{1}{\theta(x)} \, v \, \tilde{I}_v(x, v, t) \tag{8.100}$$

$\dot{s}_s$ is an entropy production rate which is always non-negative, while $\dot{s}_m$ is the entropy flux due to heat exchanges between the system and the surroundings and can have either signs.

Appendix 2: Absence of Detailed Balance Condition in the GCN

As in Appendix 1, let us consider the elementary case of a single active particle in one dimension driven by Gaussian coloured noise. The probability current, $\mathbf{I}(x, v)$, is the two dimensional vector, comprising both reversible and irreversible contribution given by Eq. (8.90). Let $p_s(x, v)$ be a steady state solution of the Fokker–Planck equation, such that

$$\text{div}\mathbf{I} = \frac{\partial I_x}{\partial x} + \frac{\partial I_v}{\partial v} = 0. \tag{8.101}$$

The detailed balance condition requires that in the steady state [30] the irreversible part of the current, represented by the terms proportional to ζ, must vanish:

$$-\frac{\Gamma(x)}{\tau} v p_s(x, v) - \frac{T}{\tau}\frac{\partial p_s}{\partial v} = 0 \tag{8.102}$$

p_s must be the product of a function of the position, $\pi(x)$, times a "local" Maxwellian whose velocity variance is also position-dependent:

$$p_s(x, v) = \pi(x) \exp\left(-\frac{\Gamma(x)}{2T} v^2\right). \tag{8.103}$$

In virtue of Eq. (8.101) the reversible part of the current vector $(vp_s, -U'p_s/(\tau\gamma))$ must fulfil the condition:

$$\left[v\frac{\partial}{\partial x} - \frac{1}{\tau\gamma}\frac{dU(x)}{dx}\frac{\partial}{\partial v}\right] p_s(x, v) = 0. \tag{8.104}$$

Plugging the distribution Eq. (8.103) into Eq. (8.104) we obtain:

$$v\left(\frac{1}{\pi(x)}\frac{d\pi(x)}{dx} + \frac{1}{\tau\gamma}\frac{dU(x)}{dx}\frac{\Gamma(x)}{T} - \frac{1}{2}\frac{d\Gamma(x)}{dx}\frac{v^2}{T}\right)$$

$$\pi(x) \exp\left(\frac{-\Gamma(x)}{2T} v^2\right) \neq 0, \tag{8.105}$$

and conclude that a function $\pi(x)$ satisfying Eq. (8.104) only exists when $\Gamma(x)$ is a constant Γ_0. This condition occurs for $\tau \to 0$, which is the equilibrium limit of the model, or when the potential is a linear or parabolic function of x. In conclusion, apart from the special case Γ_0, the Kramers equation (8.11) does not satisfy the detailed balance condition. However, we can determine an approximate steady solution consistent with the UCNA under the form:

$$\pi_{\text{trial}}(x) = \Gamma^{3/2}(x) \exp\left(-\frac{U(x) + \frac{\tau}{2\gamma}U'(x)^2}{D\gamma}\right). \tag{8.106}$$

In spite of the fact that such trial solution is not divergence-free, i.e. div$\mathbf{I} \neq 0$, the first three velocity moments, obtained by multiplying the divergence by $(1, v, v^2)$, respectively, and integrating w.r.t. v, vanish, so that in this subspace the zero divergence condition holds.

Acknowledgement A.S. acknowledges support from "Programma VALERE" of University of Campania "L. Vanvitelli".

References

1. C. Bechinger, R. Di Leonardo, H. Löwen, C. Reichhardt, G. Volpe, G. Volpe, Active particles in complex and crowded environments. Rev. Mod. Phys. **88**, 45006 (2016)
2. S. Ramaswamy, The mechanics and statistics of active matter. Ann. Rev. Condens. Matter Phys. **1**, 323–345 (2010)

3. M.C. Marchetti, J.F. Joanny, S. Ramaswamy, T.B. Liverpool, J. Prost, M. Rao, R.A. Simha, Hydrodynamics of soft active matter. Rev. Mod. Phys. **85**(3), 1143 (2013)
4. H.C. Berg, *E. coli in Motion* (Springer, Berlin, 2008)
5. M.J. Schnitzer, Theory of continuum random walks and application to chemotaxis. Phys. Rev. E **48**(4), 2553 (1993)
6. Y. Fily, M.C. Marchetti, Athermal phase separation of self-propelled particles with no alignment. Phys. Rev. Lett. **108**(23), 235702 (2012)
7. P. Hanggi, P. Jung, Colored noise in dynamical systems. Adv. Chem. Phys. **89**, 239 (1995)
8. C. Maggi, U.M.B. Marconi, N. Gnan, R. Di Leonardo, Multidimensional stationary probability distribution for interacting active particles. Sci. Rep. **5**, 10742 (2015)
9. U.M.B. Marconi, C. Maggi, S. Melchionna, Pressure and surface tension of an active simple liquid: a comparison between kinetic, mechanical and free-energy based approaches. Soft Matter **12**(26), 5727 (2016)
10. C. Cercignani, R. Illner, M. Pulvirenti, *The Mathematical Theory of Dilute Gases*, vol. 106 (Springer, Berlin, 2013)
11. M. Paoluzzi, C. Maggi, U. Marini Bettolo Marconi, N. Gnan, Critical phenomena in active matter. Phys. Rev. E **94**(5), 52602 (2016)
12. M.E. Cates, J. Tailleur, When are active Brownian particles and run-and-tumble particles equivalent? Consequences for motility-induced phase separation. Europhys. Lett. **101**(2), 20010 (2013)
13. Y. Fily, A. Baskaran, M.F. Hagan, Active particles on curved surfaces. arXiv Prepr. arXiv1601.00324 (2016)
14. C. Gardiner, *Stochastic Methods* (Springer, Berlin, 2009)
15. R.F. Fox, Correlation time expansion for non-Markovian, Gaussian, stochastic processes. Phys. Lett. A **94**(6–7), 281 (1983)
16. T.F.F. Farage, P. Krinninger, J.M. Brader, Effective interactions in active Brownian suspensions. Phys. Rev. E **91**(4), 42310 (2015)
17. R. Wittmann, C. Maggi, A. Sharma, A. Scacchi, J.M. Brader, U.M.B. Marconi, Effective equilibrium states in the colored-noise model for active matter I. pairwise forces in the fox and unified colored noise approximations. J. Stat. Mech. Theory Exp. **2017**, 113207 (2017)
18. R. Wittmann, U.M.B. Marconi, C. Maggi, J.M. Brader, Effective equilibrium states in the colored-noise model for active matter II. A unified framework for phase equilibria, structure and mechanical properties. J. Stat. Mech. Theory Exp. **2017**, 113208 (2017)
19. U.M.B. Marconi, C. Maggi, Towards a statistical mechanical theory of active fluids. Soft Matter **11**(45), 8768 (2015)
20. R.K. Pathria, *Statistical Mechanics, International Series in Natural Philosophy* (Pergamon Press, Oxford, 1986)
21. B. Widom, *Statistical Mechanics: A Concise Introduction for Chemists* (Cambridge University Press, Cambridge, 2002)
22. U.M.B. Marconi, N. Gnan, M. Paoluzzi, C. Maggi, R. Di Leonardo, Velocity distribution in active particles systems, Sci. Rep. **6**, 23297 (2016)
23. M. Joyeux, E. Bertin, Pressure of a gas of underdamped active dumbbells. Phys. Rev. E **93**(3), 32605 (2016)
24. F. Cagnetta, F. Corberi, G. Gonnella, A. Suma, Large fluctuations and dynamic phase transition in a system of self-propelled particles. Phys. Rev. Lett. **119**(15), 158002 (2017)
25. G. Szamel, Self-propelled particle in an external potential: existence of an effective temperature. Phys. Rev. E **90**(1), 12111 (2014)
26. C. Maggi, M. Paoluzzi, N. Pellicciotta, A. Lepore, L. Angelani, R. Di Leonardo, Generalized energy equipartition in harmonic oscillators driven by active baths. Phys. Rev. Lett. **113**(23), 238303 (2014)
27. S.C. Takatori, R. De Dier, J. Vermant, J.F. Brady, Acoustic trapping of active matter. Nat. Commun. **7**, 10694 (2016)

28. R.J. Riddell Jr, G.E. Uhlenbeck, On the notion of pressure in a canonical ensemble. J. Chem. Phys. **18**(8), 1066 (1950)
29. U.M.B. Marconi, A. Sarracino, C. Maggi, A. Puglisi, Self-propulsion against a moving membrane: enhanced accumulation and drag force. Phys. Rev. E **96**(3), 32601 (2017)
30. H. Risken, *Fokker–Planck Equation* (Springer, Berlin, 1984)

9

Procedure for the Construction of High Order Mixed Quadrature LB Models

Victor E. Ambruş and Victor Sofonea

9.1 Introduction

At non-negligible values of the Knudsen number Kn (defined as the ratio between the mean free path of the fluid particles in a gas and the characteristic length of the domain), the Navier–Stokes equations lose applicability [1, 2]. Such rarefied gas flows can be approached within the framework of the Boltzmann equation [3–5]. This equation describes the six-dimensional phase-space evolution of the distribution function f, where $f(t, \boldsymbol{x}, \boldsymbol{p})d^3xd^3p$ gives the number of particles at time t which are contained in an infinitesimal volume d^3x centred on $\boldsymbol{x}$, having momenta in an infinitesimal range d^3p about $\boldsymbol{p}$. Because of its complexity, the Boltzmann equation can be solved analytically only in a very limited number of cases. Alternatively, numerous well-established approaches to the numerical solutions of the Boltzmann equation are now currently used for academic or engineering purposes, of which we only mention the direct simulation Monte Carlo (DSMC) technique [6], the discrete velocity models (DVMs) [7–9], the discrete unified gas-kinetic scheme (DUGKS) [10–12] and the lattice Boltzmann (LB) models [13–20].

The LB models are a particular type of DVMs and are derived from the Boltzmann equation using a simplified version of the collision operator, as well as an appropriate discretisation of the momentum space, which ensure the recovery of the moments of the distribution function f up to a certain order N. Originally derived nearly 30 years ago from the lattice gas automata [17, 19, 20], the LB

V. E. Ambruş
Department of Physics, West University of Timişoara, Timişoara, Romania

V. Sofonea (✉)
Center for Fundamental and Advanced Technical Research, Romanian Academy, Timişoara, Romania

models were primarily designed to recover the hydrodynamics of fluid systems at the Navier–Stokes level. The LB models inherited the collision-streaming concept from their ancestors, according to which the velocities of the fluid particles are aligned along the lattice links such that after one time step δt, each particle arrives at a neighbouring node [13, 14, 17, 19, 21, 22].

One disadvantage of the collision-streaming paradigm is the increasing difficulty to approach fluid systems far from equilibrium (e.g., rarefied gases or micro/nano-scale flowing fluids) using suitable LB models. In this case, the accurate recovery of specific effects in channel flow at large values of Kn, such as the velocity slip and the temperature jump at the channel walls [1, 2], requires that higher order moments of the single particle distribution function f are ensured. Since the moments of f are derived by integration in the momentum space, their numerical computation involves the use of convenient quadrature methods. When using a quadrature method, the moments of the single particle distribution f (up to a certain order N) are exactly recovered by sums over a finite set of momentum vectors $\mathbf{p}_k$, $1 \leq k \leq K$ [13, 14, 17–19, 21–26]. As the fluid system is farther from the equilibrium and the characteristic value of the Knudsen number increases, the number K of the momentum vectors (i.e., the quadrature points) should also be increased, as it will be shown later. This task becomes more and more elaborated if one wants to keep the particles hopping from a lattice node to another one during a single time step [27–30].

An alternative to the collision-streaming paradigm is provided by the off-lattice LB models, where the distribution functions are evolved in the lattice nodes using finite-difference, finite-volume or interpolation schemes [31–34]. A fourth-order, off-lattice LB model for the simulation of thermal flows in the continuum regime (small values of the Knudsen number), where the fluid density, velocity and temperature fields are derived from a single set of distribution functions, was proposed by Watari and Tsutahara [35] for 2D flows and subsequently extended to the 3D case [36–38]. Off-lattice LB models of any order N can be easily constructed using the Gauss quadrature method in the velocity space [18, 23, 25, 26, 39, 40].

Another challenge for microfluidics simulations is due to the implementation of boundary conditions. In general, the particle–wall interaction governs the distribution of particles emerging from the wall back into the fluid. Since the distribution of particles travelling from the fluid towards the wall is essentially arbitrary, the distribution function becomes discontinuous near the wall [41, 42]. Examples of boundary conditions include the diffuse-spectral [9] and Cercignani–Lampis [43] particle–wall interaction models; however, for simplicity, we restrict the analysis to the simpler diffuse reflection model, which is a limiting case of both models mentioned above. According to the diffuse reflection paradigm, the reflected particles follow a Maxwell–Boltzmann distribution corresponding to the wall temperature and velocity. In order to accurately compute the incident and emergent fluxes required to impose diffuse reflection (kinetic) boundary conditions, it is convenient to discretise the velocity set based on half-range Gauss quadrature methods. Such techniques were also used in the frame of DVMs [44, 45] and more recently, they were adapted for the LB method [24, 26, 46, 47].

9.2 Generalities

The Boltzmann equation for a force-driven flow reads:

$$\partial_t f + \frac{\mathbf{p}}{m} \cdot \nabla f + \mathbf{F} \cdot \nabla_{\mathbf{p}} f = -\frac{1}{\tau}(f - f^{\text{eq}}), \tag{9.1}$$

where on the right-hand side we have used the Bhatnagar–Gross–Krook (BGK) single-relaxation time approximation of the collision term [48]. The distribution function $f \equiv f(\mathbf{x}, \mathbf{p}, t)$ represents the density of particles at position $\mathbf{x}$ and time t, having momentum $\mathbf{p}$, while

$$f^{\text{eq}} = n g_x g_y g_z, \tag{9.2}$$

$$g_\alpha = \frac{1}{\sqrt{2\pi m k_B T}} \exp\left[-\frac{(p_\alpha - m u_\alpha)^2}{2 m k_B T}\right], \qquad \alpha \in \{x, y, z\} \tag{9.3}$$

is the Maxwell–Boltzmann equilibrium distribution function corresponding to local thermal equilibrium. The force $\mathbf{F} \equiv \mathbf{F}(\mathbf{x})$ in Eq. (9.1) encapsulates all external forces acting on these particles.

The local quantities describing the gas flow at the macroscopic level, namely the particle number density n, the velocity $\mathbf{u}$, the stress tensor $T_{\alpha\beta}$ $(\alpha, \beta = 1, 2, 3)$ and the heat flux $\boldsymbol{q}$, can be obtained as moments of f:

$$\begin{pmatrix} n \\ \rho \mathbf{u} \\ T_{\alpha\beta} \\ \boldsymbol{q} \end{pmatrix} = \int d^3 p \begin{pmatrix} 1 \\ \mathbf{p} \\ \xi_\alpha \xi_\beta / m \\ \dfrac{\boldsymbol{\xi} \xi^2}{2m^2} \end{pmatrix} f, \tag{9.4}$$

where $\xi_\alpha = p_\alpha - m u_\alpha$ is the peculiar momentum and $\rho = mn$ is the mass density. The pressure P is defined through:

$$P = \frac{1}{3} \sum_\alpha T_{\alpha\alpha}, \tag{9.5}$$

while the temperature is obtained via $T = P/n k_B$, which represents the equation of state for an ideal gas (k_B is the Boltzmann constant). More generally, it is convenient to introduce the following notation for the moments of f and f^{eq}:

$$\begin{pmatrix} M_{s_x,s_y,s_z} \\ \\ M^{(\text{eq})}_{s_x,s_y,s_z} \end{pmatrix} = \int d^3p \begin{pmatrix} f \\ \\ f^{\text{eq}} \end{pmatrix} (p_x)^{s_x}(p_y)^{s_y}(p_z)^{s_z}. \tag{9.6}$$

Since mass m, momentum $\mathbf{p}$ and energy $\mathbf{p}^2/2m$ are collision invariant quantities, we have:

$$n = M_{0,0,0} = M^{(\text{eq})}_{0,0,0}, \tag{9.7}$$

$$\begin{pmatrix} \rho u_x \\ \\ \rho u_y \\ \\ \rho u_z \end{pmatrix} = \begin{pmatrix} M_{1,0,0} \\ \\ M_{0,1,0} \\ \\ M_{0,0,1} \end{pmatrix} = \begin{pmatrix} M^{(\text{eq})}_{1,0,0} \\ \\ M^{(\text{eq})}_{0,1,0} \\ \\ M^{(\text{eq})}_{0,0,1} \end{pmatrix}, \tag{9.8}$$

$$\begin{aligned} \frac{3}{2}nK_BT + \frac{1}{2}\rho\mathbf{u}^2 &= \frac{1}{2m}\left(M_{2,0,0} + M_{0,2,0} + M_{0,0,2}\right) \\ &= \frac{1}{2m}\left(M^{(\text{eq})}_{2,0,0} + M^{(\text{eq})}_{0,2,0} + M^{(\text{eq})}_{0,0,2}\right). \end{aligned} \tag{9.9}$$

The basic steps for the construction of an off-lattice LB model are [49]:

1. discretising the momentum space;
2. replacing the equilibrium distribution function f^{eq} in the collision term of Eq. (9.1) by a truncated polynomial with respect to the particle velocity ;
3. replacing the momentum derivative of the distribution function f in Eq. (9.1) using a suitable expression [see Eq. (9.23)];
4. choosing a numerical method for the time evolution and spatial advection;
5. implementation of the boundary conditions.

A common feature of all LB models is that the conservation equations for the particle number density n, macroscopic momentum density $\rho\mathbf{u}$ and temperature T (for thermal models) are exactly recovered. Regardless of the chosen discretisation of the momentum space, the Boltzmann–BGK equation (9.1) is replaced by a set of K equations:

$$\partial_t f_k + \frac{\mathbf{p}_k}{m} \cdot \nabla f_k + \mathbf{F} \cdot (\nabla_{\mathbf{p}} f)_k = -\frac{1}{\tau}(f_k - f_k^{\text{eq}}), \tag{9.10}$$

where f_k $(k = 1, 2, \ldots K)$ represents the distribution function corresponding to the discrete momentum $\mathbf{p}_k$. The total number K of discrete momenta are chosen such that the moments Eq. (9.6) are exactly recovered:

$$\begin{pmatrix} M_{s_x,s_y,s_z} \\ M^{(\mathrm{eq})}_{s_x,s_y,s_z} \end{pmatrix} = \sum_{k=1}^{K} \begin{pmatrix} f_k \\ f_k^{\mathrm{eq}} \end{pmatrix} (\mathbf{p}_{k,x})^{s_x} (\mathbf{p}_{k,y})^{s_y} (\mathbf{p}_{k,z})^{s_z}. \tag{9.11}$$

The order N of a given LB model is related to the maximum value of the exponents s_x, s_y, s_z for which the above equality holds.

9.3 One-Dimensional Quadrature-Based LB Models

In this section, the procedure for implementing the full-range and half-range Gauss–Hermite quadratures on a single axis of the momentum space will be discussed in Sects. 9.3.1 and 9.3.2, respectively. For convenience, in this section we will refer to the one-dimensional ($1D$) equivalent of the Boltzmann–BGK equation (9.1):

$$\partial_t f + \frac{p}{m} \partial_x f + F \partial_p f = -\frac{1}{\tau}(f - f^{\mathrm{eq}}). \tag{9.12}$$

After discretisation, this equation is replaced by a set of $K = Q$ equations, where Q is the number of quadrature points on the entire axis:

$$\partial_t f_k + \frac{p_k}{m} \partial_x f_k + F(\partial_p f)_k = -\frac{1}{\tau}(f_k - f_k^{\mathrm{eq}}). \tag{9.13}$$

The general expression of the total number K of discrete momenta employed by a D-dimensional LB model will be introduced in Sect. 9.4.

9.3.1 Full-Range Gauss–Hermite Quadrature

Let us consider integrals of f and f^{eq} along the axis of the $1D$ momentum space:

$$\begin{pmatrix} M_s \\ M_s^{(\mathrm{eq})} \end{pmatrix} = \int_{-\infty}^{\infty} dp \begin{pmatrix} f \\ f^{\mathrm{eq}} \end{pmatrix} p^s. \tag{9.14}$$

For the purpose of this section, we can consider $f^{\mathrm{eq}} = ng$, where g is expressed as in Eq. (9.3), but without using the subscript α. The function g can be expanded with respect to the full-range Hermite polynomials $\{H_\ell(\overline{p}), \ell = 0, 1, \ldots\}$ as follows [25, 26]:

$$g = \frac{\omega(\overline{p})}{p_0} \sum_{\ell=0}^{\infty} \frac{1}{\ell!} \mathcal{G}_\ell H_\ell(\overline{p}), \quad \mathcal{G}_\ell = \sum_{s=0}^{\lfloor \ell/2 \rfloor} \frac{\ell!}{2^s s! (\ell - 2s)!} \left(\frac{m K_B T}{p_0^2} - 1 \right)^s \left(\frac{mu}{p_0} \right)^{\ell - 2s}, \tag{9.15}$$

where $\lfloor \cdot \rfloor$ is the floor function, $\mathcal{G}_\ell$ is the ℓ-th expansion coefficient and $\overline{p} = p/p_0$ is the particle momentum expressed with respect to some arbitrary momentum scale p_0. The full-range Hermite polynomials [18, 26, 39, 40] satisfy the following orthogonality relation with respect to the weight function $\omega(\overline{p})$:

$$\int_{-\infty}^{\infty} d\overline{p}\, \omega(\overline{p}) H_\ell(\overline{p}) H_{\ell'}(\overline{p}) = \ell!\, \delta_{\ell,\ell'}, \qquad \omega(\overline{p}) = \frac{1}{\sqrt{2\pi}} e^{-\overline{p}^2/2}. \tag{9.16}$$

The expansion coefficients $\mathcal{G}_\ell$ given in Eq. (9.15) were obtained according to:

$$\mathcal{G}_\ell = \int_{-\infty}^{\infty} dp\, g\, H_\ell(\overline{p}). \tag{9.17}$$

Substituting Eq. (9.15) into Eq. (9.14) gives:

$$M_s^{(\mathrm{eq})} = p_0^s \sum_{\ell=0}^{\infty} \frac{1}{\ell!} \mathcal{G}_\ell \int_{-\infty}^{\infty} d\overline{p}\, \omega(\overline{p})\, H_\ell(\overline{p})\, \overline{p}^s. \tag{9.18}$$

At finite values of s and ℓ, the Gauss–Hermite quadrature can be applied to recover the integral over $\overline{p}$ on the entire momentum axis, using the following prescription:

$$\int_{-\infty}^{\infty} d\overline{p}\, \omega(\overline{p}) P_s(\overline{p}) \simeq \sum_{k=1}^{Q} w_k^H\, P_s(\overline{p}_k), \tag{9.19}$$

where $P_s(\overline{p})$ is a polynomial of order s in $\overline{p}$ and the Q quadrature points $\overline{p}_k$ ($k = 1, 2, \ldots, Q$) are the roots of the Hermite polynomial of order Q, i.e., $H_Q(\overline{p}_k) = 0$. Note that $K = \mathcal{Q} = Q$ holds only in a one-dimensional LB model based on full-range Gauss–Hermite quadratures.

Since these roots correspond to the integration over the full momentum space axis, in the case of the full-range Gauss–Hermite quadrature, the number of quadrature points on the entire axis $\mathcal{Q}$ is equal to the quadrature order Q. The quadrature weights w_k^H are given by:

$$w_k^H = \frac{Q!}{[H_{Q+1}(\overline{p}_k)]^2}. \tag{9.20}$$

The equality in Eq. (9.19) is exact if $2Q > s$. In an LB simulation, Q is fixed at runtime. Thus, in order to ensure the exact recovery of $M_s^{(\mathrm{eq})}$ in Eq. (9.18), the sum over ℓ in Eq. (9.15) must be truncated at a finite value $\ell = N$. Setting $Q > N$ ensures the exact recovery of the first $N + 1$ moments (i.e., $s = 0, 1, \ldots N$) of

$f^{\rm eq}$, since the terms of higher order in the expansion of g are orthogonal to all polynomials $P_s(\overline{p})$ of orders $0 \le s \le N$, by virtue of the orthogonality relation given by Eq. (9.16). This allows $M_s^{(\rm eq)}$ to be obtained as:

$$M_s^{(\rm eq)} = \sum_{k=1}^{Q} f_k^{\rm eq} p_k^s, \qquad f_k^{\rm eq} = n g_k^H, \qquad g_k^H = \frac{w_k^H p_0}{\omega(\overline{p}_k)} g^{H,(N)}(p_k), \tag{9.21}$$

where $p_k = p_0 \overline{p}_k$ are the discrete momenta and the notation $g^{H,(N)}(p)$ indicates that the polynomial expansion in Eq. (9.15) of $g(p)$ is truncated at order $\ell = N$ with respect to the full-range Hermite polynomials. For definiteness, we list below the expression for g_k^H [25, 26]:

$$g_k^{H,(N)} = w_k^H \sum_{\ell=0}^{N} H_\ell(\overline{p}_k) \sum_{s=0}^{\lfloor \ell/2 \rfloor} \frac{1}{2^s s! (\ell - 2s)!} \left(\frac{mT}{p_0^2} - 1 \right)^s \left(\frac{m K_B T}{p_0} \right)^{\ell - 2s}. \tag{9.22}$$

The momentum derivative $\partial_p f$ can be written as:

$$(\partial_p f)_k = \sum_{k'=1}^{Q} \mathcal{K}_{k,k'} f_{k'}, \tag{9.23}$$

where the kernel $\mathcal{K}_{k,k'}$ has the following components [49, 50]:

$$\mathcal{K}_{k,k'} = -\frac{w_k^H}{p_0} \sum_{\ell=0}^{Q-2} \frac{1}{\ell!} H_{\ell+1}(\overline{p}_k) H_\ell(\overline{p}_{k'}). \tag{9.24}$$

9.3.2 Half-Range Gauss–Hermite Quadrature

The half-range paradigm is inspired from the discontinuous nature of the distribution function due to the interaction with the channel walls. Such discontinuities naturally induce a split of the momentum space integration domain in two hemispheres, corresponding to particles travelling towards and away from the wall. In order to encompass the discontinuous nature of the distribution function in a one-dimensional LB model for confined fluid flow, it is convenient to introduce the half-range moments $M_s^{\pm}$ and $M_s^{(\rm eq),\pm}$ $(s = 0, 1, 2, \dots)$ of f and $f^{(\rm eq)}$ through:

$$\begin{pmatrix} M_s^+ \\ M_s^{(\rm eq),+} \end{pmatrix} = \int_0^\infty dp \begin{pmatrix} f(p) \\ f^{\rm eq}(p) \end{pmatrix} p^s, \qquad \begin{pmatrix} M_s^- \\ M_s^{(\rm eq),-} \end{pmatrix} = \int_{-\infty}^0 dp \begin{pmatrix} f(p) \\ f^{\rm eq}(p) \end{pmatrix} p^s. \tag{9.25}$$

The recovery of the half-range integrals in Eq. (9.25) can be achieved using the half-range Gauss–Hermite quadrature, defined with respect to the weight function $\omega(\overline{p})$:

$$\int_0^\infty d\overline{p}\,\omega(\overline{p})P_s(\overline{p}) \simeq \sum_{k=1}^{Q} w_k P_s(\overline{p}_k), \qquad \omega(\overline{p}) = \frac{1}{\sqrt{2\pi}} e^{-\overline{p}^2/2}, \tag{9.26}$$

where the equality is exact if the number of quadrature points Q satisfies $2Q > s$. The quadrature points $\overline{p}_k$ ($k = 1, 2, \ldots Q$) are the Q (positive) roots of the half-range Hermite polynomial $\mathfrak{h}_Q(\overline{p})$, while the quadrature weights w_k are given by [25, 26, 39, 40]:

$$w_k^{\mathfrak{h}} = \frac{\overline{p}_k a_Q^2}{\mathfrak{h}_{Q+1}^2(\overline{p}_k)\left[\overline{p}_k + \mathfrak{h}_{Q,0}^2/\sqrt{2\pi}\right]}, \tag{9.27}$$

where $a_Q = \mathfrak{h}_{Q+1,Q+1}/\mathfrak{h}_{Q,Q}$ and $\mathfrak{h}_{\ell,s}$ represents the coefficient of $\overline{p}^s$ in $\mathfrak{h}_\ell(\overline{p})$, i.e.,

$$\mathfrak{h}_\ell(\overline{p}) = \sum_{s=0}^{\ell} \mathfrak{h}_{\ell,s}\overline{p}^s. \tag{9.28}$$

In our convention, the half-range Hermite polynomials are normalised according to:

$$\int_0^\infty d\overline{p}\,\omega(\overline{p})\mathfrak{h}_\ell(\overline{p})\mathfrak{h}_{\ell'}(\overline{p}) = \delta_{\ell,\ell'}. \tag{9.29}$$

In order to apply the half-range Gauss–Hermite quadrature prescription, Eq. (9.26), f and $f^{\rm eq}$ must be expanded with respect to the half-range Hermite polynomials. Since the half-range Hermite polynomials are defined only on half of the momentum axis, f can be split with the help of the Heaviside step function $\theta(\overline{p})$ as follows [49]:

$$f(p) = \theta(\overline{p})f^+(p) + \theta(-\overline{p})f^-(p), \qquad \theta(\overline{p}) = \begin{cases} 1, & \overline{p} > 0, \\ 0, & \overline{p} < 0. \end{cases} \tag{9.30}$$

The functions $f^+(p)$ and $f^-(p)$ are defined only on the positive and negative momentum semiaxis, respectively, such that they can be expanded with respect to the half-range Hermite polynomials as follows:

$$f^+ = \frac{\omega(\overline{p})}{p_0}\sum_{\ell=0}^{\infty}\mathcal{F}^+\mathfrak{h}_\ell(\overline{p}), \qquad f^- = \frac{\omega(-\overline{p})}{p_0}\sum_{\ell=0}^{\infty}\mathcal{F}^-\mathfrak{h}_\ell(-\overline{p}), \tag{9.31}$$

where the coefficients $\mathcal{F}^{\pm}$ can be obtained using the orthogonality given in Eq. (9.29) of the half-range Hermite polynomials:

$$\mathcal{F}_\ell^+ = \int_0^\infty dp\, f(p)\mathfrak{h}_\ell(\overline{p}), \qquad \mathcal{F}_\ell^- = \int_{-\infty}^0 dp\, f(p)\mathfrak{h}_\ell(-\overline{p}). \tag{9.32}$$

The expansion in Eq. (9.31) with respect to the half-range Hermite polynomials $\mathfrak{h}_\ell$ can be substituted in Eq. (9.25), yielding:

$$\begin{pmatrix} M_s^+ \\ \\ M_s^- \end{pmatrix} = p_0^s \sum_{\ell=0}^{\infty} \frac{1}{\ell!} \begin{pmatrix} \mathcal{F}_\ell^+ \\ \\ \mathcal{F}_\ell^- \end{pmatrix} \int_0^\infty d\overline{p}\, \omega(\overline{p})\, \mathfrak{h}_\ell(\overline{p})(\pm\overline{p})^s. \tag{9.33}$$

Truncating the expansion in Eq. (9.31) at $\ell = Q - 1$ ensures that a quadrature of order Q can recover the moments in Eq. (9.33) for $0 \leq s \leq Q$. Since Q quadrature points are required on each semiaxis of the momentum space, the discrete momentum set of the $1D$ half-range Gauss–Hermite LB model has $K = Q = 2Q$ elements (twice as in the full-range model of the same order), which are defined as:

$$p_k = p_0\overline{p}_k, \qquad p_{k+Q} = -p_k \qquad (1 \leq k \leq Q). \tag{9.34}$$

Thus, the half-range moments in Eq. (9.25) are recovered as:

$$M_s^+ = \sum_{k=1}^{Q} f_k p_k^s, \qquad M_s^- = \sum_{k=Q+1}^{2Q} f_k p_k^s, \tag{9.35}$$

where

$$f_k = \frac{w_k^{\mathfrak{h}} p_0}{\omega(\overline{p}_k)} f(p_k), \qquad f_{k+Q} = \frac{w_k^{\mathfrak{h}} p_0}{\omega(\overline{p}_k)} f(-p_k) \qquad (1 \leq k \leq Q). \tag{9.36}$$

Let us now consider the expansion of f^{eq} with respect to the half-range Hermite polynomials, by writing $g(p) = \theta(p)g_+(p) + \theta(-p)g_-(p)$, where

$$g_\pm = \frac{\omega(|\overline{p}|)}{p_0} \sum_{\ell=0}^{\infty} \mathcal{G}_\ell^{\pm}\mathfrak{h}_\ell(|\overline{p}|). \tag{9.37}$$

The expansion coefficients $\mathcal{G}_\ell^{\pm}$ can be obtained in analogy to Eq. (9.32).

Following the convention of Eq. (9.34), the momentum space is discretised using $Q = 2Q$ elements with $p_k > 0$ (for the positive semiaxis) and $p_{k+Q} = -p_k$ (for the negative semiaxis), where $1 \le k \le Q$. The corresponding equilibrium distributions $f_k^{\rm eq} = n g_k^{\mathfrak{h},(N)}$ are constructed using

$$g_k^{\mathfrak{h},(N)} = w_k^{\mathfrak{h}} \sum_{\ell=0}^{N} \mathcal{G}_\ell^+ \mathfrak{h}_\ell(\overline{p}_k), \qquad g_{k+Q}^{\mathfrak{h},(N)} = w_k^{\mathfrak{h}} \sum_{\ell=0}^{N} \mathcal{G}_\ell^- \mathfrak{h}_\ell(\overline{p}_k), \tag{9.38}$$

where the expansion order $0 \le N < Q$ is a free parameter of the model which represents the order up to which the half-range moments of $f^{\rm eq}$ can be exactly recovered. The coefficients $\mathcal{G}_\ell^\pm$ can be found using the orthogonality relation in Eq. (9.29):

$$\mathcal{G}_\ell^+ = \int_0^\infty dp\, g\, \mathfrak{h}_\ell(\overline{p}), \qquad \mathcal{G}_\ell^- = \int_{-\infty}^0 dp\, g\, \mathfrak{h}_\ell(-\overline{p}). \tag{9.39}$$

The integrals above can be performed analytically, such that $g_k^{\mathfrak{h},(N)}$ and $g_{k+Q}^{\mathfrak{h},(N)}$ become [25, 26]:

$$\begin{aligned} g_k^{\mathfrak{h},(N)} &= \frac{w_k^{\mathfrak{h}}}{2} \sum_{s=0}^{N} \left(\frac{mT}{2p_0^2}\right)^{s/2} \Phi_s^N(\overline{p}_k) \left[(1 + \operatorname{erf}\zeta) P_s^+(\zeta) + \frac{2}{\sqrt{\pi}} e^{-\zeta^2} P_s^*(\zeta)\right], \\ g_{k+Q}^{\mathfrak{h},(N)} &= \frac{w_k^{\mathfrak{h}}}{2} \sum_{s=0}^{N} \left(\frac{mT}{2p_0^2}\right)^{s/2} \Phi_s^N(\overline{p}_k) \left[(\operatorname{erfc}\zeta) P_s^+(-\zeta) + \frac{2}{\sqrt{\pi}} e^{-\zeta^2} P_s^*(-\zeta)\right], \end{aligned} \tag{9.40}$$

where $\zeta = u\sqrt{m/2K_B T}$, $\operatorname{erf}\zeta = \frac{2}{\sqrt{\pi}} \int_0^\zeta dz\, e^{-z^2}$ is the error function, $\Phi_s^N(\overline{p}_k)$ is defined as:

$$\Phi_s^N(\overline{p}_k) = \sum_{\ell=s}^{N} \mathfrak{h}_{\ell,s} \mathfrak{h}_\ell(\overline{p}_k), \tag{9.41}$$

where $\mathfrak{h}_{\ell,s}$ is defined in Eq. (9.28), while $P_s^+(\zeta)$ and $P_s^*(\zeta)$ represent polynomials of orders s and $s-1$, respectively, defined through:

$$P_s^\pm(\zeta) = e^{\mp\zeta^2} \frac{d^s}{d\zeta^s} e^{\pm\zeta^2}, \qquad P_s^*(\zeta) = \sum_{j=0}^{s-1} \binom{s}{j} P_j^+(\zeta) P_{s-j-1}^-(\zeta). \tag{9.42}$$

The momentum derivative of f can be projected on the space of the half-range Hermite polynomials as discussed in Sect. 9.3.1. Since this projection is not relevant for the further development of this chapter, we refer the reader to Refs. [49, 50] for further details.

9.4 LB Models in the Three-Dimensional Momentum Space

In the three-dimensional ($3D$) momentum space, the discretisation procedure can be conducted using a direct product rule. On each Cartesian axis $\alpha \in \{x, y, z\}$, one can choose a specific Gauss–Hermite (full-range or half-range) quadrature of order Q_α, depending on the characteristics of the flow (e.g., the existence of a noticeable wall-induced discontinuity of the distribution function along the α axis). Let p_{α,k_α}, $1 \le k_\alpha \le \mathcal{Q}_\alpha$, be the quadrature points on the Cartesian axis α (note that $\mathcal{Q}_\alpha \in \{Q_\alpha, 2Q_\alpha\}$ as mentioned in Sect. 9.3). These quadrature points are the components of the $3D$ vectors $\mathbf{p}_k$, $k = (k_z - 1)\mathcal{Q}_x\mathcal{Q}_y + (k_y - 1)\mathcal{Q}_x + k_x$, $1 \le k \le K = \mathcal{Q}_x\mathcal{Q}_y\mathcal{Q}_z$. Following Refs. [25, 26], we generally refer to the resulting models as *mixed quadrature LB models*. The numerical solution of the discretised form Eq. (9.10) of the Boltzmann equation can be obtained following the steps described in Sect. 9.3, which will be detailed further.

In this contribution we restrict ourselves to the LB simulation of Couette and force-driven Poiseuille flows of rarefied gases between parallel plates. In these cases, the flow is homogeneous along the z axis and the computational effort can be significantly decreased by taking advantage of the reduced distribution functions introduced in Sect. 9.4.1. Section 9.4.2 discusses the construction of the mixed quadrature LB models for the investigation of rarefied Couette and force-driven Poiseuille flows using the reduced distribution functions and the resulting evolution equations are presented in Sect. 9.4.3. The section ends with a discussion of our non-dimensionalisation convention, presented in Sect. 9.4.4.

9.4.1 Reduced Distributions

In this chapter, we only consider the planar Couette and the force-driven Poiseuille flows between parallel plates. Considering that the walls are perpendicular to the x axis, these flows can be considered homogeneous with respect to the y and z axes, such that the Boltzmann equation (9.1) reduces to:

$$\partial_t f + \frac{p_x}{m}\partial_x f + F_y \nabla_{p_y} f = -\frac{1}{\tau}(f - f^{\text{eq}}). \tag{9.43}$$

The force term is present only in the case of the Poiseuille flow. Assuming that the fluid flows along the y direction, the only non-vanishing component of the force is along the y axis (see Sect. 9.5.2 for more details).

Since the flows considered in this chapter are trivial with respect to the z axis, the p_z degree of freedom can be eliminated from Eq. (9.43) [44, 51]. This helps to reduce the computational costs, especially when dealing with LB models involving high order quadratures. Defining:

$$\phi = \int_{-\infty}^{\infty} dp_z\, f, \qquad \chi = \int_{-\infty}^{\infty} dp_z\, f\, \frac{p_z^2}{m}, \tag{9.44}$$

the following two equations are obtained:

$$\partial_t \begin{pmatrix} \phi \\ \chi \end{pmatrix} + \frac{p_x}{m} \partial_x \begin{pmatrix} \phi \\ \chi \end{pmatrix} + F_y \frac{\partial}{\partial p_y} \begin{pmatrix} \phi \\ \chi \end{pmatrix} = -\frac{1}{\tau} \begin{pmatrix} \phi - \phi^{\text{eq}} \\ \chi - \chi^{\text{eq}} \end{pmatrix}, \tag{9.45}$$

where $\chi^{\text{eq}} = k_B T \phi^{\text{eq}}$ and ϕ^{eq} can be factorised using the functions g_α Eq. (9.3) as follows:

$$\phi^{\text{eq}} = n g_x g_y = \frac{n}{2\pi m k_B T} \exp\left[-\frac{(p_x - m u_x)^2 + (p_y - m u_y)^2}{2 m k_B T} \right]. \tag{9.46}$$

Note that the reduction procedure introduced above can be used also for the $3D$ pressure-driven Poiseuille flow between parallel plates, provided that there are no variations along the z axis.

9.4.2 *Mixed Quadrature LB Models with Reduced Distribution Functions*

In the mixed quadrature LB models, the momentum space is constructed using a direct product rule. This allows the quadrature on each axis to be constructed independently by taking into account the characteristics of the flow. When the gas flow is homogeneous along the z axis, the reduced distribution functions evolve in a two-dimensional space and thus, the elements of the discrete set of momentum vectors can be written as $\mathbf{p}_{ij} = (p_{x,i}, p_{y,j})$. The indices i and j run from 1 to $\mathcal{Q}_\alpha$ ($\alpha \in \{x, y\}$), where $\mathcal{Q}_\alpha = Q_\alpha$ or $\mathcal{Q}_\alpha = 2Q_\alpha$ when a full-range or a half-range quadrature of order Q_α is employed on the α axis. As shown in Refs. [25, 26], a full-range Gauss–Hermite quadrature of order $Q_y = 4$ is sufficient on the y axis in order to capture exactly the evolution of the velocity, temperature and of heat flux fields. For low Mach flows, the quadrature order Q_x can be taken to be $Q_x = 4$ in the Navier–Stokes regime, where the full-range Gauss–Hermite quadrature is efficient. As Kn is increased, Q_x must also be increased in order to retain the accuracy of the simulation results. In the case of the channel flows considered in this chapter, the discontinuity in the distribution functions ϕ and χ induced by the diffuse-reflective walls becomes significant for sufficiently large Kn. Hence, the full-range Gauss–Hermite quadrature on the x axis becomes inefficient compared to the half-range Gauss–Hermite quadrature, as demonstrated in Refs. [25, 26]. In this chapter, we only consider the half-range Gauss–Hermite quadrature of order Q_x on the x axis. The resulting models are denoted HHLB(Q_x) × HLB(4) following the convention in Ref. [26], employing $8Q_x$ velocities and $16Q_x$ distinct populations (ϕ_{ij} and χ_{ij}), as discussed below.

9.4.3 The Lattice Boltzmann Equation

The reduced distribution functions ϕ_{ij} and χ_{ij} corresponding to the momentum vector $\mathbf{p}_{ij} = (p_{x,i}, p_{y,j})$ are linked to ϕ and χ through the direct extension of Eq. (9.36):

$$\begin{pmatrix} \phi_{ij} \\ \chi_{ij} \end{pmatrix} = \left(\frac{w_i^x \, p_{0,x}}{\omega(\overline{p}_{x,i})} \right) \left(\frac{w_j^y \, p_{0,y}}{\omega(\overline{p}_{y,j})} \right) \begin{pmatrix} \phi^{Q_x,Q_y}(p_{x,i}, p_{y,j}) \\ \chi^{Q_x,Q_y}(p_{x,i}, p_{y,j}) \end{pmatrix}. \tag{9.47}$$

The weights w_i^x and w_j^y are given by Eqs. (9.27) and (9.20), respectively. After the discretisation of the momentum space, Eq. (9.45) becomes:

$$\partial_t \begin{pmatrix} \phi_{ij} \\ \chi_{ij} \end{pmatrix} + \frac{p_{x,i}}{m} \partial_x \begin{pmatrix} \phi_{ij} \\ \chi_{ij} \end{pmatrix} + F_y \sum_{j'=1}^{Q_y} \mathcal{K}_{j,j'} \begin{pmatrix} \phi_{i,j'} \\ \chi_{i,j'} \end{pmatrix} = -\frac{1}{\tau} \begin{pmatrix} \phi_{ij} - \phi_{ij}^{\rm eq} \\ \chi_{ij} - \chi_{ij}^{\rm eq} \end{pmatrix}, \tag{9.48}$$

where the kernel $\mathcal{K}_{j,j'}$ is given in Eq. (9.24). In particular, for the case $Q_y = 4$ considered in this chapter, $\mathcal{K}_{j,j'}$ has the following elements:

$$\mathcal{K}_{j,j'} = \frac{1}{p_{0,y}} \begin{pmatrix} \frac{1}{2}\sqrt{3+\sqrt{6}} & \frac{\sqrt{3+\sqrt{3}}}{2(3+\sqrt{6})} & -\frac{\sqrt{3-\sqrt{3}}}{2(3+\sqrt{6})} & \frac{1}{2}\sqrt{1-\sqrt{\frac{2}{3}}} \\ -\sqrt{\frac{5+2\sqrt{6}}{2(3-\sqrt{3})}} & \frac{1}{2}\sqrt{3-\sqrt{6}} & \frac{1}{2}\sqrt{1+\sqrt{\frac{2}{3}}} & -\frac{\sqrt{27+11\sqrt{6}}-\sqrt{3+\sqrt{6}}}{2\sqrt{6}} \\ \frac{\sqrt{27+11\sqrt{6}}-\sqrt{3+\sqrt{6}}}{2\sqrt{6}} & -\frac{1}{2}\sqrt{1+\sqrt{\frac{2}{3}}} & -\frac{1}{2}\sqrt{3-\sqrt{6}} & \frac{\sqrt{27+11\sqrt{6}}+\sqrt{3+\sqrt{6}}}{2\sqrt{6}} \\ -\frac{\sqrt{3-\sqrt{6}}}{2\sqrt{3}} & \frac{\sqrt{3-\sqrt{3}}}{2(3+\sqrt{6})} & -\frac{\sqrt{3+\sqrt{3}}}{2(3+\sqrt{6})} & -\frac{1}{2}\sqrt{3+\sqrt{6}} \end{pmatrix}. \tag{9.49}$$

Numerically, the above expression reduces to:

$$\mathcal{K}_{j,j'} \simeq \frac{1}{p_{0,y}} \begin{pmatrix} 1.1672 & 0.1996 & -0.1033 & 0.2142 \\ -1.9757 & 0.3710 & 0.6739 & -1.0227 \\ 1.0227 & -0.6739 & -0.3710 & 1.9757 \\ -0.2142 & 0.1033 & -0.1996 & -1.1672 \end{pmatrix}. \tag{9.50}$$

The equilibrium distribution $\phi_{ij}^{\rm eq} = n g_i^{\mathfrak{h},(N_x)} g_j^{H,(N_y)}$ is obtained as the product between the expansions $g_i^{\mathfrak{h},(N_x)}$ from Eq. (9.40) and $g_j^{H,(N_y)}$ from Eq. (9.22), performed with respect to the half-range and full-range Hermite polynomials, respectively. For this particular case, the orders of the expansions are $N_x = 3$ and $N_y = 3$. For definiteness, we list below the exact expression for $g_j^{H,(3)}$:

$$g_j^{H,(3)} = w_j^H \left[1 + \overline{p}_{y,j}\mathfrak{U}_y + \frac{1}{2}(\overline{p}_{y,j}^2 - 1)(\mathfrak{U}_y^2 + \mathfrak{T}_y) + \frac{1}{6}\left(\overline{p}_{y,j}^3 - 3\overline{p}_{y,j}\right)\mathfrak{U}_y(\mathfrak{U}_y^2 + 3\mathfrak{T}_y)\right], \tag{9.51}$$

where $\mathfrak{U}_y$ and $\mathfrak{T}_y$ are defined as [26]:

$$\mathfrak{U}_y = \frac{m u_y}{p_{0,y}}, \qquad \mathfrak{T}_y = \frac{m K_B T}{p_{0,x}^2} - 1. \tag{9.52}$$

Similarly, $g_i^{\mathfrak{h},(3)}$ is given by Ambruș and Sofonea [25]:

$$\begin{aligned} g_i^{\mathfrak{h},(3)} = \frac{w_i^{\mathfrak{h}}}{2} \Big\{ &(1 + \operatorname{erf}\zeta_{x,i}) \Big[\Phi_0^3(|\overline{p}_{x,i}|) + 2\zeta_{x,i}\mathcal{T}_x \Phi_1^3(|\overline{p}_{x,i}|) \\ &+ 2\mathcal{T}_x^2(2\zeta_{x,i}^2 + 1)\Phi_2^3(|\overline{p}_{x,i}|) + 4\zeta_{x,i}\mathcal{T}_x^3(2\zeta_{x,i}^2 + 3)\Phi_3^3(|\overline{p}_{x,i}|) \Big] \\ &+ \frac{2e^{-\zeta_\alpha^2}}{\sqrt{\pi}}\mathcal{T}_x \Big[\Phi_1^3(|\overline{p}_{x,i}|) + 2\zeta_{x,i}\mathcal{T}_x\Phi_2^3(|\overline{p}_{x,i}|) \\ &+ 4\mathcal{T}_x^2(\zeta_{x,i}^2 + 1)\Phi_3^3(|\overline{p}_{x,i}|) \Big] \Big\}, \end{aligned} \tag{9.53}$$

where $\zeta_{x,i} = u_x \sigma_{x,i}\sqrt{m/2K_B T}$, $\sigma_{x,i}$ is the sign of $p_{x,i}$ and $\mathcal{T}_x = \sqrt{mK_BT/2p_{0,x}^2}$, while the functions $\Phi_s^3(z)$ are given below:

$$\Phi_0^3(z) = \frac{2\pi(9\pi - 28) - z\sqrt{2\pi}(21\pi - 64) + 2\pi z^2(10 - 3\pi) - z^3\sqrt{2\pi}(16 - 5\pi)}{32 - 29\pi + 6\pi^2},$$

$$\Phi_1^3(z) = \frac{2\pi z(15\pi - 44) - \sqrt{2\pi}(21\pi - 64) - z^2\sqrt{2\pi}(16 - 3\pi) + 2\pi z^3(10 - 3\pi)}{32 - 29\pi + 6\pi^2},$$

$$\Phi_2^3(z) = \frac{2\pi(10 - 3\pi) - z\sqrt{2\pi}(16 - 3\pi) + 2\pi z^2(3\pi - 7) - z^3\sqrt{2\pi}(3\pi - 8)}{32 - 29\pi + 6\pi^2},$$

$$\Phi_3^3(z) = \frac{-\sqrt{2\pi}(16-5\pi) + 2\pi z(10-3\pi) - z^2\sqrt{2\pi}(3\pi-8) + 2\pi z^3(\pi-3)}{32-29\pi+6\pi^2}. \tag{9.54}$$

Finally, the macroscopic moments Eq. (9.4) can be written in terms of ϕ_{ij} and χ_{ij} as follows:

$$n = \sum_{i,j} \phi_{ij} = \sum_{i,j} \phi_{ij}^{\text{eq}}, \qquad \begin{pmatrix} \rho u_x \\ \rho u_y \end{pmatrix} = \sum_{i,j} \phi_{ij} \begin{pmatrix} p_{x,i} \\ p_{y,j} \end{pmatrix} = \sum_{i,j} \phi_{ij}^{\text{eq}} \begin{pmatrix} p_{x,i} \\ p_{y,j} \end{pmatrix},$$

$$\frac{3}{2} n K_B T + \frac{1}{2}\rho \mathbf{u}^2 = \sum_{i,j} \left[\phi_{ij} \frac{p_{x,i}^2 + p_{y,j}^2}{2m} + \frac{1}{2}\chi_{ij} \right]$$

$$= \sum_{i,j} \left[\phi_{ij}^{\text{eq}} \frac{p_{x,i}^2 + p_{y,j}^2}{2m} + \frac{1}{2}\chi_{ij}^{\text{eq}} \right]. \tag{9.55}$$

It can be seen that χ_{ij} appears only in the definition of the temperature field. It is essential to track the evolution of ϕ_{ij} and χ_{ij} simultaneously in order to correctly compute the temperature field appearing in the definition of ϕ^{eq} given in Eq. (9.46), as well as in the definition of χ^{eq}.

9.4.4 Non-Dimensionalisation Procedure

In order to perform numerical simulations, we non-dimensionalise all quantities with respect to the following parameters:

- The wall temperature, such that $T_w = 1$.
- The particle mass, such that $m = 1$.
- The reference speed $c_{\text{ref}} = \sqrt{k_B T_w / m}$.
- The channel width, such that $L = 1$.
- The average particle number density, such that the total number of particles obeys:

$$N_{\text{tot}} = \int_{-1/2}^{1/2} dx\, n(x) = 1. \tag{9.56}$$

The reference time is $t_{\text{ref}} = L/c_{\text{ref}}$ and we set $p_{0,x} = p_{0,y} = 1$ for the rest of this chapter. With the above conventions, the relaxation time τ is set to

$$\tau = \frac{\mathrm{Kn}}{nT}, \tag{9.57}$$

which ensures that the viscosity $\mu = \tau n T = \mathrm{Kn}$ is constant throughout the simulation domain.

9.5 Simulation Results

The advantage of the quadrature-based approach to LB modelling quickly becomes apparent when considering rarefied flows. An excellent arena for this type of tests is represented by channel flows. In particular, we will restrict the discussion to the Couette and the force-driven Poiseuille flows between parallel plates, which have become canonical benchmark problems in the microfluidics community. In the context of these flows, the distribution function becomes discontinuous due to the diffuse reflection interaction with the boundary. Thus, at large values of Kn, half-range quadratures are much more efficient than the more traditional full-range ones [25, 26, 52, 53]. More complex flows, where the application of half-range Gauss–Hermite quadrature is essential are investigated in Refs. [50, 54, 55].

9.5.1 Couette Flow Between Parallel Plates

In this section, we consider the Couette flow between parallel plates. The geometry of this flow can be seen in Fig. 9.1. The system consists of two parallel plates at

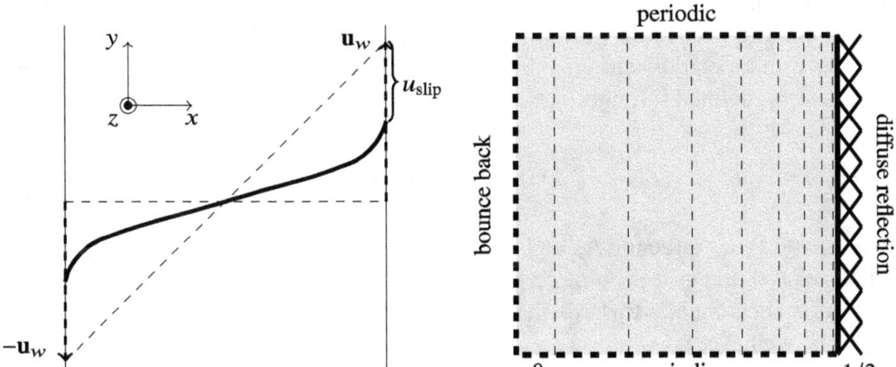

Fig. 9.1 *Left:* Setup for the Couette flow problem, highlighting the slip velocity $u_{\text{slip}} = u_{\text{w}} - u(1/2)$. *Right:* Boundary conditions and grid characteristics. The fine dotted lines show a grid comprised of $S = 8$ cells, stretched according to Eq. (9.58) with $A = 0.95$. Only one cell is used along the y direction

rest located at $x = \pm 1/2$.[1] The gas between these plates is initially in thermal equilibrium at the wall temperature $T_w = 1$. At $t = 0$, the left and right plates are set into motion with velocities $-\boldsymbol{u}_w = (0, -u_w, 0)$ and $\boldsymbol{u}_w = (0, u_w, 0)$, respectively, as shown in Fig. 9.1 (left). The evolution of the fluid is simulated using the LB algorithm described in Sect. 9.4, until the stationary state is reached. The analysis presented in this section is restricted to the stationary state.

In the stationary state of Couette flow, rarefied gases exhibit a non-linear velocity profile in the proximity of the moving walls. This nonlinearity originates from the wall-induced discontinuity of the particle distribution function and its spatial extension (i.e., the width of the so-called Knudsen layer, where the discontinuity induced through interparticle collisions) is of the order of the mean free path of the fluid particles [1, 2]. Diffuse reflection boundary conditions are used to capture this wall-induced discontinuity [24–26], as shown in Fig. 9.1 (right).

Mathematically, diffuse reflection boundary conditions entail that the distribution functions for the particles emerging from the walls back into the fluid satisfy $f(x = \pm L/2, \boldsymbol{p}, t) = f^{\text{eq}}(\boldsymbol{p}; \pm \boldsymbol{u}_w)$, which is valid for $\pm p_x < 0$, respectively. Noting that $f^{\text{eq}}(-\boldsymbol{p}; \boldsymbol{u}) = f^{\text{eq}}(\boldsymbol{p}; -\boldsymbol{u})$, it can be seen that the solution of the Boltzmann equation (9.12) possesses the symmetry $f(-x, \boldsymbol{p}, t) = f(x, -\boldsymbol{p}, t)$. This symmetry allows only the right half of the channel to be considered, provided that bounce back boundary conditions are implemented at the channel centre-line [i.e., $f(0, -\boldsymbol{p}, t) = f(0, \boldsymbol{p}, t)$]. This simplification effectively halves all computation times. Moreover, since the system is homogeneous along the y axis, no advection is performed in this direction and a discretisation using a single node is sufficient. In fact, this corresponds to implementing periodic boundary conditions along the y axis. The p_z degree of freedom is integrated out, as explained in Sect. 9.4.1, and no advection is performed along the z direction. The right panel of Fig. 9.1 presents schematically the implementation of the Couette flow geometry.

In order to capture the Knudsen layer, it is convenient to use a grid which is more refined in the vicinity of the wall. This can be achieved by employing a standard grid-stretching procedure [56, 57]. In this chapter, we follow Refs. [50, 54, 55] and perform an equidistant grid discretisation with respect to the non-dimensional parameter η, defined through:

$$x(\eta) = \frac{1}{2A} \tanh \eta, \tag{9.58}$$

where $0 \le \eta \le \operatorname{arctanh}(A)$ and $0 < A < 1$ controls the stretching such that when $A \to 0$, the grid becomes equidistant with respect to x, while as $A \to 1$, the grid points accumulate towards the right boundary. For a discretisation employing S points, we have:

[1] All quantities presented in this section are non-dimensionalised according to the conventions presented in Sect. 9.4.4.

$$\eta_s = \frac{1}{S}\left(s - \frac{1}{2}\right)\operatorname{arctanh}(A), \qquad x_s = \frac{1}{2A}\tanh\eta_s, \tag{9.59}$$

where the points with $1 \leq s \leq S$ lie within the flow domain. For the simulations presented in this section, we found that $S = 16$ points with $A = 0.98$ are sufficient to yield accurate results. The stretching procedure is illustrated in Fig. 9.1 (right) for a grid with $S = 8$ cells, when $A = 0.95$.

In order to employ the finite-difference scheme described in the Appendix, three ghost nodes are required on either side of the simulation domain. The bounce back boundary conditions [17, 19, 20] employed on the left side of the domain can be written as:

$$\phi_{0;ij} = \phi_{1;\tilde{\imath}\tilde{\jmath}}, \qquad \phi_{-1;ij} = \phi_{2;\tilde{\imath}\tilde{\jmath}}, \qquad \phi_{-2;ij} = \phi_{3;\tilde{\imath}\tilde{\jmath}}, \tag{9.60}$$

and similarly for $\chi_{s;ij}$. The notation $\tilde{\imath}$ ($\tilde{\jmath}$) refers to the component $p_{x,\tilde{\imath}}$ ($p_{y,\tilde{\jmath}}$) defined through:

$$p_{x,\tilde{\imath}} = -p_{x,i}, \qquad p_{y,\tilde{\jmath}} = -p_{y,j}. \tag{9.61}$$

On the right boundary, the diffuse reflection concept [24–26] is imposed. This requires that the flux of particles coming from the boundary cell at $s = S + \frac{1}{2}$ towards the first fluid node at $s = S$ is Maxwellian:

$$\begin{pmatrix} \Phi_{S+\frac{1}{2};ij} \\ X_{S+\frac{1}{2};ij} \end{pmatrix} = \frac{p_{x,i}}{m}\begin{pmatrix} \phi^{\rm eq}_{w;ij} \\ \chi^{\rm eq}_{w;ij} \end{pmatrix}, \qquad p_{i,x} < 0, \tag{9.62}$$

where $\phi^{\rm eq}_{w;ij}$ is the reduced equilibrium distribution Eq. (9.46) corresponding to the wall parameters n_w, $\boldsymbol{u}_w$ and $T_w = 1$ and $\chi^{\rm eq}_{w;ij} = \phi^{\rm eq}_{w;ij}$. In the above, the notations $\Phi_{S+\frac{1}{2};ij}$ and $X_{S+\frac{1}{2};ij}$ represent the fluxes corresponding to the reduced distributions ϕ_{ij} and χ_{ij}, which can be computed using Eq. (9.76) by replacing p_x with $p_{x,i}$ and F_s with $\phi_{ij;s}$ and $\chi_{ij;s}$, as required. Equation (9.62) can be achieved in the frame of the WENO-5 scheme [50] described in the Appendix, when

$$\phi_{S+1;ij} = \phi_{S+2;ij} = \phi_{S+3;ij} = \phi^{\rm eq}_{w;ij}, \qquad p_{x,i} < 0. \tag{9.63}$$

Similar relations hold also for $\chi_{s;ij}$. The distributions of the particles travelling from the fluid towards the wall are obtained by quadratic extrapolation with respect to the equidistant η coordinate from the fluid towards the wall:

$$\phi_{S+1;ij} = 3\phi_{S;ij} - 3\phi_{S-1;ij} + \phi_{S-2;ij}, \quad \phi_{S+2;ij} = 6\phi_{S;ij} - 8\phi_{S-1;ij} + 3\phi_{S-2;ij}. \tag{9.64}$$

The same relations are valid for $\chi_{s;ij}$. The wall density n_w can be obtained by imposing mass conservation:

$$\sum_{i,j} \Phi_{S+\frac{1}{2};ij} = 0 \Rightarrow n_w = -\frac{\displaystyle\sum_{i,j,p_{x,i}>0} \Phi_{S+\frac{1}{2};ij}}{\displaystyle\sum_{i,j,p_{x,i}<0} \frac{\phi^{\text{eq}}_{w;ij}}{n_w} \frac{p_{x,i}}{m}}. \tag{9.65}$$

It can be seen that the accurate computation of n_w requires the recovery of half-space quadrature sums, which is the reason why we choose the half-range Gauss–Hermite quadrature on the x axis.

We illustrate the capabilities of our models by considering the velocity profile for a low Mach number flow ($u_w = 0.1$) and perform simulations at various values of Kn. While the Navier–Stokes equations predict a straight-line velocity profile $u_y = 2xu_w/L$ [60, 61], the kinetic analysis shows that in the vicinity of the boundary, there is always a Knudsen layer, having an extension of the order of the particle mean free path, where the velocity profile curves along the wall [42]. Figure 9.2 (left) shows the excellent agreement between our LB results and the benchmark linearised Boltzmann–BGK results reported in Ref. [58]. These results are also presented in Ref. [53], but with less accuracy and for a smaller range of values of Kn. The dependence of the slip velocity with respect to Kn is shown in Fig. 9.2 (right), where our results are compared with the linearised Boltzmann–BGK results reported in Refs. [58, 59]. Excellent agreement is found in both cases. In order to compare our simulation results with those reported in Ref. [58], we employed the relation $\text{Kn} = k/\sqrt{2}$ between the Knudsen number defined in Eq. (9.57) and the parameter k employed in Ref. [58]. The quadrature orders used in these simulations were $Q_x = 4$ ($k \le 0.1$), 5 ($k = 0.3$), 10 ($k = 1$), 11 ($k = 2$), 20

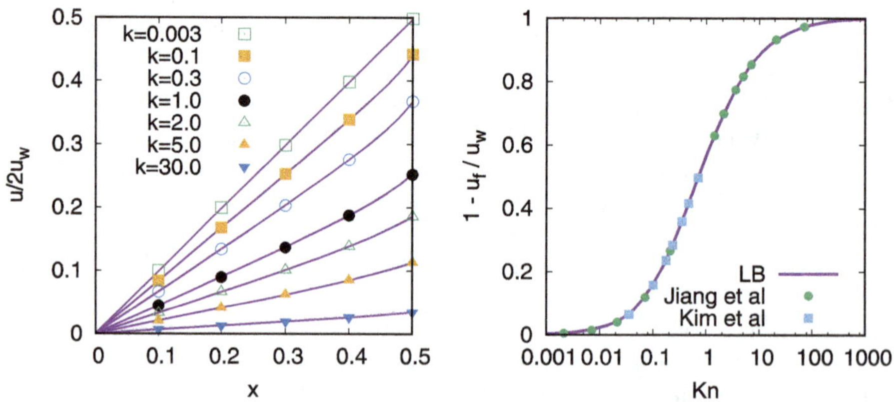

Fig. 9.2 Validation of the LB results (lines) in the context of the Couette flow. *Left:* Comparison of the velocity profile in the half-channel $0 \le x \le 1/2$ with the benchmark results reported in Ref. [58] at various values of $k = \text{Kn}\sqrt{2}$ (points). *Right:* Comparison of the slip velocity as a function of Kn with the linearised Boltzmann–BGK results reported in Refs. [58] (circles) and [59] (squares)

($k = 5$) and 40 ($k = 30$). As k is increased, the fluid velocity at the wall approaches 0 (its free-streaming value). Since the slip velocity can be recovered accurately even if the velocity profile presents visible deviations with respect to the benchmark data, the results presented in the right panel of Fig. 9.2 were obtained using a quadrature order $Q_x = 21$ for all values of Kn.

9.5.2 Force-Driven Poiseuille Flow Between Parallel Plates

In this section, we consider the force-driven Poiseuille flow between parallel plates. The geometry of this flow can be seen in Fig. 9.3 (left). The system consists of two parallel plates at rest which are taken to be perpendicular to the x axis. The gas between these plates is initially in thermal equilibrium at the wall temperature $T_w = 1$. At $t = 0$, a constant force $\boldsymbol{F} = (0, ma, 0)$ is applied throughout the fluid domain. According to the non-dimensionalisation discussed in Sect. 9.4.4, the acceleration a is expressed in units of c_{ref}^2/L and $m = 1$. The evolution of the fluid is simulated using the LB algorithm presented in Sect. 9.4.

The flow geometry, the boundary conditions and the Boltzmann equation (9.12), possess the symmetry property $f(-x, p_x, p_y, t) = f(x, -p_x, p_y, t)$. As was the case for the Couette flow, this symmetry allows only half of the channel to be simulated ($0 \le x \le \frac{1}{2}$), while the symmetry $f(0, p_x, p_y, t) = f(0, -p_x, p_y, t)$ is ensured using specular boundary conditions [17, 19, 20], as shown in Fig. 9.3 (right). In order to implement specular boundary conditions, the distribution functions in the nodes to the left of the flow domain, having indices $s = 0, -1, -2$, are populated according to:

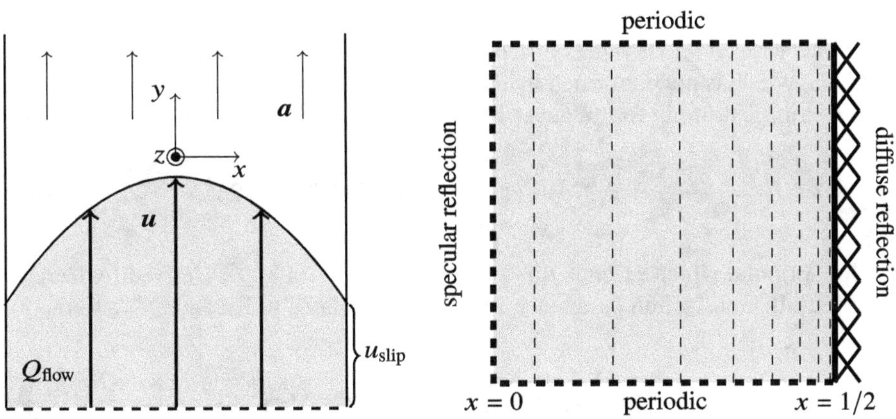

Fig. 9.3 *Left:* Setup for the force-driven Poiseuille flow problem, highlighting the slip velocity $u_{\text{slip}} = u(1/2)$. The mass flow rate is shown in the shaded area. *Right:* Boundary conditions and grid characteristics. The fine dotted lines show a grid comprised of $S = 8$ cells, stretched according to Eq. (9.58) with $A = 0.95$. Only one cell is used along the y direction

$$\phi_{0;ij} = \phi_{1;\bar{\imath}j}, \qquad \phi_{-1;ij} = \phi_{2;\bar{\imath}j}, \qquad \phi_{-2;ij} = \phi_{3;\bar{\imath}j}, \tag{9.66}$$

where only the x component of the momentum is reversed in the right-hand side of the above equations, as shown in Eq. (9.61). On the right boundary, the diffuse reflection concept is imposed, as discussed in Sect. 9.5.1. Furthermore, the grid is stretched using Eq. (9.58), with $A = 0.98$. In order to accurately capture the main features of the flow, we employed $S = 32$ nodes along the x axis, distributed in the right half of the channel.

In the case of the force-driven Poiseuille flow, we discuss two features which manifest at non-negligible values of Kn. The first one refers to the Knudsen paradox, according to which the flow rate through the channel decreases with Kn from its value in the Navier–Stokes limit down to a minimum, after which it increases towards infinity as the ballistic regime settles in. The non-dimensionalised mass flow rate Q_{flow} can be computed as follows:

$$Q_{\text{flow}} = \frac{\sqrt{8}}{a\sqrt{\pi}} \int_{-1/2}^{1/2} dx\, \rho(x) u_y(x). \tag{9.67}$$

For small values of Kn, Cercignani [62] derived the following approximation for Q_{flow}:

$$Q_{\text{flow}} \simeq \frac{1}{6\widetilde{\text{Kn}}} + s + (s^2 - 1)\widetilde{\text{Kn}}, \tag{9.68}$$

where $s = 1.01615$ and $\widetilde{\text{Kn}}$ is defined as:

$$\widetilde{\text{Kn}} = \text{Kn}\sqrt{\frac{\pi}{2}}. \tag{9.69}$$

While accurate at small values of Kn, Eq. (9.68) predicts a linear increase of Q_{flow} with $\widetilde{\text{Kn}}$, which is not confirmed by experiments or numerical simulations.

An empirical fitting formula was given by Sharipov in Eq. (11.136) of Ref. [9]:

$$G_P^* = -\frac{\ln \delta}{\sqrt{\pi}} + 0.376 - (1.77 \ln \delta + 0.584)\delta + 2.12\delta^2. \tag{9.70}$$

In this formula, which extends the asymptotic term $-\ln \delta/\sqrt{\pi}$ derived by Cercignani [62], the rarefaction parameter δ and G_P^* are related to Kn and Q_{flow} through:

$$\delta = \frac{1}{\text{Kn}\sqrt{2}} = \frac{\sqrt{\pi}}{2\widetilde{\text{Kn}}}, \qquad Q_{\text{flow}} = G_P^* \sqrt{\frac{4}{\pi}}. \tag{9.71}$$

Our numerical results for Q_{flow}, together with the approximations Eq. (9.68) and Eq. (9.71), as well as various other semi-analytical or numerical results are shown in Fig. 9.4 (left). These results were obtained using the mixed LB model described in

Sect. 9.4.2 with the order of the half-range Gauss–Hermite quadrature set to $Q_x = 21$. Since the velocity profile, and hence the mass flow rate, do not scale linearly with a at large values of a, we used $a = 0.01$ throughout the simulations in order to ensure good agreement with the validation data.

The second remarkable microfluidics specific effect occurring in the force-driven Poiseuille flow refers to the development of a dip (local minimum) in the temperature profile $T(x)$ at the centre $x = 0$ of the channel. The dip occurrence was predicted by the kinetic theory at the super-Burnett level and observed by DSMC simulations [1, 65–71].

Using a moments method approach, Mansour et al. [67, 68] derived analytically the following dependence of the temperature profile on the distance x from the centre of the channel:

$$T(x) = T_0 + \alpha x^2 + \beta x^4. \tag{9.72}$$

Using a numerical fit, we found an excellent match between the above functional form and our simulation results. For clarity, Fig. 9.4 (right) shows the half-channel profile of $[T(x) - 1]/(T_0 - 1)$, where $T_w = 1$ is the wall temperature and T_0 represents the temperature at the centre of the channel, as determined by fitting Eq. (9.72) to the numerical data. The values of the parameters T_0, α and β for the values of Kn considered in Fig. 9.4 (right) are given in Table 9.1. In these simulations

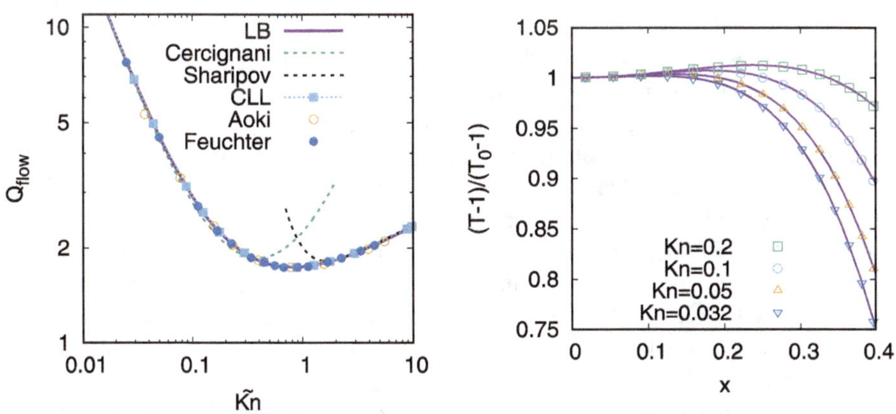

Fig. 9.4 Validation of the LB results in the context of the force-driven Poiseuille flow. *Left:* Comparison between the LB results (continuous line) for the flow rate Q_{flow}, defined in Eq. (9.67), and the asymptotic formulae in Eqs. (9.67) and (9.68) due to Cercignani [62] and Sharipov [9] (dashed lines), the results of Cercignani, Lampis and Lorenzani (CLL) [63] (dashed line with filled squares), the DVM results from Ref. [64] (hollow circles), as well as the DSMC results reported by Feuchter and Scheifenbaum in Ref. [52] (filled circles). The results are represented with respect to $\widetilde{\text{Kn}}$, defined in Eq. (9.69). *Right:* Illustration of the dip in the temperature profile at various values of Kn. The lines represent the best fits of the analytic expression, Eq. (9.72) to the LB results (points), as described in Sect. 9.5.2 and in Table 9.1

Table 9.1 Values of the coefficients T_0, a and b obtained by fitting Eq. (9.72) to the simulation results shown in Fig. 9.4

Kn	T_0	a	b
0.032	1.00778861	0.0020561	−0.0887483
0.05	1.00397973	0.00143734	−0.0393261
0.1	1.00171142	0.000749203	−0.0117178
0.2	1.000997806	0.000434946	−0.00385886

Only the points inside the domain $0 < x/L < 0.4$ are taken into account when performing the fit

we used $a = 0.05$ in order to enhance the development of the temperature dip. We employed the mixed quadrature LB model described in Sect. 9.4.2 where we set $Q_x = 4$ for Kn $\in \{0.032, 0.05, 0.1\}$, while at Kn $= 0.2$, we used $Q_x = 7$.

9.6 Conclusions

In this chapter, we presented a systematic procedure for the construction of high order mixed quadrature LB models based on the full-range and half-range Gauss–Hermite quadratures. A particular attention was given to the case when the flow is homogeneous along the z axis, when reduced distribution functions can be used in order to minimise the computational effort. The capabilities of these models are demonstrated in the context of the Couette and force-driven Poiseuille flows between parallel plates at various values of Kn. Excellent agreement is found between our results and benchmark data available in the literature from the Navier–Stokes level up to the transition regime.

The success of our models relies on the accurate recovery of half-range integrals required for the implementation of diffuse reflection. Such integrals are exactly recovered by employing the half-range Gauss–Hermite quadrature.

Our numerical method for solving the LB evolution equations employs finite-difference techniques. In particular, we implemented the advection using the fifth-order weighted essentially non-oscillatory (WENO-5) scheme and the time-stepping was performed using the third-order Runge–Kutta (RK-3) method. This allowed us to obtain accurate results using a small number of nodes (16 for the Couette flow and 32 for the force-driven Poiseuille flow).

Taking advantage of the homogeneity of the flows studied in this chapter with respect to the z axis, we eliminated the z axis degree of freedom by integrating the Boltzmann–BGK equation with respect to p_z. In order to correctly track the evolution of the temperature and heat flux fields, we employed two reduced distribution functions, ϕ and χ, obtained by integrating with respect to p_z the Boltzmann distribution f multiplied by 1 and p_z^2/m, respectively. The extension of the methodology presented in this chapter to more complex flow domains is straightforward since the mixed quadrature paradigm allows the type of quadrature and the quadrature orders to be adjusted for each axis separately. The treatment of complex boundaries can be performed by using the standard staircase approximation [72, 73] or the more recent vielbein approach [50]. Finally, more complex relaxation time models, such as the Shakhov model [74, 75], can be implemented as described in, e.g., Refs. [54, 55].

We conclude that the models described in this chapter can be used to obtain numerical solutions of the Boltzmann–BGK equation for channel flows at arbitrary values of the Knudsen number.

Appendix: Numerical Scheme

The simulation results presented in this chapter were obtained using an explicit third-order total variation diminishing (TVD) Runge–Kutta (RK-3) time marching procedure [76–79], together with the fifth-order weighted essentially non-oscillatory (WENO-5) scheme [80, 81] for computing the advection.

In order to implement the time-stepping algorithm, it is convenient to cast the Boltzmann–BGK equation (9.1) in the following form:

$$\partial_t f = L[f], \qquad L[f] = -\frac{\boldsymbol{p}}{m}\cdot\nabla f - \boldsymbol{F}\cdot\nabla_{\boldsymbol{p}} f - \frac{1}{\tau}(f - f^{\rm eq}). \tag{9.73}$$

Following the discretisation of the time variable using equal time steps δt, the distribution function at time step l is $f_l \equiv f(t_l)$, when the time coordinate has the value $t_l = l\delta t$, taken with respect to the initial time $t_0 = 0$. For simplicity, the dependence of the distribution function on the spatial coordinates and on the momentum degrees of freedom was omitted. The third-order Runge–Kutta TVD integrator described using the Butcher tableau summarised in Table 9.2 gives the following algorithm for computing the value f_{l+1} of the distribution function at time t_{l+1}:

$$\begin{aligned} f_l^{(1)} &= f_l + \delta t\, L[f_l], \\ f_l^{(2)} &= \frac{3}{4} f_l + \frac{1}{4} f_l^{(1)} + \frac{1}{4}\delta t\, L[f_l^{(1)}], \\ f_{l+1} &= \frac{1}{3} f_l + \frac{2}{3} f_l^{(2)} + \frac{2}{3}\delta t\, L[f_l^{(2)}]. \end{aligned} \tag{9.74}$$

For more information regarding the Butcher tableaux representation, we refer the reader to Ref. [82].

Table 9.2 Butcher tableau for the third-order Runge–Kutta time-stepping procedure described in Eq. (9.74)

0			
1	1		
1/2	1/4	1/4	
	1/6	1/6	2/3

The advection term is computed, as follows:

$$\frac{\boldsymbol{p}}{m} \cdot \nabla f = \frac{p_x}{m} \partial_x f + \frac{p_y}{m} \partial_y f + \frac{p_z}{m} \partial_z f. \tag{9.75}$$

Since the flows considered in this chapter are effectively one-dimensional (being homogeneous with respect to the y and z axes), the discussion on the implementation of the WENO-5 scheme for the computation of the above derivatives will cover only the derivative with respect to the x coordinate. Considering that the spatial domain is discretised equidistantly with respect to the η coordinate Eq. (9.59), the derivative with respect to x can be written as:

$$\left(\frac{p_x}{m} \partial_x f\right)_s = \frac{\mathcal{F}_{s+1/2} - \mathcal{F}_{s-1/2}}{x_{s+1/2} - x_{s-1/2}}. \tag{9.76}$$

The flux $\mathcal{F}_{s+1/2}$ corresponding to the interface between the cells centred on $x_s \equiv x(\eta_s)$ and x_{s+1} is computed in an upwind-biased approach using the WENO-5 algorithm [50, 80, 83], which we summarise below for the case when the advection velocity $p_x/m > 0$:

$$\mathcal{F}_{s+1/2} = \overline{\omega}_1 \mathcal{F}^1_{s+1/2} + \overline{\omega}_2 \mathcal{F}^2_{s+1/2} + \overline{\omega}_3 \mathcal{F}^3_{s+1/2}. \tag{9.77}$$

The interpolating functions $\mathcal{F}^q_{s+1/2}$ ($q = 1, 2, 3$) are given by:

$$\begin{aligned}
\mathcal{F}^1_{s+1/2} &= \frac{p_x}{m}\left(\frac{1}{3} f_{s-2} - \frac{7}{6} f_{s-1} + \frac{11}{6} f_s\right), \\
\mathcal{F}^2_{s+1/2} &= \frac{p_x}{m}\left(-\frac{1}{6} f_{s-1} + \frac{5}{6} f_s + \frac{1}{3} f_{s+1}\right), \\
\mathcal{F}^3_{s+1/2} &= \frac{p_x}{m}\left(\frac{1}{3} f_s + \frac{5}{6} f_{s+1} - \frac{1}{6} f_{s+2}\right),
\end{aligned} \tag{9.78}$$

while the weighting factors $\overline{\omega}_q$ are defined as:

$$\overline{\omega}_q = \frac{\widetilde{\omega}_q}{\widetilde{\omega}_1 + \widetilde{\omega}_2 + \widetilde{\omega}_3}, \qquad \widetilde{\omega}_q = \frac{\delta_q}{\sigma_q^2}. \tag{9.79}$$

The ideal weights δ_q are:

$$\delta_1 = 1/10, \qquad \delta_2 = 6/10, \qquad \delta_3 = 3/10, \tag{9.80}$$

while the indicators of smoothness σ_q are given by:

$$\sigma_1 = \frac{13}{12}\left(f_{s-2} - 2 f_{s-1} + f_s\right)^2 + \frac{1}{4}\left(f_{s-2} - 4 f_{s-1} + 3 f_s\right)^2,$$

Table 9.3 The values of the weighting factors $\overline{\omega}_q$ defined in Eq. (9.79) when one, two or all three of the σ_i ($i = 1, 2, 3$) functions have vanishing values

	$\overline{\omega}_1$	$\overline{\omega}_2$	$\overline{\omega}_3$
$\sigma_1 = \sigma_2 = \sigma_3$	0.1	0.6	0.3
$\sigma_2 = \sigma_3 = 0$	0	2/3	1/3
$\sigma_3 = \sigma_1 = 0$	1/4	0	3/4
$\sigma_1 = \sigma_2 = 0$	1/7	6/7	0
$\sigma_1 = 0$	1	0	0
$\sigma_2 = 0$	0	1	0
$\sigma_3 = 0$	0	0	1

$$\sigma_2 = \frac{13}{12}\left(f_{s-1} - 2f_s + f_{s+1}\right)^2 + \frac{1}{4}\left(f_{s-1} - f_{s+1}\right)^2,$$

$$\sigma_3 = \frac{13}{12}\left(f_s - 2f_{s+1} + f_{s+2}\right)^2 + \frac{1}{4}\left(3f_s - 4f_{s+1} + f_{s+2}\right)^2. \tag{9.81}$$

In the case when one, two or all three of the σ_q indicators vanish, the computation of the functions $\widetilde{\omega}_q$ using Eq. (9.79) implies illegal division by zero operations. In this case, the weighting factors $\overline{\omega}_q$ can be computed directly as shown in Table 9.3. Alternatively, a small quantity $\varepsilon \simeq 10^{-6}$ can be added to the σ_q functions. A more thorough discussion on the side effects of this approach can be found in Ref. [77].

References

1. G. Karniadakis, A. Beskok, N. Aluru, *Microflows and Nanoflows: Fundamentals and Simulation* (Springer, Berlin, 2005)
2. Y. Sone, *Molecular Gas Dynamics: Theory, Techniques and Applications* (Birkhäuser, Boston, 2007)
3. C. Cercignani, *The Boltzmann Equation and Its Applications* (Springer, New York, 1988)
4. S. Harris, *An Introduction to the Theory of the Boltzmann Equation* (Holt, Rinehart and Winston, New York, 1971)
5. R.L. Liboff, *Kinetic Theory: Classical, Quantum and Relativistic Descriptions*, 3rd edn. (Springer, New York, 2003)
6. G.A. Bird, *Molecular Gas Dynamics and the Direct Simulation of Gas Flows* (Oxford University Press, Oxford, 1994)
7. J.E. Broadwell, Study of rarefied shear flow by the discrete velocity method. J. Fluid Mech. **19**, 401 (1964)
8. M.T. Ho, I. Graur, Heat transfer through rarefied gas confined between two concentric spheres. Int. J. Heat Mass Transf. **90**, 58 (2015)
9. F. Sharipov, *Rarefied Gas Dynamics: Fundamentals for Research and Practice* (Wiley, Weinheim, 2016)
10. W. Peng, M.T. Ho, Z.L. Guo, Y.H. Zhang, A comparative study of discrete velocity methods for low-speed rarefied gas flows. Comput. Fluids **161**, 33 (2018)
11. L.H. Zhu, Z.L. Guo, K. Xu, Discrete unified gas kinetic scheme on unstructured meshes. Comput. Fluids **127**, 211 (2016)
12. L.H. Zhu, S.Z. Chen, Z.L. Guo, dugksFoam: an open source OpenFOAM solver for the Boltzmann model equation. Comput. Phys. Commun. **213**, 155 (2017)

13. C.K. Aidun, J.R. Clausen, Lattice-Boltzmann method for complex flows. Annu. Rev. Fluid Mech. **42**, 439 (2010)
14. S. Chen, G.D. Doolen, Lattice Boltzmann method for fluid flows. Annu. Rev. Fluid Mech. **30**, 329 (1998)
15. Z. Guo, C. Shu, *The Lattice Boltzmann Method and its Applications in Engineering* (World Scientific, Singapore, 2013)
16. H. Huang, M.C. Sukop, X.Y. Lu, *Multiphase Lattice Boltzmann Methods: Theory and Application* (Wiley Blackwell, Chichester, 2005)
17. T. Krüger, H. Kusumaatmaja, A. Kuzmin, O. Shardt, G. Silva, E.M. Viggen, *The Lattice Boltzmann Method, Principles and Practice* (Springer, Oxford, 2017)
18. X.W. Shan, X.F. Yuan, H.D. Chen, Kinetic theory representation of hydrodynamics: a way beyond the Navier–Stokes equation. J. Fluid Mech. **550**, 413 (2006)
19. S. Succi, *The Lattice Boltzmann Equation for Fluid Dynamics and Beyond* (Clarendon Press, Oxford, 2001)
20. D.A. Wolf-Gladrow, *Lattice-Gas Cellular Automata and Lattice Boltzmann Models* (Springer, Berlin, 2000)
21. X. He, L.S. Luo, A priori derivation of the lattice Boltzmann equation. Phys. Rev. E **55**, R6333 (1997)
22. X. He, L.S. Luo, Theory of the lattice Boltzmann method: from the Boltzmann equation to the lattice Boltzmann equation. Phys. Rev. E **56**, 6811 (1997)
23. V.E. Ambruş, V. Sofonea, High-order thermal lattice Boltzmann models derived by means of Gauss quadrature in the spherical coordinate system. Phys. Rev. E **86**, 16708 (2012)
24. V.E. Ambruş, V. Sofonea, Implementation of diffuse-reflection boundary conditions using lattice Boltzmann models based on half-space Gauss–Laguerre quadratures. Phys. Rev. E **89**, 041301(R) (2014)
25. V.E. Ambruş, V. Sofonea, Application of mixed quadrature lattice Boltzmann models for the simulation of Poiseuille flow at non-negligible values of the Knudsen number. J. Comput. Sci. **17**, 403 (2016)
26. V.E. Ambruş, V. Sofonea, Lattice Boltzmann models based on half-range Gauss–Hermite quadratures. J. Comput. Phys. **316**, 760 (2016)
27. S.S. Chikatamarla, I.V. Karlin, Lattices for the lattice Boltzmann method. Phys. Rev. E **79**, 46701 (2009)
28. M. Namburi, S. Krithivasan, S. Ansumali, Crystallographic lattice Boltzmann method. Nat. Sci. Rep. **6**, 27172 (2016)
29. P.C. Philippi, J. L. A. Hegele, L.O.E.d. Santos, R. Surmas, From the continuous to the lattice Boltzmann equation: the discretization problem and thermal models. Phys. Rev. E **73**, 56702 (2006)
30. M. Sbragaglia, R. Benzi, L. Biferale, S. Succi, K. Sugiyama, F. Toschi, Generalized lattice Boltzmann method with multirange pseudopotential. Phys. Rev. E **75**, 26702 (2007)
31. N.Z. Cao, S.Y. Shen, S. Jin, D. Martinez, Physical symmetry and lattice symmetry in the lattice Boltzmann method. Phys. Rev. E **55**, R21 (1997)
32. H. Chen, Volumetric formulation of the lattice Boltzmann method for fluid dynamics: basic concept. Phys. Rev. E **58**, 3955 (1998)
33. X.Y. He, Error analysis for the interpolation-supplemented lattice-Boltzmann equation scheme. Int. J. Mod. Phys. C **8**, 737 (1997)
34. R.W. Mei, W. Shyy, On the finite difference-based lattice Boltzmann method in curvilinear coordinates. J. Comput. Phys. **143**, 426 (1998)
35. M. Watari, M. Tsutahara, Two-dimensional thermal model of the finite-difference lattice Boltzmann method with high spatial isotropy. Phys. Rev. E **67**, 36306 (2003)
36. M. Watari, M. Tsutahara, Possibility of constructing a multispeed Bhatnagar–Gross–Krook thermal model of the lattice Boltzmann method. Phys. Rev. E **70**, 16703 (2004)
37. M. Watari, M. Tsutahara, Supersonic flow simulations by a three-dimensional multispeed thermal model of the finite difference lattice Boltzmann method. Phys. A **364**, 129 (2006)

38. M. Watari, Velocity slip and temperature jump simulations by the three-dimensional thermal finite-difference lattice Boltzmann method. Phys. Rev. E **79**, 66706 (2009)
39. F.B. Hildebrand, *Introduction to Numerical Analysis*, 2nd edn. (Dover, New York, 1987)
40. B. Shizgal, *Spectral Methods in Chemistry and Physics: Applications to Kinetic Theory and Quantum Mechanics* (Springer, Berlin, 2015)
41. E.P. Gross, E.A. Jackson, S. Ziering, Boundary value problems in kinetic theory of gases. Ann. Phys. **1**, 141 (1957)
42. S. Takata, H. Funagane, Singular behaviour of a rarefied gas on a planar boundary. J. Fluid Mech. **717**, 30 (2013)
43. C. Cercignani, M. Lampis, Kinetic models for gas-surface interactions. Transp. Theory Stat. Phys. **1**, 101 (1971)
44. Z.H. Li, H.X. Zhang, Study on gas kinetic unified algorithm for flows from rarefied transition to continuum. J. Comput. Phys. **193**, 708 (2004)
45. M.T. Ho, L. Zhu, L. Wu, P. Wang, Z. Guo, Z.H. Li, Y. Zhang, A multi-level parallel solver for rarefied gas flows in porous media. Comput. Phys. Commun. **234**, 14 (2019)
46. A. Frezzotti, L. Gibelli, B. Franzelli, A moment method for low speed microflows. Contin. Mech. Thermodyn. **21**, 495 (2009)
47. Y. Shi, Y.W. Yap, J.E. Sader, Linearized lattice Boltzmann method for micro- and nanoscale flow and heat transfer. Phys. Rev. E **92**, 13307 (2015)
48. P.L. Bhatnagar, E.P. Gross, M. Krook, A model for collision processes in gases. I. Small amplitude processes in charged and neutral one-component systems. Phys. Rev. **94**, 511 (1954)
49. V.E. Ambrus, V. Sofonea, R. Fournier, S. Blanco, Implementation of the force term in half-range lattice Boltzmann models. arXiv preprint arXiv:1708.03249 (2017)
50. S. Busuioc, V.E. Ambruş, Lattice Boltzmann models based on the vielbein formalism for the simulation of flows in curvilinear geometries. Phys. Rev. E. **99**(3), 033304 (2019)
51. J. Meng, L. Wu, J.M. Reese, Y. Zhang, Assessment of the ellipsoidal-statistical Bhatnagar–Gross–Krook model for force-driven Poiseuille flow. J. Comput. Phys. **251**, 383 (2013)
52. C. Feuchter, W. Schleifenbaum, High-order lattice Boltzmann models for wall-bounded flows at finite Knudsen numbers. Phys. Rev. E **94**, 13304 (2016)
53. Y.W. Yap, J.E. Sader, High accuracy numerical solutions of the Boltzmann Bhatnagar–Gross–Krook equation for steady and oscillatory Couette flows. Phys. Fluids **24**, 32004 (2012)
54. V.E. Ambruş, V. Sofonea, Half-range lattice Boltzmann models for the simulation of Couette flow using the Shakhov collision term. Phys. Rev. E **98**(6), 063311 (2018)
55. V.E. Ambruş, F. Sharipov, V. Sofonea, Lattice Boltzmann approach to rarefied gas flows using half-range Gauss–Hermite quadratures: comparison to DSMC results based on ab initio potentials. arXiv preprint arXiv:1810.07803 (2018)
56. Z. Guo, T.S. Zhao, Explicit finite-difference lattice Boltzmann method for curvilinear coordinates. Phys. Rev. E **67**, 66709 (2003)
57. R. Mei, W. Shyy, On the finite difference-based lattice Boltzmann method in curvilinear coordinates. J. Comput. Phys. **143**, 426 (1998)
58. S. Jiang, L.S. Luo, Analysis and accurate numerical solutions of the integral equation derived from the linearized BGKW equation for the steady Couette flow. J. Comput. Phys **316**, 416 (2016)
59. S.H. Kim, H. Pitsch, I.D. Boyd, Accuracy of higher-order lattice Boltzmann methods for microscale flows with finite Knudsen numbers. J. Comput. Phys. **227**, 8655 (2008)
60. P.K. Kundu, I.M. Cohen, D.R. Dowling, *Fluid Mechanics*, 6th edn. (Academic Press, New York, 2015)
61. M. Rieutord, *Fluid Mechanics: An Introduction* (Springer, Berlin, 2015)
62. C. Cercignani, *Theory and Application of the Boltzmann Equation* (Scottish Academic Press, Edinburgh, 1975)
63. C. Cercignani, M. Lampis, S. Lorenzani, Variational approach to gas flows in microchannels. Phys. Fluids **16**, 3426 (2004)
64. K. Aoki, S. Takata, T. Nakanishi, Poiseuille-type flow of a rarefied gas between two parallel plates driven by a uniform external force. Phys. Rev. E **65**, 26315 (2002)

65. M. Tij, A. Santos, Perturbation analysis of a stationary nonequilibrium flow generated by an external force. J. Stat. Phys. **76**, 1399 (1994)
66. M. Tij, M. Sabbane, A. Santos, Nonlinear Poiseuille flow in a gas. Phys. Fluids **10**, 1021 (1998)
67. M.M. Mansour, F. Baras, A.L. Garcia, On the validity of hydrodynamics in plane Poiseuille flow. Phys. A **240**, 255 (1997)
68. S. Hess, M.M. Mansour, Temperature profile of a dilute gas undergoing a plane Poiseuille flow. Phys. A **272**, 481 (1999)
69. K. Aoki, S. Takata, T. Nakanishi, Poiseuille-type flow of a rarefied gas between two parallel plates driven by a uniform external force. Phys. Rev. E **65**, 26315 (2002)
70. Y. Zheng, A.L. Garcia, B.J. Alder, Comparison of kinetic theory and hydrodynamics for Poiseuille flow. J. Stat. Phys. **109**(3/4), 495 (2002)
71. K. Xu, Super-Burnett solutions for Poiseuille flow. Phys. Fluids **15**(7), 2077 (2003)
72. X. Yin, J. Zhang, An improved bounce-back scheme for complex boundary conditions in lattice Boltzmann method. J. Comput. Phys. **231**, 4295 (2012)
73. L. Li, R. Mei, J.F. Klausner, Boundary conditions for thermal lattice Boltzmann equation method. J. Comput. Phys. **237**, 366 (2013)
74. E.M. Shakhov, Generalization of the Krook kinetic relaxation equation. Fluid Dyn. **3**, 95 (1968)
75. E.M. Shakhov, Approximate kinetic equations in rarefied gas theory. Fluid Dyn. **3**, 112 (1968)
76. S. Gottlieb, C.W. Shu, Total variation diminishing Runge–Kutta schemes. Math. Comput. **67**, 73 (1998)
77. A.K. Henrick, T.D. Aslam, J.M. Powers, Mapped weighted essentially non-oscillatory schemes: achieving optimal order near critical points. J. Comput. Phys **207**, 542 (2005)
78. C.W. Shu, S. Osher, Efficient implementation of essentially non-oscillatory shock-capturing schemes. J. Comput. Phys. **77**, 439 (1988)
79. J.A. Trangenstein, *Numerical Solution of Hyperbolic Partial Differential Equations* (Cambridge University Press, New York, 2007)
80. Y. Gan, A. Xu, G. Zhang, Y. Li, Lattice Boltzmann study on Kelvin–Helmholtz instability: roles of velocity and density gradients. Phys. Rev. E **83**, 56704 (2011)
81. G.S. Jiang, C.W. Shu, Efficient implementation of weighted ENO schemes. J. Comput. Phys. **126**, 202 (1996)
82. J.C. Butcher, *Numerical Methods for Ordinary Differential Equations*, 2nd edn. (Wiley, Chichester, 2008)
83. V.E. Ambruş, R. Blaga, High-order quadrature-based lattice Boltzmann models for the flow of ultrarelativistic rarefied gases. Phys. Rev. C **98**, 035201 (2018)

Permissions

All chapters in this book were first published by Springer; hereby published with permission under the Creative Commons Attribution License or equivalent. Every chapter published in this book has been scrutinized by our experts. Their significance has been extensively debated. The topics covered herein carry significant information for a comprehensive understanding. They may even be implemented as practical applications or may be referred to as a beginning point for further studies.

The contributors of this book come from diverse backgrounds, making this book a truly international effort. We would like to thank all the contributing authors for lending their expertise to make the book truly unique. They have played a crucial role in the development of this book. Without their invaluable contributions this book wouldn't have been possible. They have made vital efforts to compile up to date information on the varied aspects of this subject to make this book a valuable addition to the collection of many professionals and students.

This book was conceptualized with the vision of imparting up-to-date and integrated information in this field. To ensure the same, a matchless editorial board was set up. Every individual on the board went through rigorous rounds of assessment to prove their worth. After which they invested a large part of their time researching and compiling the most relevant data for our readers.

The editorial board has been involved in producing this book since its inception. They have spent rigorous hours researching and exploring the diverse topics which have resulted in the successful publishing of this book. They have passed on their knowledge of decades through this book. To expedite this challenging task, the publisher supported the team at every step. A small team of assistant editors was also appointed to further simplify the editing procedure and attain best results for the readers.

Apart from the editorial board, the designing team has also invested a significant amount of their time in understanding the subject and creating the most relevant covers. They scrutinized every image to scout for the most suitable representation of the subject and create an appropriate cover for the book.

The publishing team has been an ardent support to the editorial, designing and production team. Their endless efforts to recruit the best for this project, has resulted in the accomplishment of this book. They are a veteran in the field of academics and their pool of knowledge is as vast as their experience in printing. Their expertise and guidance has proved useful at every step. Their uncompromising quality standards have made this book an exceptional effort. Their encouragement from time to time has been an inspiration for everyone.

The publisher and the editorial board hope that this book will prove to be a valuable piece of knowledge for students, practitioners and scholars across the globe.

Index

U

V

www.ingramcontent.com/pod-product-compliance
Lightning Source LLC
LaVergne TN
LVHW051048220926
848485LV00001BA/48
* 9 7 8 1 6 4 7 2 5 4 4 7 6 *